W9-ADH-372

EDUCATION LIBRARY
UNIVERSITY OF KENTUCKY

Research in Collegiate Mathematics Education. VII

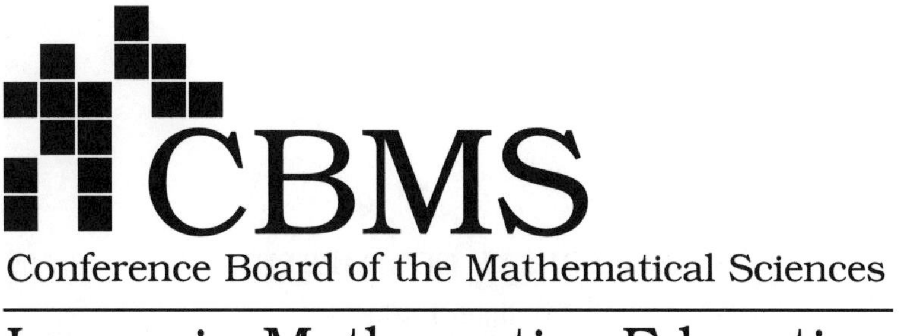

Conference Board of the Mathematical Sciences

Issues in Mathematics Education

Volume 16

Research in Collegiate Mathematics Education. VII

Fernando Hitt
Derek Holton
Patrick W. Thompson
Editors

Shandy Hauk, *Production Editor*

American Mathematical Society
Providence, Rhode Island
in cooperation with
Mathematical Association of America
Washington, D. C.

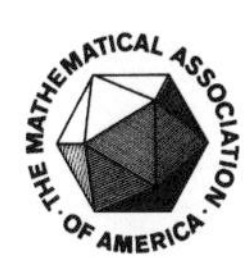

EDUC
QA
11
.A1
M277X
2010

EDITORIAL COMMITTEE

William Barker
Carl Cowen
Solomon Friedberg
W. James Lewis (Chair)

2000 *Mathematics Subject Classification.* Primary 97Axx, 97Cxx, 97Dxx, 97Fxx, 97Ixx, 97Uxx, 97–XX, 00–XX.

ISBN-13: 978-0-8218-4996-5
ISBN-10: 0-8218-4996-4
ISSN: 1047-398X

Copying and reprinting. Material in this book may be reproduced by any means for educational and scientific purposes without fee or permission with the exception of reproduction by services that collect fees for delivery of documents and provided that the customary acknowledgment of the source is given. This consent does not extend to other kinds of copying for general distribution, for advertising or promotional purposes, or for resale. Requests for permission for commercial use of material should be addressed to the Acquisitions Department, American Mathematical Society, 201 Charles Street, Providence, Rhode Island 02904-2294, USA. Requests can also be made by e-mail to `reprint-permission@ams.org`.

Excluded from these provisions is material in articles for which the author holds copyright. In such cases, requests for permission to use or reprint should be addressed directly to the author(s). (Copyright ownership is indicated in the notice in the lower right-hand corner of the first page of each article.)

© 2010 by the American Mathematical Society. All rights reserved.
The American Mathematical Society retains all rights
except those granted to the United States Government.
Copyright of individual articles may revert to the public domain 28 years
after publication. Contact the AMS for copyright status of individual articles.
Printed in the United States of America.

♾ The paper used in this book is acid-free and falls within the guidelines
established to ensure permanence and durability.
Visit the AMS home page at `http://www.ams.org/`

10 9 8 7 6 5 4 3 2 1 15 14 13 12 11 10

Contents

Preface

Welcome to the seventh volume of *Research in Collegiate Mathematics Education (RCME VII)*. The present volume, like previous volumes in this series, reflects the importance of research in mathematics education at the collegiate level. The editors in this series encourage communication between mathematicians and mathematics educators and, as pointed out by the International Commission on Mathematics Instruction (ICMI), much more work is needed in concert by these two groups. As is true in any research field, over time researchers in collegiate mathematics education have developed a variety of theoretical approaches. As a natural part of the development of ways to investigate the learning and teaching of mathematics, researchers have constructed a specialized terminology that mathematicians may find challenging. In fact, the editors of the first volume of this series wrote about this (Dubinsky, Schoenfeld, & Kaput, 1994):

> As Alan Schoenfeld's opening chapter makes clear, mathematicians who are not familiar with educational research are likely to be in for some surprises. The field is not what you might think it is, and its methods are not what you might expect. This should come as no great shock. Compare people's preconceptions of what a mathematician does with the reality of being a mathematical researcher. The same applies here. (pp. vii-viii)

Overview of this Volume

Nine papers constitute this volume. The first two examine problems students experience when converting a representation from one particular system of representations to another. One paper is on the role of fractional and decimal representations in relation to determining the rationality of a number and the second is about moving among numeric, algebraic, and graphic representation in the context of the definite integral. The next three papers investigate student learning about proof. The first one looks at this issue in terms of professors' points of view about proof and its teaching while the second examines two students' experiences with proving. The third paper in this group focuses on students making and testing conjectures about infinite processes. In the next two papers, the focus is instructor knowledge for teaching calculus. The first examines how graduate teaching assistants gain knowledge of student thinking and the second explores the perceptions and practices of professors teaching calculus. The final two papers in the volume address the nature of "conception" in mathematics. The first of these proposes a model that leads to a definition of "conception" and the last paper reports on the availability of strategies for problem solving in connection with the idea of "conception."

Conversions Among Representations

There are two papers related to theoretical approaches with representations: one from Zazkis and Sirotic, and the other from Camacho, Depool and Santos. Zaskis and Sirotic investigated prospective teachers' understanding of irrational number representations by surveying 46 future teachers and through semi-structured interviews with 16 volunteers from that group. From a theoretical point of view of representations, they assume two things. First, they take into account the fundamental fact that there is no direct access to mathematical objects but only to their representations, and secondly that any representation captures only partially the mathematical object. Zazkis and Sirotic use the notions of *transparent* and *opaque* representation, with respect to a certain property of the represented object, as if the property can be "seen" (i.e., a maximum or minimum in a graph) or is "hidden" (i.e., a maximum or minimum in the algebraic representation of a function). Zazkis and Sirotic show us some ways that students' identifications of irrational numbers may be correct, but that their mathematical reasoning about the representations may not be appropriate. The results of the study support, in general, the importance of promoting articulation about representations among students. In particular, Zazkis and Sirotic provide us with pedagogical elements to improve the teaching of rational and irrational number concepts.

In a semester long study conducted by Camacho, Depool and Santos, the researchers observed the problem-solving efforts of 31 first-year engineering students enrolled in a reform calculus course, interviewing a subset of 6 students, to explore students' learning of the definite integral. The instructional approach embraced the "Rule of Four," including algebraic, numeric, graphic, and technologically rich representations (see, e.g., Schwarz, 1989). Camacho, Depool and Santos' perspective is based on Duval's (1993) theoretical approach for learning through both distinguishing between mathematical concepts and negotiating among multiple representations of a concept. Particularly, the authors were aware of the difficulties that students have when passing from one representation to another in a mathematical task related to definite integrals. Therefore, they designed and had students work on a number of non-routine problems related to geometrical representations and algebraic representations. Some of these problems asked students to decide whether given statements were true or false. The results of the study suggest that, regardless of the representations or technology used, students tended to give primacy to algebraic over graphic and numeric representations. The results also indicate that students were able to analyze data given in a graphic form though the authors further state that students experienced serious difficulties in constructing examples and counter-examples that could help them in the resolution process of tasks. From a general point of view, the results obtained by the authors – in a technology rich setting – are similar to observations by Eisenberg and Dreyfus (1991) and Vinner (1989) about the challenges of learning to use visualization in mathematics.

Teaching and Learning Proof

In relation to exploring and proving conjectures, we have three papers in this volume: one written by Alcock, another by Alcock and Weber, and the last by Brown, McDonald and Weller. Alcock's paper is related to the teaching and learning of mathematical proof. From discussions with five mathematicians who taught courses about mathematical reasoning, Alcock distilled four important aspects of student experience when learning proof: *instantiation*, *structural thinking*, *creative thinking*, and *critical thinking*. In the first part of Alcock's report, she shares various comments from professors about how these aspects are exhibited. In the second part, Alcock provides evidence from professors' explanations about the types of strategies used in the mathematics classroom to describe the four types of thinking among students. The professors play an important role in demonstrating the use of examples and counter-examples. In her analysis, through the voices of the professors, Alcock provides examples of professors who were directing their approach to promote structural thinking, while others directed their approach towards instantiation and creative thinking. In the third part of the article, Alcock asserts that the four modes of thinking can be viewed as interdependent and that those who are successful at producing correct proofs will switch flexibly from one to another mode of thinking according to changing goals during the proving process. Alcock concludes with some suggestions for teaching proof, including a balancing of attention among the four modes of thinking.

The paper by Alcock and Weber reports some of the problems students have in the process of proving mathematical statements. They analyze two interviews from two different students, Brad and Carla, around the verification or falsification of four mathematical statements. In two of the situations it is necessary to provide a counter-example, in one a proof by contradiction is efficacious, and in the fourth it is possible to construct a direct proof. Examination of the techniques used by the two students suggest they had two distinct approaches: (1) a *syntactic* approach to proving, focused on logical syntax, without much attention to variety in the representation of mathematical concepts, and (2) a *referential* approach to proving, focused on the use of representations of mathematical concepts. Literature has provided us with mathematicians who can be classified under this theoretical approach (Hadamard, 1945/1975): Hermite was a mathematician who used a syntactic approach, and Poincaré was a mathematician who used a referential approach. Alcock and Weber show us the importance of the two approaches in the learning of mathematics.

Brown, McDonald, and Weller address one of the major problem areas for mathematics learning at university level: the intersection between the concept of proof and the concept of infinity. The investigation carried out by Brown, McDonald, and Weller included 12 students learning set theory. The principal focus was: Prove or disprove: $\bigcup_{k=1}^{n} P(1,2,\ldots,k) = P(\mathbb{N})$, where $\mathbb{N}$ denotes the set of natural numbers and P "the power set of" (the set of all subsets of the given set). This task was difficult for students and, for them, involved both potential and actual infinity; these ideas are a major challenge documented by authors in the didactics of mathematics and by historians of mathematics. Grounded in Action, Process, Object, Schema (APOS) theory, the researchers conducted their study in two phases. First, they interviewed five students using an individual semi-structured protocol; and one year later they talked with seven students, using small groups and a expanded version of the original semi-structured interview. The design of the questions in the second phase, and the subsequent answers given by students, presented interesting information about the students' difficulties and how they addressed them. The authors have suggested a teaching approach that takes into account APOS theory-based instruction to help undergraduates in the construction of an understanding of actual infinity. One of the implications of their analysis is that even if the results obtained with their study are consistent with the Basic Metaphor of Infinity offered by Lakoff and Núñez (2000), the results go beyond what is suggested by Lakoff and Núñez; new structures are needed to interpret the Brown, McDonald and Weller results.

Knowledge for Teaching

Two papers in this volume address aspects of the knowledge needed for teaching calculus. In the first, Kung interviewed current and former teaching assistants (TAs) about student thinking in calculus and how the TAs learned about it. The author connects the ideas of college instructor learning about student thinking to the literature on the knowledge needed for teaching in grades K-12 and to research on college calculus learning. As Kung points out, the first teaching experiences of many future college faculty occur when they are TAs, during graduate school. And, like the induction of K-12 teachers, the early experiences of TAs may be quite formative. A main result of his interviews with TAs about their on-the-job learning around student thinking was that the participants reported learning different things about student thinking, depending on the setting. Kung discusses this duality between types of knowledge about student thinking and types of interactions participants had with students (e.g., watching students work in groups or grading students' written work), then offers a framework to use in understanding and developing TAs' instructional practice. Where Kung's paper explores the complexities of the knowledge needed for teaching by a careful examination of the experiences of novices (TAs), Sofronas and DeFranco report on beliefs about calculus instruction held by more experienced instructors: college mathematics professors.

Sofronas and DeFranco suggest that the efforts made by ICMI to develop communication between mathematicians and mathematics educators has had little impact. Taking into account research on pedagogical content knowledge and the Knowledge Base for Teaching (KBT) at the K-12 level, through interviews with seven mathematicians teaching calculus, Sofronas and DeFranco propose a KBT model for collegiate mathematics instruction. As noted by a participant in their work, it is true that "most empirical research on the effectiveness of collaborative learning was concerned with a small scale [groups]" (see, Dillenbourg, 1999), so if we think that collaborative learning is important, we need to continue doing and communicating more research on the subject (see, e.g., Star & Smith III, 2006; Hitt, 2007). One of the results of the Sofronas and DeFranco study was that the instructional methods of the mathematician participants were grounded in their own learning styles without taking into consideration the knowledge base in pedagogy. It is interesting to see that for mathematicians, the history of mathematics is influential for some in their teaching, and some of the participants also thought that a knowledge of this history permitted them to have a better approach to teaching mathematics. The papers by Kung and Sofronas and DeFranco illustrate the complexity of the KBT in higher education.

Mathematical Conception

The final two papers in the volume explore the structure of mathematical conception. The theoretical foundations of Balacheff and Gaudin's paper are the works of the French tradition, mainly using the theoretical approaches and epistemological language of authors like Bachelard (1938), Brousseau (1997), and Vergnaud (1991). For some time the term *conception* was used in the didactics of mathematics in a vague way without settling on a characterization of the term so that researchers in mathematics education and professors of mathematics might share it (Artigue, 1991). One of the first attempts at characterization was provided by Duroux in connection with epistemological obstacles (as cited by Brousseau, 1997, pp. 99–100) and developed further by Vergnaud. Balacheff and Gaudin's work includes another fundamental element not considered by Verngaud, which they have named *control structure*. Indeed, in other theoretical approaches this notion of control structure has already been suggested, for example, as control processes in problem-solving (Schoenfeld, 1985), or as an epistemic frame (Perkins & Simmons, 1988), or as a verification process (Margolinas, 1989, 1993). Balacheff and Gaudin's extension of Vergnaud's theoretical approach adds a control structure to result in a conception C consisting of a quadruplet (P, R, L, Σ) in which: P is a set of problems, R is a set of operators, L is a representation system, and Σ is a control structure. Using this approach, Balacheff and Gaudin provide examples of conceptions related to function such as the Conception of Table: $C_T = (P_T, R_T, \text{Table}, \Sigma_T)$. Their analysis about C_T is made from a historical point of view, affirming that C_T is essentially formed from empirical foundations. Like the mathematicians in Sofronas and DeFranco's study, Balacheff and Guadin turn to the historical genesis of concept. They also conducted an experiment to understand the complexity of notions that students associate with the concept of function. Their experiment was carried out in a technological environment using the geometry software Cabri with 12th grade students. Balacheff and Gaudin illustrate their theoretical framework with two conceptions: Curve-Algebraic and Algebraic-Graph.

The volume concludes with the paper by Mesa. Mesa's theoretical frame is provided in an early version of Balacheff's work on what a conception is (see above and the article of Balacheff and Gaudin in this volume). According to Balacheff and Gaudin's theoretical approach, explicit attention to control structures is very important in problem-solving. Mesa takes into consideration the structure of control related to the conception of Initial Value Problems (IVPs) and makes a compelling argument that IVPs are a kind of problem that permits the development of a control structure. Related to the development of this control structure, Mesa analyzes calculus textbooks' treatment of IVPs to examine the richness of strategy discourse and explicitness of control structures offered in the texts. Mesa's results portray the textbooks' focus on solution verification as a primary strategy and illuminate the limited nature of strategy discourse about control structures supported by the textbooks in her study.

Finally, we would like to thank all those who have given of their time and expertise to review manuscripts for this and previous *RCME* volumes. Also, the editors express deep gratitude and appreciation to Annie Selden who oversaw the editing of several of the included manuscripts and to Shandy Hauk for her professional work and the special attention she gave to the present issue of *RCME* for which she served as Production Editor. Again, welcome and we hope you enjoy reading the content in this volume.

Fernando Hitt
Derek Holton
Patrick W. Thompson

References

Artigue, M. (1991). Épistémologie et didactique. *Recherches en Didactique des Mathématiques*, *10*(2/3), 241–285.

Bachelard, G. (1938). *La formation de l'esprit scientifique.* Paris: Librairie Philosophique J. Vrin.

Brousseau, G. (1997). *Theory of didactical situations in mathematics.* Dordrecht, The Netherlands: Kluwer.

Dillenbourg, P. (1999). What do you mean by collaborative learning? In P. Dillenbourg (Ed.), *Collaborative learning: Cognitive and computational approaches* (pp. 1–19). Amsterdam: Pergamon.

Dubinsky, E., Schoenfeld, A. H., & Kaput, J. (Eds.). (1994). *Research in collegiate mathematics education. I* . Providence, RI: American Mathematical Society.

Duval, R. (1993). Registres de représentation sémiotique et fonctionnement cognitif de la pensée. *Annales de Didactique et de Sciences Cognitives*, *5*, 37-65.

Eisenberg, T., & Dreyfus, T. (1991). On the reluctance to visualize in mathematics. In W. Zimmermann & S. Cunningham (Eds.), *Visualization in teaching and learning mathematics* (pp. 26–37). Washington, DC: Mathematical Association of America.

Hadamard, J. (1945/1975). *Essai sur la psychologie de l'invention dans le domaine mathématique.* Gauthier-Villars.

Hitt, F. (2007). Utilisation de calculatrices symboliques dans le cadre d'une méthode d'apprentissage collaboratif, de débat scientifique et d'auto-réflexion. In M. Baron, D. Guin, & L. Trouche (Eds.), *Environnements informatisés et ressources numériques pour l'apprentissage, conception et usages, regards croisés* (pp. 65–88). Paris: Éditorial Hermes.

Lakoff, G., & Núñez, R. (2000). *Where mathematics comes from.* New York: Basic Books.

Margolinas, C. (1989). *Le point de vue de la validation: Essai de synthèse et d'analyse en didactique des mathématiques.* Unpublished doctoral dissertation, Université Joseph Fourier, Grenoble I.

Margolinas, C. (1993). *De l'importance du vrai et du faux dans la classe de mathématiques.* Grenoble: La Pensée Sauvage.

Perkins, D., & Simmons, R. (1988). Patterns of misunderstanding: An integrative model for science, math, and programming. *Review of Educational Research*, *58*, 303–326.

Schoenfeld, A. H. (1985). *Mathematical problem solving.* Orlando, FL: Academic.

Schwarz, B. (1989). *The use of a microworld to improve the concept image of a function: The triple representation model curriculum.* Unpublished doctoral dissertation, Weizmann Institute of Science, Rehovot, Israel.

Star, J. R., & Smith III, J. P. (2006). An image of calculus reform: Students' experiences of Harvard Calculus. In A. Selden, F. Hitt, & G. Harel (Eds.), *Research in collegiate mathematics education. VI* (pp. 1–26). Providence, RI: American Mathematical Society.

Vergnaud, G. (1991). La théorie des champs conceptuels. *Recherches en Didactique des Mathématiques*, *10*(2/3), 133-169.

Vinner, S. (1989). The avoidance of visual considerations in calculus students. *Focus on Learning Problems in Mathematics*, *11*, 149-156.

CBMS Issues in Mathematics Education
Volume **16**, 2010

Representing and Defining Irrational Numbers: Exposing the Missing Link

Rina Zazkis and Natasa Sirotic

ABSTRACT. This article reports on a study of prospective secondary teachers' understanding of the irrationality of numbers. Specifically, we focused on how different representations influenced participants' responses with respect to irrationality. As a theoretical perspective we used the distinction between transparent and opaque representations, that is, representations that "show" some features of numbers and representations that "hide" some features. The results suggest that often participants did not rely on a given transparent representation (e.g., 53/83) in determining whether a number is rational or irrational. Further, the results indicate participants' tendencies to rely on a calculator, preference towards decimal over common fraction representation, and confusion between irrationality and infinite decimal representation regardless of the structure of this representation. As a general recommendation for teaching practice, we suggest a tighter focus on representations and conclusions that can be derived from considering them.

Definitions of irrational numbers provided at a school level are strongly linked to representations. This report gives insight into the extent to which representational features of numbers are attended to when rationality or irrationality is considered, and into the ways in which different representations relate to one another, as perceived by preservice teachers. This report is part of ongoing research on understanding of irrational numbers by prospective secondary school teachers. Specifically, we focus here on how irrational numbers can be (or cannot be) represented and how different representations influenced participants' responses with respect to irrationality.

On Representations and Irrational Numbers

There is an extensive body of research on representations in mathematics and their role in mathematical learning (see, for example, Cuoco, 2001 and Goldin and Janvier (1998)). The role of representations is recognized in manipulating mathematical objects, communicating ideas, and assisting in problem solving. Researchers draw strong connections between the representations students use and their understanding (Lamon, 2001). Janvier (1987) describes understanding as a "cumulative process mainly based upon the capacity of dealing with an 'ever-enriching' set of representations" (p. 67). Furthermore, representations are considered as a means in the formation of conceptual understanding. The ability to

©2010 American Mathematical Society

move smoothly between various representations of the same concept is seen as an indication of conceptual understanding and also as a goal for instruction (Lesh, Behr, & Post, 1987). Moreover, according to Kaput (1991), possessing an abstract mathematical concept "is better regarded as a notationally rich web of representations and applications" (p. 61). Only a small part of the work on representation addresses representation of numbers, and it focuses primarily on fractions and rational numbers (e.g., Lamon, 2001; Lesh et al., 1987)

In contrast, research on irrational numbers is rather slim. Fischbein, Jehiam, and Cohen (1994, 1995) offer research reports that treat the issue explicitly. The main objective of these studies was to examine the knowledge of irrational numbers of high school students and prospective teachers. Based on historical and psychological grounds, Fischbein and colleagues assumed that the concept of irrational number presented two major obstacles: incommensurability and nondenumerability. Contrary to expectations, their studies found that these intuitive difficulties did not manifest in participants' reactions. Instead, they found that participants at all levels were not able to define correctly rational and irrational numbers or place given numbers as belonging to either of these sets. It was concluded that the expected obstacles are not of a primitive nature – they imply a certain mathematical maturity that the participants in these studies did not possess.

In a study by Arcavi, Bruckheimer, and Ben-Zvi (1987), the authors focused on the development of instructional materials related to the history of mathematics in the design of courses for pre-service and in-service teachers. One of their stated objectives was to strengthen teachers' mathematical knowledge related to the concept of irrationals. In their formative assessment of teachers' previous knowledge, they found, consistent with the work of Fischbein et al., that many teachers had trouble recognizing numbers as being rational or irrational. They suggested that "one of the sources of confusion between rational and irrational numbers is the common use of rational approximation to an irrational as the irrational itself" (p. 19). Their example of such approximation was the use of 3.14 or 22/7 in the case of π.

A study by Peled and Hershkovitz (1999), which involved 70 prospective teachers in their second or third year of college mathematics, focused on the difficulties that prevent student teachers from integrating various knowledge pieces of the concept into a flexible whole. Contrary to the Fischbein et al. study, these researchers found that student teachers knew the definitions and characteristics of irrational numbers but failed in tasks that required a flexible use of different representations. They identified the misconceptions related to the limit process as the main source of difficulty and argued the need for creating tasks that facilitate integration of different pieces of knowledge.

The research findings of the above studies were based on written response to questionnaires where students were asked to: (a) solve a problem involving the application of irrational measure (Peled & Hershkovitz, 1999), (b) recall a definition of irrational number (Arcavi et al., 1987), and – common to all three of the above-mentioned studies, (c) classify numbers as rational or irrational (Arcavi et al., 1987; Fischbein et al., 1995; Peled & Hershkovitz, 1999). However, in literature to date there has been no comprehensive attempt to probe in-depth to investigate the thinking that guided students' responses. Previous findings call for examining the understanding of irrationality and attending to issues of concern that were identified. In school mathematics, definitions of rational and irrational numbers rely on number representations. Therefore, investigation of understanding of irrational numbers from the perspective of representations and ways in which these relate to one another is the focus of research reported herein.

Representations: Transparent and Opaque

As a theoretical perspective, we use the distinction between transparent and opaque representations introduced by Lesh et al. (1987). According to these researchers, a transparent representation has no more and no less meaning than the represented idea(s) or structure(s). An opaque representation emphasizes some aspects of the ideas or structures and de-emphasizes others. Borrowing this terminology in drawing the distinction between transparent and opaque representations, Zazkis and Gadowsky (2001) focused on representations of numbers and introduced the notion of relative transparency and opaqueness. Namely, they suggested that all representations of numbers are opaque in the sense that they always hide some of the features of a number, although they might reveal others, with respect to which they would be "transparent." For example, representing the number 784 as 28^2 emphasizes that it is a perfect square, but de-emphasizes that it is divisible by 98. Representing the same number as $13 \times 60 + 4$ makes it transparent that the remainder of 784 in division by 13 is 4, but de-emphasizes its property of being a perfect square. In general, we say that a representation is transparent with respect to a certain property, if the property can be "seen" or derived from considering the given representation.

Subsequently, Zazkis and Liljedahl (2004) used the notions of transparency and opaqueness in their investigation of prospective elementary school teachers' understanding of prime numbers. They mentioned that while the "evenness" or "oddness" of a number is transparent in its representation as $2k$ or $2k + 1$, respectively, for a whole number k, there is no transparent representation for a prime number. Following Dubinsky (1991), they claimed that the ability to manipulate or perform actions on concepts contributes to object construction. However, performing actions often relies on the availability of a representation. For example, in order to act on odd numbers, the odd numbers need to be first represented in a transparent way. Therefore, Zazkis and Liljedahl concluded, the nonexistence of a transparent representation was one of the causes for difficulties participants had in treating prime numbers as objects. In this report, we explore irrational numbers in terms of their transparent or opaque representations and discuss to what degree the use of certain representations assists or obscures students' understanding.

Irrational Numbers: Definitions and Representations

The definition of rational number relies on the existence of a certain representation: a rational number is a number that can be represented as a/b, where a is an integer and b is a nonzero integer. When a real number cannot be represented in this way, it is called irrational. Until exposure to a formal construction of irrational number using, for instance, Dedekind cuts, this distinguishing representational feature is used as a working definition of irrational number. That is to say, an irrational number is a number that cannot be represented as a quotient of integers. An alternative definition of irrational number refers to the infinite non-repeating decimal representation. Applying the notions of opaqueness and transparency to these definitions, we suggest that representation as a common fraction is a transparent representation of a rational number (that is, rationality is embedded in the representation), while infinite non-repeating decimal representation (such as 0.010011000111..., for instance) is a transparent representation of an irrational number (that is, irrationality can be derived from this representation). These two definitions provide an introductory description for the concept of irrationality as introduced at a high school level and are used interchangeably by teachers and students alike. However, are they equivalent? In fact, the reason that we can use the decimal characterization as a "definition" is that a decimal number represents a rational number if and only if it terminates or repeats. Ironically, this characterization has nothing to do with the original motivation for distinguishing between the rational and irrational numbers.

Any fraction whose denominator is not a factor of a power of ten – and only the products of powers of 2 and 5 are – cannot be written as a finite decimal. Consequently, a great number of fractions that students see in school mathematics are those that have infinite decimal expansions. Using the decimal representation it might seem that it would be difficult to distinguish such rational numbers from irrational numbers, except for one interesting thing about infinite decimals that come from fractions – they repeat. It is possible to see, by long division, that the decimal expansion of any fraction a/b with a and b integers, $b \neq 0$, necessarily repeats. For a more unified perspective, we can say that the "terminating decimals" are also infinite repeating, having bi-unique infinite expansions. For example, the terminating decimal 2.3 can be seen as an infinite repeating decimal either as 2.2999... or as 2.3000.... The possible remainders on dividing a by b can only be 0, 1, 2, ..., $(b-1)$, so with only $b-1$ possible choices for remainder other than zero, the calculations in the long division must eventually start repeating. The number of digits in the period will therefore be no greater than 1 less than the denominator. For example it could take up to 16 places for the decimal expansion of $1/17$ to repeat, but no more than that. This is because when dividing out $1/17$ there are only 16 possible remainders (excluding the remainder of zero).

The converse, that any repeating decimal is a fraction, is much more subtle and difficult to grasp. The general proof of this notion requires the summing of an infinite decreasing geometric progression. Although a formula for the sum of such progression is given in high school, the derivation of the result requires the use of the limit process (usually done in a first year university course). In high school, and sometimes as early as in Grade 7, a type of symbolic "juggling" where operations are conducted on infinite decimal expansions is presented to students to convince them that repeating decimal representations can be turned into common fractions. Here is a typical example:

> *Problem:* Convert 0.12121212... to a fraction.
> *Solution:* Let $x = 0.12121212\ldots$, then $100x = 12.121212\ldots$, so $100x - x = 12$. But $100x - x$ is also $99x$, so $99x = 12$. Dividing both sides by 99, we get $x = 12/99 = 4/33$.

Although resourceful, this juggling is a bit contrived and can possibly leave the student with an impression that there might exist another trick that will turn a non-repeating decimal into a fraction. Compounding this conflict are the fractions which can be "seen" to have no repeating pattern, such as $1/257$ or even $1/17$ when displayed on a calculator. Furthermore, there remains the danger of conflict between the theoretical requirement for infinite decimals and the practical experience that finite decimals are both convenient and sufficient.

Research Setting

Situating the Study. In this report we share part of a larger study on prospective secondary school teachers' (PSTs') understanding of irrational numbers. The purpose of this larger study was to provide an account of PSTs' knowledge related to irrational numbers within the system of real numbers, to interpret how their understanding of irrationality is acquired, to analyze the difficulties that occur, and to consider implications for teaching practice. Facets of this research have included investigation of PSTs' intuitions and beliefs related to the relationship between the sets of rational and irrational numbers with respect to size and operations (Sirotic & Zazkis, 2007b) and geometric interpretation of irrational numbers and locating them on a real number line (Sirotic & Zazkis, 2007a).

The purpose of the facet presented here was to examine PSTs' understanding related to different representations of irrationals. Specifically, we were interested in addressing the following questions: How does the availability of certain representations influence participants' decisions with respect to irrationality? Are prospective teachers aware of the two definitions for irrational numbers mentioned above? Are they able to apply them in a flexible manner? How do they perceive the connection between the two definitions of irrational numbers? What is the effect of this perception on their overall inferences related to irrational numbers? We further explored the ways in which the absence of the link that renders the two definitions equivalent – referred herein as "the missing link" – impeded the prospective teachers' overall understanding of irrationality.

The Tasks. For this investigation, we designed the following items:

Item 1: Consider the following number 0.12122122212... (there is an infinite number of digits where the number of 2s between the 1s keeps increasing by one). Is this a rational or irrational number? How do you know?

Item 2: Consider 53/83. Let's call this number M. In performing this division, the calculator display shows 0.63855421687. Is M a rational or an irrational number? Explain.

Note that the numbers in Item 2 were carefully chosen so that the repeating is "opaque" on a calculator display. The length of the period in this case is 41 digits. Using the theoretical framework presented above, we say that in Item 1 there is a transparent representation of an irrational number (infinite non-repeating decimal), whereas in Item 2 there is a transparent representation of a rational number (ratio of two integers).

Participants and Data Collection. As the first step in our data collection, the two items indicated above were presented to a group of 46 prospective secondary school mathematics teachers as part of a written questionnaire. These participants were mathematics and science majors in their final course of the teacher education program. Their mathematical backgrounds varied but all had at least two calculus courses in their background. The time for completing the questionnaire was not limited.

TABLE 1. Quantification of results for Items 1 and 2 ($n = 46$)

Response category	Item 1 0.121221222...	Item 2 53/83
Correct answer with correct justification	27 (58%)	31 (67%)
Correct answer with incorrect justification	7 (15%)	7 (15%)
Correct answer with no justification	1 (2%)	2 (4%)
Incorrect answer	6 (13%)	5 (11%)
No answer	5 (11%)	1 (2%)

As the next step, upon completion of the questionnaire, 16 volunteers from the group participated in a clinical semi-structured interview. These participants represented a wide range of responses to the written questionnaire items, including incorrect decisions, correct answers followed by incorrect justifications, and correct answers followed by correct justifications. The items of the questionnaire served as a starting point in the clinical interview, where the participants had an opportunity to clarify and extend upon their responses and arguments and at times change their judgment or their justification. Based on initial responses of participants, the interviewer presented additional probing questions to identify the "strength of belief" (Ginsburg, 1997), that is, to identify whether a participant's response to a presented task was an arbitrary choice or whether it was based on persistent and robust knowledge or belief. Participants' responses were analyzed with specific attention to the role that representation of a number played in their decision, and their reliance or non-reliance on the given representation.

Results and Analysis

Using Items 1 and 2 we analyzed participants' reliance on the representational features of numbers affecting their decisions with respect to irrationality. We first present a quantitative summary of written responses for the two items. We then focus on the details of several interviews, identifying some common trends in participants' approaches to the presented questions. We also discuss some common erroneous beliefs found amongst the participants and we attempt to identify their sources.

As shown in Table 1, over 40% of the participants did not recognize the non-repeating decimal representation as a representation of an irrational number. Further, over 30% of the participants either failed to recognize a number represented as a common fraction as being rational or provided incorrect justifications for their claim. It is evident that for a significant number of participants, the definitions of rational and irrational numbers were not a part of their active repertoire of knowledge.

Introductory Excerpt. We start with presenting an excerpt from the interview with one participant, Steve, as his responses shed light on the possible sources of errors and misconceptions and foreshadow the forthcoming data analysis (italics added).

Steve: (claiming 0.121221222... is irrational) Um hm, I would say because it's not a common, there's not a *common element* repeating there that it would make it a rational...

Interviewer: How about this one, 0.0122222... with 2 repeating endlessly, is this rational or irrational?

Steve: Okay, I would have to say that's irrational real number.

Interviewer: Irrational or rational, I couldn't hear you.

Steve: Irrational. Well oh, the 2 repeats, no but it has to be, then it repeats, even though the 2 repeats, it has to be a *common pattern*, so I would say it's irrational.

Interviewer: Okay, so 0.01222... repeating infinitely is irrational.

Steve: I think so, but I forget if the fact that that, if the 1 there changes, I would have thought it would have to be *012, 012,...* but if it starts repeating later, yeah I can't remember if it starts repeating later, I'm pretty sure it's irrational, but I could be mistaken.

Interviewer: How about the second question, when you consider 53 divided by 83...

Steve: Um hm...

Interviewer: And let's call this quotient M, if you perform this division on the calculator the display shows this number, 0.63855421687.

Steve: And I assume it keeps going, that's just what fits on your calculator...

Interviewer: Yeah, that's what the calculator shows, that's right. So is M rational or irrational?

Steve: So this is the *quotient* M, yeah I would say it's *irrational.*

Interviewer: Because?

Steve: Because *we can't see a repeating decimal.*

Interviewer: But maybe later, down the road it starts repeating.

Steve: Well that's true, it's possible...

Interviewer: So we can't really determine?

Steve: Well I guess we don't, we *wouldn't know* for sure just *from looking at that number on the calculator, but chances are that if it hasn't repeated that quickly, then it would be irrational.* I haven't seen a lot of examples where they start repeating with 10 digits or more. I'm sure there are some but....

Interviewer: Okay, and the fact that it comes from dividing 53 by 83, does that not qualify it as rational?

Steve: Oh so that is a fraction, it's 53/83?

Interviewer: Yeah we, that's how we got this number, so we divided 53 by 83 and called this M.

Steve: *53/83 as it's written would be rational,* but yeah, I see what you mean, *if you took that decimal, yeah, I guess that's a good point. I see what you're, you're saying that fact that it's 53/83 that is A/B, so that is rational, but then when you take, if you started dividing... It would just go on and on and on and on, so that you would think it is irrational.* Yeah, I must say I don't know the answer to that.

In the beginning of the interview Steve claims correctly that an infinite non-repeating decimal fraction represents an irrational number. However, his use of the words "common element" prompts an inquiry into his perception of "common." This perception is clarified in Steve's incorrect claim that 0.0122222... is also irrational. Steve is looking for a common pattern, and the repeating digit of 2 does not seem to fit his perception of a pattern. For the next question Steve is presented with a fraction 53/83 and distracted by its display on a calculator. Focusing on the decimal representation rather than the common fraction representation, his first response – this quotient is irrational – presents an oxymoron. It is based on the inability to "see" the repeating pattern. The underlying assumption here is that a repeating pattern, if it exists, has a short and easily detectable repeating cycle. This perception is confronted by the interviewer in directing Steve's attention to the number representation as a fraction, 53/83. From his reply it appears that Steve believes that whether the number is rational or irrational depends on how it is written; that is, a common fraction represents a rational number, but its equivalent decimal representation could be irrational. Perceptions and beliefs demonstrated by Steve were echoed in responses of many other participants and are discussed in further detail in what follows.

Identifying Dispositions. In analyzing the interviews it was immediately observed that some participants referred to irrational numbers using either their decimal or fractional representation, while others persistently relied on one particular representation or definition. We identified these tendencies as either balanced, fractional, or decimal disposition. Among the 16 interviewees, 7 exhibited the balanced disposition, that is, relied on fractional or decimal disposition interchangeably, 2 exhibited the fractional disposition and 7 participants exhibited the decimal disposition. Though it is impossible to determine a participant's disposition accurately based on the written responses alone, the written arguments suggested that the decimal disposition prevailed. In what follows we exemplify these dispositions and illustrate in what way they support or impede PSTs' understanding of irrationality.

Decimal Disposition

In the arguments provided by participants to justify their decisions, we found that decimal representation was the preferred representation in deciding the rationality or irrationality of a number for seven interviewees. We refer to this preference as decimal disposition. As in the case of Steve, there was a tendency to ignore the transparency of rationality inherent in the fractional representation, and instead rely on the truncated decimal representation offered by the calculator display. Noting that reliance on decimal approximation was consistent with findings of Arcavi et al. (1987), we further investigated what might be the reasons for this state of affairs. We suggest that a possible reason lies in participants' purely operational approach to rationals. That is to say, the division of integers was still only a process for them, so 53/83 was not perceived as one number, as an object. We learned from one of the interviewees, Amy, why it was so unnatural for her to consider the representation as a common fraction in deciding irrationality or rationality of a number. After claiming 0.121221222... is irrational on the first item, and after the interviewer directs her attention to the number representation as a fraction on the second item, she says:

> *Amy:* Well okay, what I see from, thinking about it that way, because *I don't see this as being a number*, like if you're going to decide about something. Like this isn't a number, *53/83 isn't a number*, but 0.638.., and does this continue? Is that ..., or is that?

Amy cannot decide simply by looking at 53/83; she thinks she must divide it out first. Amy interprets a/b as an instruction for division. Unlike Steve, however, Amy is not convinced that periods should always be short. She considers the conflict that arises from judging rationality of a number based on the digits displayed on the calculator. She recognizes that it may lead her to a wrong conclusion, because "the decimals could start repeating."

> *Amy:* It seems like a contradiction (laugh). But how do you know, like how far, like maybe the pattern happens and it's harder to see. Like maybe there's something that's, maybe this does repeat at some point, I don't know. Yeah, I don't know... I mean you could always long divide it out with long division, and keep going and going and going and look for a pattern, um, because you get more numbers, or with a computer that you can set up to look at more decimal places and look for a pattern.

It is most likely that Amy has not had much experience with division in her life, because if she had, she could have known that the decimals must start repeating. We designate this as the case of the "missing link," and discuss it further in the section on definitions.

Students with persistent decimal dispositions have exhibited several misconceptions or confusions related to the features entailed in the decimal representations that can be used to determine rationality or irrationality of a number. We discuss below confusions related to different kinds of decimal representations – immediately repeating versus eventually repeating, repeating versus patterned non-repeating, infinite versus terminating. Moreover, we focus on the interpretation of the two words "pattern" and "irrational" that played a role in participants' inferences with respect to irrational numbers.

Infinite Decimal Representation and Patterns. If one is to determine rationality or irrationality of a number from its decimal representation, one must have a reliable way to do this. The way we speak about things influences the way we think about things. In this section we explore how participants' interpretation of the word "pattern" influenced or guided their responses.

The word "pattern" with respect to decimal expansion of a number is not used as a guiding principle in discerning rationality or irrationality in any of the standard mathematics textbooks; however, we find it to be a common reference in discussion of irrational numbers. Nearly all the participants who exhibited decimal disposition, such as Steve and Amy, used this word, either in the stricter sense of "repeating pattern" or in the looser sense without the crucial repeating part. Regardless of the sense in which they used it, we find this language to be vague and prone to personal interpretation. We thus consider it a verbal obstacle.

According to Brousseau (1997), an obstacle has its domain of validity but in another domain it is false, ineffective, and a source of errors – which makes it so resistant. All participants who relied on "pattern" successfully identified as rational the immediately repeating decimals (such as 0.121212...) but not necessarily the eventually repeating decimals (such as 0.012222...). The following excerpt demonstrates this.

> *Interviewer:* How about this one, 0.0122222 with 2 repeating endlessly, is this number rational or irrational?

Amy: Like why is it irrational, why did I say that?

Interviewer: Yeah...

Amy: Um, because there isn't um, *it repeats without a pattern. I think that the pattern should have to start right after the decimal point, not anywhere in the middle...*

Interviewer: Okay...

Amy: Because .01 wouldn't come up again, just the 2, if that's written the way, like the way that I would see it is like the low line over the 2, to say only that the 2 repeats.

Interviewer: Okay, so if you have 0.012 with a little line indicating that only 2 repeats endlessly then you would slot it as irrational?

Amy: Right...

Interviewer: Because the repeating doesn't have the starting right from the decimal point...

Amy: Yeah, yeah, I think the *whole decimal has to be part of the pattern for it to be rational.*

This is the same misconception as presented earlier; however, in contrast with Steve, Amy is very sure of her rule while Steve admits he could be mistaken. The point is, why would one ever want to memorize such strange and insignificant rules, to think that if the decimal starts repeating right away then it is a rational number, while if it starts repeating later it is irrational? This has nothing to do with rationality or irrationality of numbers. Cases such as these speak of why it is, for so many, that mathematics is nothing but a collection of arbitrary rules to be memorized. We have demonstrated how some PSTs applied their notion of "pattern" to conclude that the eventually repeating decimals are irrational. Interestingly, this verbal obstacle works the other way as well. It allows people to claim that irrational numbers are rational. More specifically, the transcendental irrationals with patterned decimal expansions, such as the one in Item 1, were identified by some PSTs as rational numbers. The interview with Matthew exemplifies this.

Interviewer: So can you please tell me how you recognize an irrational number?

Matthew: It's a number that um *doesn't have any sort of pattern*, or doesn't, or it doesn't terminate ever, or as far as you can see. And it *doesn't have any sort of pattern...*

Interviewer: So number like 0.12 122 1222, where the number of 2s between every successive 1s keeps increasing by 1, would that be a rational or irrational number?

Matthew: Rational...

Interviewer: So to you, what is the definition of a rational number?

Matthew: A number that either does terminate, or doesn't terminate and *has a definite pattern.*

Interviewer: Well, so you consider this to be a pattern...

Matthew: Yes, it's definitely a pattern... So it is rational.

We witnessed the omission of the requirement for *repeating* a number of times in our interviews, despite the fact that such patterned transcendental numbers are often given in textbooks to exemplify that it is the periodicity, and not the pattern, that is the distinguishing characteristic of rational number. The unfortunate misinterpretation of repeating pattern as simply "pattern," and the reliance on this interpretation, has a limited domain of validity and, as we have demonstrated, serves only to complicate matters.

Irrationals as "Unreasonable" Numbers. In the following excerpt Ed also uses the requirement for "pattern" as a distinguishing feature between rational and irrational numbers. Moreover, he is influenced by the colloquial meaning of the words "rational" and "irrational" and attempts to link this meaning to mathematics and to the existence of a pattern.

Interviewer: So tell me what does it mean to you for a number to be irrational?

Ed: Well I mean, it's, *the word irrational obviously means it can do something that we can't understand*, right, it's kind of like you have to shake your head and you can't figure it out right...

Interviewer: The word? And as a number?

Ed: As a, yeah, so it's the same kind of thing right, so you think of it as like *a number that you can't really explain*, it's just sort of, you know, it goes on and on and *there's no system to it*...

Interviewer: How do you identify them though? How do you distinguish them from rational numbers?

Ed: Um, well my understanding is that like *with a rational number you always have some kind of a pattern*, [...] because we know the rule that the number follows, no matter how long it is, whatever, we know what they're going to be, all the digits, whereas irrational number it's like you can't really, you can't know its digits. You have to actually sit down like with pi right, *you don't know ahead of time* like, okay what's the 1,000th digit, I can't just quickly tell you right. Of course the computer will tell you right, so it's a complicated process to find out what that would be, there's no like systematic way of finding the answer right.

Interviewer: Okay, and you can tell easily what the 1,000th digit will be for any rational number, no problem?

Ed: That's my feeling yeah.

Interviewer: How exactly do you tell for a rational?

Ed: Uh, well like I'm saying, my understanding is that *there is a system* so I guess um like if the number, for example, goes 1, 2 and it goes, like because of the decimal places right, 0.121221222..., right, there's one 2 and there's two 2s and there's three 2s and four 2s and five 2s um I mean I'm assuming there's a quick way to turn that into sort of a formula or whatever where you can actually establish what that 1,000th digit, whether it would be a 2 or would it be a 1 right,...

Interviewer: So is this number that you just mentioned a rational or irrational?

Ed: I think it would be rational...

Interviewer: Rational...

Ed: Because *there seems to be a system to it* like, like um, you know, you could, I mean it is interminable, like it goes on forever, but at the same time like *you can easily explain how it's working* right.

Like that's my feeling, so in an irrational number it's impossible to pin it down like that, I mean *you can't really explain the number itself*, why are these digits the way they are (laugh).

Interviewer: Okay, so what would you say for 53 divided by 83? Is this number rational or irrational?

Ed: Yeah, I think it's irrational...

Interviewer: You think it's irrational...

Ed: Just by looking, I mean *I have no idea what the result would be just by looking at the two, 53 and 83 right*, by looking at the answer right, probably doing it on your calculator, I mean, it just looks like it would be an irrational number, right. I mean that's, like the thing is, you know, my understanding of math isn't huge, so I just look at a number and like "that looks irrational to me" right, then I was thinking at the time like if I could, if I really want to *I could turn on my computer* and put it in there, one of those it just displays all the numbers, it doesn't limit you to 10 digits.

Interviewer: So you're aware that you're limited by 10 digits here...

Ed: Yeah, yeah... Yeah, I *mean I just have to assume one or the other and it looks like a disorganized number so (laugh), it looks an irrational to me.*

As evident, Ed thinks rational numbers are organized, they have a system to them, they are reasonable. His reference to the ability to predict digits and explain the "system" of their appearance is a clear indication of a decimal disposition. Further in the interview he describes rational numbers as "nice" and "having certain kind of appealing nature." On the other hand, irrational numbers are something we cannot explain, they are disorganized and unpredictable.

Ed's conception of irrational number hinges on the lay meaning of the word irrational as "unreasonable" to which he adds a somewhat more mathematical connotation as "disorganized." From the perspective of "rational" as "reasonable" instead of "ratio," one is doomed to rely on a calculator to judge the reasonableness of digits. That is, the terms "rational" and "irrational" themselves present a verbal obstacle as their colloquial meaning may intertwine with and overshadow the mathematical one.

Infinite Decimals versus Terminating Decimals. Although we were aware that there exists a belief among students that terminating decimals are rational and infinite decimals are irrational, we did not expect to find this belief among our group of participants, given their educational background. One prospective teacher, Katie, who had been out of school for a while and decided to take her teaching degree, held this belief.

Interviewer: Okay, and how do you distinguish rational numbers from irrational numbers?

Katie: Um, (pause) Irrational numbers are something which you don't know the exact value. If you divide something and then it gives $0.3333\ldots$, and *you don't know what the exact number is*, that would be irrational, but .5 would be rational.

Interviewer: Because?

Katie: It terminates and then it's a, then I can see if that's the real number and otherwise if it's let's say 0.3333... and *it just repeats and then there's no end, then I can't really tell what the actual number is.*

Interviewer: So if it repeats and there's no end to it, you would say that it is irrational...

Katie: Um hm...

Interviewer: Okay. Um, now here on question 2 if you remember, when you divide 53 by 83 and we said let's call this quotient M, but when we perform this division on the calculator, this is what the display shows, so the question is, is M rational or irrational? What would you say?

Katie: It ends there right? So the number ends there, and when I multiplied, I don't know what I was thinking, but I thought that when I multiplied this number with 83 I do get 53, so I think that's why I put down it's a rational number.

Interviewer: Oh, okay I see, because it's a decimal that terminates...

Katie: This is what I was thinking, *it's infinite numbers and it's finite numbers...*

Interviewer: So am I right if I say that you distinguish rationals from irrationals just by looking at whether it is...

Katie: Whether they terminate or not, that's what I was looking at really. If, let's say *if it's a ratio, and I try to express it in decimal form and if it doesn't terminate, then I would say it's an irrational and if it terminates, then I would say it's rational.*

From her response we see that Katie is troubled by the infinite decimals, she feels the number is somehow constructed in time and she cannot know what the exact number is since there is no end to its digits. From her perspective, the number is still in the making; therefore, impossible to be conceived of as an object.

Despite her erroneous belief, Katie answered both Items 1 and 2 correctly, and from the justifications on the written response it would be hard to tell that she maintains this view. For example, for Item 1 she wrote that 0.12122122212... is irrational because "it cannot be expressed as a ratio and has infinite number of entries after the decimal point." The way she used this language was not clear until the probing of her thinking occurred during the clinical interview where we found out that whether a number can be expressed as a ratio of two integers is in fact irrelevant for Katie.

On Item 2, her thinking was that M and the number on the calculator display were identical, and since the number on the display was a terminating decimal, M would have to be rational. In the written response, she justified this by writing "$M \times 83 = 53$." We found out during the interview that she was in fact aided by the calculator in confirming her thinking that $M =$ "calculator display."

Katie: (Explaining her decision that 53/83 is rational) Okay, *there's another way I was testing whether I should get irrational or not...*

Interviewer: How is that?

Katie: Uh huh, so because then I was looking at this one *and I was trying to test whether it would be rational or not. I was trying to cross multiply M with 83 and see if I can come up with 53...*

Interviewer: And you did?

Katie: And I did, so I figured this is rational because, this is another thing that I remembered that *if you could multiply and then come up with the numerator then it's rational.*

In fact, Katie was using the calculator to test out whether the digits of the resulting quotient continue beyond what is seen on the display or whether they terminate. When she used the finite decimal and multiplied it by the denominator, and got the numerator – this confirmed that the division of 53 by 83 yields a finite decimal, thus M was concluded to be rational. As we can see, Katie had a good reason, not just a plain guess, to believe that what the calculator showed was indeed equal to M. However, had she entered the digits of M directly on a calculator, rather than having arrived at them as a result of division, the result could have been different. (In fact, the result will depend on the "round-off" ability of a calculator. In a simple calculator, multiplying 0.6385542 – which is the displayed result of division of 53 by 83 – by 83 will result in 52.999999).

From the perspective of a decimal disposition, it seems reasonable to distinguish between infinite and finite decimals. It is certainly the most prominent feature of a number when looking at its decimal form. But to bundle the infinite repeating decimals with terminating decimals and to call that rational, and to call all the other infinite decimals that are left (the non-repeating) irrational, does not make much sense for many students.

In a three-way distinction – terminating, infinite-repeating, and infinite non-repeating – Katie simply associated with rationality the first one, rather than the first two. However, as shown below, this view may have been adopted by her due to the confusion between π and one of its approximations.

π as a Special Case. In our inquiry into how Katie acquired her understanding of infinite and finite decimal expansions to correspond to irrational and rational numbers respectively, we found that it was the result of an adaptation of her concept image to fit the evidence concerning the number π. Namely, Katie believes that π is irrational. Of course she would, everyone knows that – this fact does not escape anyone who was schooled. The problem is, Katie *also* believes that π *equals* 22/7. Necessarily, these two "pieces of knowledge" fuse together to bring about the kind of understanding that Katie holds.

Interviewer: With pi, your teacher said it's...

Katie: Irrational, I had heard that. Like I was trying to, when talking real and rational, irrationals, I had to close my eyes and reflect back so many years, and one of the numbers that came from my memory bank as irrational was pi and I based the whole questionnaire on that. I was trying to answer based on that limited knowledge, because I didn't have the depth of knowledge. I had two strings that I was hanging on to, one was... Pi and yes pi is there...

Interviewer: So you remembered that pi is irrational...

Katie: *Yeah, and pi is normally expressed as 22/7 and I knew pi is irrational, and then I thought when I divide 22/7 the number repeats and never ends, so I thought a number that never ends is irrational.*

The excerpt with Katie exemplifies one of the findings of the Fischbein's et al. (1995) study mentioned earlier – namely, that one of the sources of confusion between rational and irrational numbers are the commonly used rational approximations of irrational numbers. We note further that Katie's distinction between rational and irrational numbers is shaped by the need to ensure consistency in her cognitive schema; unfortunately, this consistency is achieved by coordinating faulty information.

Summary of Responses Stemming from the Decimal Disposition. As mentioned above, seven participants demonstrated decimal dispositions in the interview; however, arguments based on decimal representations were most common in the written responses of this group of PSTs. In what follows we demonstrate several frequent erroneous beliefs expressed by the participants in their written questionnaires, some of which have been exemplified in the excerpts from various interviews. Given that several of the PSTs we interviewed operated from a decimal disposition, we see that conflicts arise from applying incorrect or incomplete characterization of decimal expansions in deciding the irrationality of numbers. However, the source of the conflict is poor understanding of the relationship between fractions and their decimal representations. It seems that the present day didactical choice is that somehow we can short-circuit the need for understanding this relationship and give the student a substitute for understanding instead — a recipe for how to look at the decimal expansion of a number to decide whether it is rational ("if the digits terminate or have a repeating pattern") or irrational ("if the digits are non-repeating and non-terminating"). The following examples summarize the major themes that were brought to light with respect to incorrect usage and over-reliance on this recipe.

(1) If there is a pattern, then the number is rational. Therefore, the number 0.12122122212... is rational, (similarly, 0.100200300... is rational, but 0.745555... is not, because there is no pattern).
(2) 53/83 is irrational because there is no pattern in the decimal 0.63855421687.
(3) 53/83 is rational because it terminates (the calculator shows 0.63855421687).
(4) 53/83 could be rational or irrational – I cannot tell whether digits will repeat because too few digits are shown. They might repeat or they might not.
(5) There is no way of telling if 53/83 is rational – unless you actually do the division, which could take forever. Digits might terminate after a million places or they might start repeating after a million places.
(6) It is possible that a number is rational and irrational at the same time. For example, there are fractions that have non-repeating non-terminating decimals, yet they can be represented as a/b.

The first illustration above echoes Amy's and Matthew's reliance on a personal interpretation of "pattern" which ignores the requirement for the repetition of digits. The other responses demonstrate participants' dependence on a calculator and a preference towards decimal representation, which is misinterpreted as either terminating or having no repeating pattern, or treated as ambiguous. These themes also emerged in the interviews with Ed and Katie. The last response involves a contradiction in terms ("irrational fractions"), resulting from a warped understanding of the two competing definitions of irrational number used in school mathematics.

Fractional Disposition

We now consider the reactions of those participants who had a primarily fractional disposition. We consider a person exhibiting fractional disposition as someone who primarily bases understanding of rational number on the idea of a number that can be expressed as a ratio of two integers with a nonzero denominator, and an irrational number as a number that cannot be expressed in this way. Fractionally disposed participants had an advantage in that they automatically concluded rationality of common fractions; it was natural for them to attend to the transparent feature of a number represented in this form. However, some fractionally disposed participants had difficulty concluding rationality of infinite repeating decimals, beyond simple cases such as 0.33333... The interview with Paul demonstrates this.

Interviewer: (After Paul claims that 0.12122122212... is irrational because its digits go on forever) Yeah, okay. So what about if the question was about 0.12121212..., like this infinitely, is that also, is that a rational or not a rational number?

Paul: I would say it's (pause) not a rational number because it... Rational number can be written one over the other... as the ratio of two whole numbers...

Interviewer: That's right. So you're saying that 0.121212 with one-two repeating cannot be written as a ratio of two whole numbers?

Paul: Hmm, the way I see it is like this, you'd have to go... (pause) you have, I can't imagine this number, *I can't imagine 0.121212... being multiplied by any number to get a whole number...*

Interviewer: I see, that's the way you think about it, okay. Um, hmm, interesting point. But how about 0.3333 repeating, is that rational?

Paul: 0.333...

Interviewer: Yeah...

Paul: (pause) This, this number *0.333... it comes up like a lot, when, when you just do your math... over the years, and I can see it's, when multiplied by 3, you get 1...*

Among the participants that we interviewed we found only two who were fractionally disposed. Although it may seem that this view would be a more desirable one than the decimal disposition, we find that simply having this disposition does not immunize one to erroneous beliefs.

Rational Number as a Ratio: Ignoring the Requirement for Integers. When asked about how she distinguishes between rational and irrational numbers, this is what Anna offers:

Anna: Well like I've heard before that, like the way that I was taught in school is how that *rational numbers can be written as a fraction, and when they're put into a decimal, they either, they repeat in some way, or else they terminate and so irrational are all the non-repeating, continuous decimals.*

On the surface we would judge this as a balanced disposition; that is, it would seem that Anna would be able to coordinate the two representations as needed. Also, she is successful on both Items 1 and 2. Only later in the interview, after Anna repeatedly makes statements such as: "any number divided by any other number is how, is like where we find our rational number," or "an irrational number is never uh the result of an operation that we do, like if the operation is dividing, the result wouldn't be, an irrational" that we learn she is missing the essential part of the definition, namely the requirement that the two numbers must be integers. She bases all her reasoning on the "nonexistence of a representation as a ratio" as the defining characteristic of an irrational; therefore, we would consider this to be a case of fractional disposition. In the excerpt that follows, the interviewer probes further Anna's conception of ratio and her idea that division is the identifying feature of a rational number.

Anna: So what I was saying , like dividing 12 by 75 would never produce, any division, if the *operation is operation of division, that will never produce an irrational number...*

Interviewer: Okay...

[...]

Anna: Because *if you divide something by something else, that means you can put it in a fraction, because of what a fraction is, something divided by something...*

[...]

Interviewer: So, for example, taking a square root of 5 and dividing it by a square root of 2 would that be in your opinion a rational or an irrational?

Anna: Oh (laugh) I think that would be rational, yeah...

Interviewer: By your argument...

Anna: That the, by the fraction argument, yeah. I just think *the fraction argument extends to be able to say that, because a fraction argument produces a quotient right, ...*

Interviewer: Um hm...

Anna: *So the quotient came from a fraction, so the quotient can never be irrational...*

Compounding the problem of Anna's incomplete definition is that fraction is seen as an instruction for division, the result of which is the quotient that "can never be irrational." We see that Anna's concept of rational number is still very much tied to its operational origins. Based on her own conception, Anna is driven to conclude that π is rational, when reminded that it is defined as a ratio of circumference and diameter. This realization and the dissonance it creates "destroys her whole theory," as she acknowledges herself. It is interesting that Anna had not run into these inconsistencies before, because if she had, she would have been forced to adapt her concept image and refine her definition so that it would not admit numbers such as π into the set of rationals.

In the interview with Anna we learned that one has to be aware that what we have access to are only symbols and representations (verbal, mostly), and it is hard to know how close or distant the conceptions might be from their formal counterparts. At first all seemed to be in order, with the exception of somewhat loose wording when casting her definitions. However, we did not question that in much detail at the onset of the interview, given that she correctly identified the numbers in Items 1 and 2. However, as we scratched a little through these symbols and verbal representations, we got to the conceptions that were hidden behind. Anna's case demonstrates that conceptions stemming from the incomplete and unquestioned definition can survive for a very long time despite the many contradictions that this may cause.

Balanced Disposition

A balanced disposition assumes a flexible use of both fractional and decimal representation of a number. We consider it the most desirable one. Seven of the 16 participants who were interviewed demonstrated this disposition. We are interested in how a balanced disposition is acquired, and how it manifests. To relate to the reader what we consider a balanced disposition, we present an excerpt from the interview with Dave.

Interviewer: How do you distinguish rationals from irrationals?

Dave: Just the, you know, it's rational if you can *write it in the form* A/B where A and B are integers, and B is not 0, (laugh) but I didn't start using that till like, I saw it in the number theory course up here...

Interviewer: And when you see a number that is in a decimal form, do you have a way of telling?

Dave: Um, in the number theory course we talk about how you can predict like, um, when you see a bunch of decimals, there *should be a section that repeats* and then you can calculate how long the section would be and then, it boils down if you see a pattern and it's probably like a pattern that repeats, not the one that keeps getting bigger, then it's rational.

Interviewer: Okay. Could you tell me what do you think is the connection between the repeating and the being able to express it as A/B, what's the connection here?

Dave: Oh, like... If you see something repeating and if it's a rational number... What would the connection be, oooh (laugh), good question, um...

Interviewer: Are these two equivalent? Because people use them, yeah people use both of these as definitions of a rational number, so I was just wondering, are they two separate definitions, or is there a connection between them...

Dave: Well I think *there probably would be a connection*, but I'm not sure how to describe it, but you, if there's something that looks like it's there, it probably is (laugh). I think, like if you say yeah, *if there is a pattern, there is a very good chance that it would be rational...*

Interviewer: Very good chance...

Dave: Yeah, (laugh)...

Interviewer: But sometimes it might happen that it's not...

Dave: Yeah, like maybe you just didn't look far enough right, and actually it does repeat or something...

Interviewer: Oh okay, so if it repeats, you are sure that every time it's going to be a rational number...

Dave: Pretty sure (laugh), *I wouldn't say all the time*, I'm not that confident in it, but... yeah...

Although Dave is able to attend to the transparent properties of a number to conclude its rationality or irrationality, regardless of how it is represented, he is unsure of what it is that makes the two definitions equivalent. It appears that the number theory course that Dave refers to shaped his understanding of irrational numbers. However, number theory is often an elective course for mathematics majors. The majority of the participants did not have a similar exposure. In fact, Dave was the only interviewee who mentioned a number theory course in his responses. This course aided in Dave's conviction that "almost always" a repeating decimal would be a rational number. In our analysis of moving between the two representations, we would consider this to be a case of a "leap of faith." That is to say, knowing that a decimal number represents a fraction if and only if that decimal number terminates or repeats is based on some authority and not on one's own understanding. We found that all of the 7 interviewees who held a balanced disposition held it more or less on the basis of a "leap of faith." We wish to point out that "balanced" does not necessarily imply the understanding of the connection between fractions and decimals; rather, we use the term to describe the situation where both views are employed. In fact, we did not find a single PST who would be able to explain the connection between repeating decimals and a ratio of two integers. We analyzed why this is so and we present our findings in greater detail in the following section.

Definitions and Coordination of Definitions

As demonstrated in the previous section, transparent features of the given representations were often either not recognized or not attended to. We suggest that the main reason for this is that the equivalence of the two definitions of irrational numbers given in school mathematics – the nonexistence of representation as a/b, where a is an integer and b is a nonzero integer and the infinite non-repeating decimal representation – is not recognized. We consider this as a missing link that is rooted in understanding of rational numbers, that is, the understanding of how and when the division of whole numbers gives rise to repeating decimals, and conversely, that every repeating decimal can be represented as a ratio of two integers. With this in mind, we examine the role of the two competing definitions in conceptualizing irrationality.

We initially assumed that the 7 participants who demonstrated a flexible use of both definitions did so because they understood their equivalency. This assumption was probed during the clinical interviews (on the basis of written responses alone this is impossible to judge). Our assumption was challenged time and again during the interviews. We were forced to conclude, on the basis of evidence, that the equivalence of the two definitions is not common knowledge even among those prospective teachers who displayed a balanced disposition and responded correctly to both items. Rather, we observed that a balanced disposition demonstrated a kind of opportunistic thinking which is not concerned that both definitions yield a consistent result: As long as the number fits one or the other definition, that definition can be applied to conclude rationality or irrationality.

One Direction in the Missing Link: From Fraction to Repeating Decimal. There are three reasons for the difficulty in reaching a general conclusion on why any ratio of two integers results in a repeating decimal number (taken broadly). The extension of the common experience of division of two whole numbers is needed. First, part of the problem, we believe, lies in the lack of algorithmic experience with division. Most likely this is because calculators are commonly used – before this insight is given a chance to develop. Second, fractions that have long periods further complicate the issue, as not many people have ever taken the time to reach the full expansion of the entire period of such fractions. Third, the separation into terminating and repeating decimals further compounds the issue. While finite decimals are indeed the first kind met, it would be beneficial later on to support the acquiring of a perspective that terminating decimals can be seen as repeating decimals, as we indicated earlier. Consequently, there would be no need to consider terminating decimals as a separate case, which intuitively people are likely to do. Instead, we could point out that terminating decimals are just a special case of repeating decimals. Furthermore, most participants could not even tell how one would know whether a decimal that came from a fraction repeated or terminated. Neither could they tell why it should repeat at all. Next, we consider the response of Kathryn, who correctly identified 53/83 as rational considering the "ratio" in 53/83, but was probed on whether this number could have a non-repeating non-terminating decimal expansion.

> *Kathryn:* It might not be repeating, but we can't tell from, from this... *we could never tell because we can never see every single digit in the number and see if it repeated some millions of digits down the line.*

In fact, we did not find a single participant who could explain this direction of the missing link, that is, that every division of two whole numbers must yield a repeating decimal. However, it should be noted that we found two participants who firmly believed that this must be the case. Not because they came to understand the connection, but because they trusted their correct recall of both definitions and they believed that in mathematics there is no place for inconsistencies of this sort to happen. In these two cases we would say that the equivalence of the two definitions was reached not by genuine understanding of the underlying concepts, but by a "leap of faith." Stephanie, who we found to have the most complete understanding of irrational number of all the participants, exemplified this kind of rendering of the two competing definitions as equivalent via a leap of faith.

> *Stephanie:* (Considering why 53/83 is rational and what does that have to do with repeating or terminating decimals) Because it's a ratio of integers. There is a pattern, it might just be down here, or it might be a repeating of 50 digits and you just don't have the whole repeating thing...
>
> *Interviewer:* Oh so we just have a partial view because of the calculator display?
>
> *Stephanie:* Yes.
>
> *Interviewer:* But are you sure this is going to repeat?
>
> *Stephanie:* Yes I am.
>
> *Interviewer:* What makes you say that?

Stephanie: I don't know, I don't know, it's, anyway I'm pretty sure it will repeat, I'm not, I don't think...
Interviewer: But there's no chance it will just keep going randomly?
Stephanie: No. No. So it might for the first 300 digits and then start repeating. So... it either terminates or repeats...

Stephanie is sure that the decimals of $53/83$ either terminate or repeat, but she does not have a mathematical explanation for this. However, her faith is extremely stable, she does not need to abandon either of the definitions to assure consistency of her thinking and a flexible usage of both definitions. We found this kind of faith very rare. We are not suggesting that learners should be able to reproduce the proof instantly. In mathematics we often use the results that we have once been convinced of without having to go back to first principles. However, the equivalence of the two definitions together with the reasons for it should be clear to an undergraduate mathematics student in general and to a secondary school mathematics teacher in particular.

The Other Direction in the Missing Link: From Repeating Decimal to Fraction. The converse, that a repeating decimal represents a fraction, was also not commonly found in the knowledge repertoire of the participants of this study. Most common responses regarding this were:

(1) It is easy to turn a fraction into a decimal. But there is no easy, general way of turning a decimal into a fraction. Looking at a decimal, unless it is a terminating decimal, you cannot tell if it is rational or not.
(2) 0.012222... is not rational. I cannot think of any two numbers to divide to get that decimal.

Furthermore, not a single participant reproduced the symbolic juggling on infinite decimals to show how a repeating decimal can be transformed into a fraction (e.g., see the Problem and Solution presented earlier in the section on Irrational Numbers: Definitions and Representations). Often, upon requesting that a repeating decimal number be converted into a fraction, we received the outright claim that this cannot be done. An excerpt from our interview with Erica exemplifies this.

Erica: See it's, I think it's for a student and for me, it's virtually impossible to look at a decimal and put it into a fraction, like to go that direction. To go from a fraction to a decimal no problem, but to go the other way, is impossible...

In the local curriculum, the method of converting a repeating decimal into a fraction is taught as early as in Grade 7 (for decimals with a single digit repeating) and again in Grade 9 (for any repeating decimal). Although the intention behind the inclusion of this juggling method in the curriculum is probably to convince students that every repeating decimal can be turned into a fraction, we wonder whether it serves the purpose. Amongst the participants of this study, we found only three individuals who believed that every repeating decimal can be represented as a fraction. Again, this belief was largely based on faith in the consistency of mathematics and not on an actual ability to perform this conversion.

Furthermore, only two participants in our study, Dave and Stephanie, referred to some "method" of turning a repeating decimal into a fraction. Neither could recall or reconstruct the particulars of the method. In our assessment of the accessibility, portability, and transferability of the knowledge of this symbolic juggling method, we found that more than anything the method tends to leave people mystified and is difficult to retrieve. On top of this, as evidenced in the following excerpt with Dave, participants saw no practical benefit for knowing it; that is to say, it is not perceived as critical piece of knowledge for the development of further concepts.

Dave: (Considering the question of how we can know that every infinite repeating decimal has a representation as a/b where a and b are integers, and b is nonzero)...
Yeah, yeah, from what I remember, like *if I remember correctly from my number theory course*, like the repeating part will, it doesn't matter where it starts, as long as there is one... I remember um our professor, he, like say there's a decimal then a bunch of random, looks like random numbers and then the pattern, he was able to do it, to convert it to a fraction...

Interviewer: Okay, every time?

Dave: Yeah pretty, yeah pretty, like for examples he had, you could do it, like it took, *I remember I did some work and I couldn't really understand how he did it, like I followed it, but I couldn't remember...*

Interviewer: Yeah...

Dave: But I didn't look at it long enough to really learn what he was doing, I just go okay, it works, I wasn't, because *I was thinking oh, after this course, you know, I'm not going to use this too often. I couldn't see it being applied to anything*, except for, and just doing those kind of problems, although I thought they were very interesting, because it took a while for the brain power to try to make sense of it (laugh), so...

For Dave the method of symbolic juggling on infinite repeating decimals succeeded in convincing him that every repeating decimal can be represented as a ratio of two integers. His understanding is more robust and it does not suffer from the instability caused by the "missing link." In his reply we note that he had been recently exposed to this method in his (elective) number theory course at the university. Similarly, the only other participant who had any reference to the method, Stephanie, was also a recent mathematics major graduate in the honors program. She referred to the method as a "trick" and could not reproduce it; however, the "trick" did the job of convincing her that every repeating decimal can be transformed into a fraction.

Interviewer: Why repeating, what's so special about repeating?

Stephanie: Um, it helps us put them in a ratio, if digits repeat, we can manipulate them and put into a ratio of integers which is a rational number.

Interviewer: How, is there a technique or something that you are referring to?

Stephanie: Um, uh there's a technique, I know they teach it in grade 10 or 11 or something and you can manipulate it in some way, but um, that's... yeah and it doesn't matter where it starts repeating, as long as it would repeat.

Interviewer: Any number that repeats can be manipulated in that way and put into a ratio?

Stephanie: Well there's a trick that you can show that it is a ratio of integers ...

Undoubtedly, the quality and depth of understanding are affected if the equivalence of the two definitions and the reasons for the equivalence are not recognized. When directly confronted with the issue, the majority of PSTs recognized that their notions of these matters were inextricably muddled. We present Kyra's response, provoked by drawing her explicit attention to the problem of the "missing link."

Kyra: Umm, I just thought like, by saying that a rational can be expressed as a fraction means or implies that every rational is a quotient, right, it's the answer of something divided by something else. And it's just really difficult for kids to go back, or take a quotient and find out what the two dividends would have been, so I think that's why we use the second criteria, but I have no idea who figured that out, or why that rule, does that rule support the other one. Is there proof...

Interviewer: Does it always support it?

Kyra: And that's something I've, yeah, I have never even considered it.

Confronting the Two Definitions: Exposing the Missing Link. It is interesting to note that people can be largely unaware of the problem of the "missing link," particularly if they have a balanced disposition. This is probably the best possible outcome of the present didactical choice – to have people fluently use either of the definitions as the situation warrants, never questioning their equivalency. In the next excerpt, we present Erica's confrontation with the missing link, provoked by the use of an interviewing technique referred to as the "probing of the strength of belief" (Ginsburg, 1997).

Erica: (Referring to 53/83) *I said it was rational because it's a ratio, I'm, I don't know whether I even considered whether it terminates or not. Which means my definitions aren't very strong, are they.* Um, yeah, I don't know. If it, looking at it again, if it terminates then I would say that it's a rational number, but if it doesn't and it doesn't repeat, then it would be considered an irrational number.

Interviewer: See I had a student who had similar kind of reasoning, he said um, this, he believed, continues, and it doesn't show any repeats so this decimal, in his opinion, would be an irrational number, yet 53/83 looking at another definition of what a rational number is, would be considered rational, so he said well there are some numbers obviously, that can be both, rational and irrational, I mean both at the same time. They can be rational and irrational, do you think that it is possible...

Erica: (laugh) So I think *my definition maybe isn't clear*, um, because that, my understanding is, it's either going to be a rational number or an irrational number, not both...

Interviewer: So it can't be both...

Erica: So I would say that this would be an example of my definition not being strong enough for either one. Um, so maybe my understanding of one or the other isn't correct, because obviously there needs to be some way of deciding...

Interviewer: One or the other, what do you mean by one or the other...

Erica: Rational or irrational, like my definition of what I, *how I define a rational number or an irrational number is not complete perhaps*, which is why I'm not identifying, like, because if I simply say a rational number is any number that's a ratio, yet *a ratio can give a number that has a non-repeating decimal* and I say a non-repeating decimal is irrational. *There's got to be something else that's going* to distinguish between the two, because otherwise it's going to overlap too much, so I'm thinking that my definition is missing a part, rather than say that some are both, is my fence.

Only after Erica's attention is drawn to the conflicts arising under the condition of an unresolved missing link, her mathematical notions of rational and irrational numbers are cast into shades of doubt. While Erica experienced this consciously, we believe a similar kind of insecurity is experienced subconsciously in many learners. What is more, we believe that this problem could be avoided, or at least reduced by a more prudent didactical choice.

Summary and Pedagogical Considerations

We investigated the understanding of irrational numbers of a group of prospective secondary mathematics teachers. In this report we focused on the role that representations play in concluding rationality or irrationality of a number. The main contribution of our study can be seen in two arenas. It is the first comprehensive study of prospective secondary teachers' understanding of irrational numbers from the perspective of their representations. As such, it contributes to the detailed research accounts of learners' understanding of a particular mathematical content as well as to the research on teachers' content knowledge. Further, it enriches the research on representations, highlighting the difficulties in connecting different representations of irrational numbers and comprehending their equivalence. Our findings call for reconsideration of instructional and pedagogical approaches to irrational numbers. In what follows we summarize our findings and discuss several pedagogical suggestions.

Though the majority of participants provided correct and appropriately justified responses and attended to the transparency in the given representation, the incorrect responses of the minority are troublesome – especially taking into account participants' formal mathematical background and the fact that they were close to the certification to teach mathematics at the secondary level. For this significant minority:

(1) The definitions of irrational, as well as rational, numbers were not in the active repertoire of their knowledge.
(2) There was a tendency to rely on a calculator and participants expressed preference towards decimal representation over the common fraction representation.
(3) There was a confusion between irrationality and infinite decimal representation, regardless of the structure of this representation.

(4) The idea of "repeating pattern" in the decimal representation of numbers was at times overgeneralized to mean any pattern.

From our theoretical perspective, we say that the transparent features of the given representations were either not recognized or not attended to. A possible obstacle to students' understanding is that students do not recognize the equivalence of the two definitions of irrational numbers given in school mathematics (i.e., the nonexistence of representation as a/b, where a is an integer and b is a nonzero integer and the infinite non-repeating decimal representation). We consider this a missing link between ideas that is rooted in understanding of rational numbers. In particular, it is connected to the understanding of how and when the division of whole numbers gives rise to repeating decimals, and conversely, that every repeating decimal can be represented as a ratio of two integers. As stated above, even participants who responded correctly and justified their responses appropriately could not explain the link.

We showed that the same definitions as introduced in school mathematics are used by mature university students who are en route to becoming teachers very soon. Their knowledge in this area seems to get stifled, and this seems to happen very early on, as we detect no further growth in mathematical maturation of the concept of irrational numbers. This warrants an investigation, or at least a consideration of how this content is being taught, especially in looking for reasons why understanding in this area is relatively poor. Further, given that understandings acquired in school seem not to be influenced by "advanced" mathematical training, we wonder what experiences at the undergraduate level may have an effect on students' understanding of irrationality.

In the interviews it was revealed that the very essence of mathematics – consistency of related ideas – may be compromised if the two competing definitions are used without giving explicit attention to how and why they are equivalent. Is it possible to link the two interpretations of irrationals with the aim of supporting a balanced and connected perception? We consider two pedagogical suggestions: (a) postponing the link and (b) exposing the link. The former is most appropriate for secondary school, the latter could take place in school, but is essential to be reinforced at the undergraduate level.

We prefer to defer the description of irrational numbers in terms of decimal representation. From the perspective of opaqueness and transparency applied to the decimal representation as displayed on a calculator screen, it is obvious that all displays of irrational numbers are opaque with respect to irrationality and that the majority of displays of rational numbers are opaque with respect to rationality. Note that, on the calculator, the only transparent representations are those of "short" terminating rational numbers; that is, those that use up fewer digits than the calculator display allows.

Even those decimal numbers that can be "seen" as repeating are not necessarily rational. To illustrate this, we can create examples of irrational numbers that, when entered into the calculator can be seen as "infinite repeating" and therefore indeed seem rational. Addressing Peled's (1999) call for creating tasks that integrate different pieces of knowledge related to irrational numbers, we invite the reader to consider the following example, using an 8-digit calculator.

Consider the following number, m:

$$m = \sqrt{10000000} \div \sqrt{91} \div \sqrt{989011}$$

Entering it in the calculator, we end up with 0.3333333 on the display, which could be interpreted as a demonstration that m is rational. However, this is false because each one of the three roots above is irrational: both 91 and 989,011 are prime, so their square roots are irrational, also $\sqrt{10000000} = 1000\sqrt{2}\sqrt{5}$ is irrational; since none of the factors combine to form a perfect square, m is irrational. The same calculation performed on a 10-digit calculator, yields 0.333333331, which at least casts some doubt. Therefore, using the calculator display to decide on anything is open to errors. The problem is that this method "works" with most if not all school examples.

This highlights the problematics associated with using the decimal characterization of a number in deciding whether or not that number is irrational. We cannot possibly tell from a calculator display whether a decimal repeats or even whether it terminates. As the results of this study show, a natural side effect – and a realistic danger – of introducing this characterization is that many people revert to it as their only working definition for deciding whether or not a number is rational or irrational.

However, reasonable as it may sound, in an era of obsessive reliance on calculators the suggestion to avoid or even to postpone description of irrationals in terms of their decimal representation may appear unfeasible. Therefore, a more practical approach can be in reinforcing the algorithmic experience to attend to the connection between the fractional and decimal representations of numbers. Explicit attention to the link between two representations, where both exist, would reinforce students' understanding of rational numbers and serve well in preparing them for the encounter with the counterpart – the irrationals. Simply put, we suggest that directing students' explicit attention to representations and to mathematical connections that render the two representations equivalent is helpful in acquiring a more profound understanding of number.

A more general suggestion for teaching practice calls for a tighter emphasis on representations and conclusions that can be derived from considering them. It is a "connected-balanced" disposition that we aim for, rather than "opportunistic-balanced," as was held by several participants in this research. In particular, attending to the connections between decimal (binary, etc.) and other representations (geometric, symbolic, common fraction, and even continued fractions) of a number can be an asset for teachers and students alike.

References

Arcavi, A., Bruckheimer, M., & Ben-Zvi, R. (1987). History of mathematics for teachers: The case of irrational numbers. *For the Learning of Mathematics*, 7(2), 18–23.

Brousseau, G. (1997). *Theory of didactical situations in mathematics.* Boston: Kluwer.

Cuoco, A. (Ed.). (2001). *The roles of representation in school mathematics.* Reston, VA: National Council of Teachers of Mathematics.

Dubinsky, E. (1991). Reflective abstraction in advanced mathematical thinking. In D. O. Tall (Ed.), *Advanced mathematical thinking* (pp. 95–123). Boston: Kluwer.

Fischbein, E., Jehiam, R., & Cohen, C. (1994). The irrational numbers and the corresponding epistemological obstacles. In J. da Ponte & J. Matos (Eds.),

Proceedings of the 18th conference of the International Group for the Psychology of Mathematics Education (Vol. 2, pp. 352–359). Lisbon, Portugal: University of Lisbon.

Fischbein, E., Jehiam, R., & Cohen, C. (1995). The concept of irrational number in high school students and prospective teachers. *Educational Studies in Mathematics*, *29*, 29–44.

Ginsburg, H. (1997). *Entering the child's mind: The clinical interview in psychological research and practice.* Cambridge, UK: Cambridge University Press.

Goldin, G., & Janvier, C. E. (Eds.). (1998). *Representations and the psychology of mathematics education, parts I and II.* Special issue of the *Journal of Mathematical Behavior, 17*(1&2).

Janvier, C. (1987). Representation and understanding: The notion of function as an example. In C. Janvier (Ed.), *Problems of representation in the teaching and learning of mathematics* (pp. 67–72). Hillsdale, NJ: Erlbaum.

Kaput, J. (1991). Notations and representations as mediators of constructive processes. In E. von Glasersfeld (Ed.), *Radical constructivism in mathematics education* (pp. 53–74). Dordrecht, the Netherlands: Kluwer.

Lamon, S. J. (2001). Presenting and representing: From fractions to rational numbers. In A. Cuoco (Ed.), *The roles of representation in school mathematics* (pp. 41–52). Reston, VA: National Council of Teachers of Mathematics.

Lesh, R., Behr, M., & Post, M. (1987). Rational number relations and proportions. In C. Janvier (Ed.), *Problems of representation in the teaching and learning of mathematics* (pp. 41–58). Hillsdale, NJ: Erlbaum.

Peled, I., & Hershkovitz, S. (1999). Difficulties in knowledge integration: Revisiting Zeno's paradox with irrational numbers. *International Journal of Mathematical Education in Science and Technology*, *30*(1), 39–46.

Sirotic, N., & Zazkis, R. (2007a). Irrational numbers on the number line – where are they? *International Journal of Mathematical Education in Science and Technology*, *38*(4), 477–488.

Sirotic, N., & Zazkis, R. (2007b). Irrational numbers: The gap between formal and intuitive knowledge. *Educational Studies in Mathematics*, *65*, 49–76.

Zazkis, R., & Gadowsky, K. (2001). Attending to transparent features of opaque representations of natural numbers. In A. Cuoco (Ed.), *The roles of representation in school mathematics* (pp. 146–165). Reston, VA: National Council of Teachers of Mathematics.

Zazkis, R., & Liljedahl, P. (2004). Understanding primes: The role of representation. *Journal for Research in Mathematics Education*, *35*(3), 164–186.

Faculty of Education, Simon Fraser University, Burnaby, British Columbia, Canada
E-mail address: zazkis@sfu.ca

Faculty of Education, Simon Fraser University, Burnaby, British Columbia, Canada
E-mail address: nsirotic@telus.net

CBMS Issues in Mathematics Education
Volume **16**, 2010

Students' Use of *Derive* Software in Comprehending and Making Sense of Definite Integral and Area Concepts

Matías Camacho Machín, Ramón Depool Rivero, and Manuel Santos-Trigo

ABSTRACT. The study documents representations, resources and strategies that first-year engineering students demonstrate as a result of using *Derive* software to comprehend and solve problems that involve the concept of definite integral. Three student profiles appear to emerge from analysis of their problem-solving approaches: (a) students who relied on the use of the software as a means to validate their paper and pencil work, (b) those who used the software to represent graphically and calculate approximated areas and (c) a group of students who combined both paper and pencil and software approaches to solve problems but often did not connect concepts that appeared in the study of the definite integral with basic ideas (and procedures) previously studied (e.g., area of simple figures). There is evidence that the use of *Derive* software helped students elicit ideas that needed to be discussed in order to refine their initial approaches to problems.

Introduction

This study investigates how first year engineering university students performed after they had taken a calculus course in which they systematically used *Derive* software to work on a series of tasks that involve numerical, graphic and algebraic approaches. Problem solving activities that included the study of definite integral were designed in accordance with research results identified in the literature review. During problem-solving sessions, students had the opportunity to use a specially designed Utility File as a means to approximate areas of bounded curves (through the use of rectangles, trapezoids, and parabolic regions). Thus, the study focused on documenting the extent to which students were able to utilize *Derive* software in their problem solving approaches. In particular, we were interested in analyzing the approaches used by the students to understand and solve different types of problems that involve concepts of area and definite integral.

The research questions that were used to guide the development of the study included: To what extent do students display relationships between graphic, algebraic, and numerical representations to understand and apply concepts associated with the definite integral in their problem solving approaches? What types of difficulties do students experience as a result of using *Derive* and the Utility File to work on problems? To what extent does the use of *Derive* software help students understand concepts involved in the study of the definite integral? And, what

©2010 American Mathematical Society

type of reasoning do students develop regarding the definite integral concept and problem solving activities as a result of using *Derive* software?

Conceptual Framework

It is recognized that the use of computational tools plays an important role in developing and understanding mathematical ideas. Artigue (2002) states that for some mathematicians the use of these tools has changed not only the methods used in mathematical practice but also the themes and problems they investigate. With different tools, students are likely to learn different mathematics and learn them differently (Heid, 2002, p. 109). How do students develop and employ a working method or work habits in the context of using a new tool? To what extent does the use of computational tools help students construct their mathematical knowledge? To what extent does the use of the tools enhance students' problem solving strategies? What type of mathematical understanding do students develop as a result of using Computer Algebra Systems? What type of mathematical reasoning do students exhibit while working on mathematical tasks with the use of computational tools? How do students transform the use of a particular artifact (*Derive* in this case) into a problem-solving tool or instrument? Discussing these questions provided relevant information to organize and structure elements of a conceptual framework that supported and guided the development of the study.

A central aspect in analyzing the impact of using Computer Algebra Systems (CAS) in students' construction or development of mathematical understanding involves a characterization of what the process of understanding entails. As Heid (2002) noted:

> It is interesting to examine the impact of that choice on the role of the CAS in mathematics instruction. ... One who believes that learning occurs through one's deliberate action and reflection on that action will find it necessary not just to have students generate instances of a pattern, but to reflect on the meaning of those results. (p. 101)

In this context, we recognize the importance for students to conceptualize the discipline as problem-solving activities in which they need to constantly pose and pursue relevant questions. Thus, understanding a concept or solving a mathematical problem involves thinking of the problem or concept in terms of dilemmas or questions that need to be examined in multiple ways or from diverse perspectives. Thus, substantial questions associated with understanding the concept of definite integral may include: How is the concept of area under a curve related to the definite integral concept? What does it mean for a function to be nonnegative and continuous on one given interval? What does the limit of the sum of the areas of the inscribed rectangles mean? How is the concept of definite integral defined in terms of the limit of Riemann sums? How is the limit process represented geometrically? In the same vein, Thurston (1994) stated that the concept of derivative can be thought of as:

> (1) Infinitesimal: the ratio of the infinitesimal change in the value of a function to the infinitesimal change in a function.
> (2) Symbolic: the derivative of x^n is nx^{n-1}, the derivative of $\sin(x)$ is $\cos(x)$, the derivative of $f \circ g$ is $f' \circ g * g'$, etc.

(3) Logical: $f'(x) = d$ if and only if for every ϵ there is a δ such that when $0 < |\Delta x| < \delta$,

$$\left| \frac{f(x + \Delta x) - f(x)}{\Delta x} - d \right| < \epsilon.$$

(4) Geometric: the derivative is the slope of a line tangent to the graph of the function, if the graph has a tangent.
(5) Rate: the instantaneous speed of $f(t)$ when t is time.
(6) Approximation: The derivative of a function is the best linear approximation to the function near a point.
(7) Microscopic: The derivative of a function is the limit of what you get by looking at it under a microscope of higher and higher power. (p. 163)

Thus, Thurston's ideas shed light on the importance of recognizing that mathematical learning is a continuous process in which students need to represent and examine mathematical concepts in different ways. In particular, learning a mathematical concept demands that students construct a web of relations and meaning associated with that concept. Thurston suggested that understanding the concept of derivative implies that students can relate and transit through, in terms of meaning, the ideas and representations associated with each way of thinking about this concept. Similarly, in order for students to develop a clear understanding of the definite integral and area concepts, they need to identify, examine, connect, and operate on various relationships as well as develop ways to represent those concepts. Here, the use of CAS may be of importance to generate representations of problems that involve the concept of definite integral. Discussing those representations may allow students to access and use basic knowledge to understand and develop meaning associated with that concept.

A relevant aspect in mathematical understanding is the need for students to develop a web of connections between distinct representations and operations associated with a concept or problem solving activity. Thompson and Saldanha (2003) indicated:

> We see understanding as, fundamentally, what results from a person's interpreting signs, symbols, interchanges, or conversations – assigning meaning according to a web of connections the person builds over time through interactions with his or her own interpretations of settings and through interactions with other people as they attempt to do the same. (p. 99)

And, regarding the use of CAS, Ruthven (2002) stated that:

> The availability of calculating tools facilitates a numerical treatment of derivative in terms of the infinitesimal ratio sense; likewise, graphing tools assist work on the tangent slope and linear approximation senses. Equally, the new possibility of 'zooming' in and out on a graph permits an operationalisation of the microscopic image sense through tasks and techniques focusing directly on the local slope of the graph itself, without appeal to secant and tangent constructions. (p. 280)

How are aspects of mathematical practice that involve the use of techniques or algorithms and concepts reconciled with the use of CAS? Artigue (2002) argued that "[techniques] have also an *epistemic value*, as they contribute to the understanding of the objects they involve, and thus techniques are a source of questions about mathematical knowledge" (p. 248). Thus, analyzing and transforming results produced with the use of CAS becomes important for students in order to understand the meaning of operations and concepts. That is, "technique – whether mediated by technology or not – fulfills not only a pragmatic function in accomplishing mathematical tasks, but an epistemic function in building mathematical concepts" (Ruthven, 2002, p. 283). Heid (2002) suggested that results from CAS studies challenge the assumption that students' abilities to perform procedures must precede the development of conceptual understanding: "[CAS studies] have provided evidence that, prior to developing related by-hand routines, students can learn at a greater depth than in a traditional skills-before-concepts curriculum" (p. 98).

Guin and Trouche (2002) introduced the idea of instrumental genesis to explain the process of transforming an artifact (a material object) into an instrument (when students use it as a tool to solve problems). They recognized that this process is complex and involves aspects related both to the actual design features of the tool and also to the cognitive process involved in students' appropriation of the instrument to solve problems (the development of instrumentation schema). That is, it is crucial to document the way students develop and use a style of work in the context of a new tool (Cuoco, 2002). In this sense, it becomes important to pay attention to the design limitations associated with the use of the software that might interfere with students' work. Drijvers (2002) identified local and global obstacles that students often face while using a computer algebra environment. In particular, we have taken the idea that obstacles can be seen as an opportunity for students to reflect on their own learning, rather than as a barrier to achieving understanding of mathematical ideas.

To document students' process for transforming an artifact into an instrument, that is making the use of CAS functional to comprehend or solve mathematical problems, we recognize that students need to develop cognitive schemes to transform a physical device or material into a mathematical tool. Artigue (2002) has stated that:

> For a given individual, the artefact at the outset does not have an instrument value. It becomes an instrument through a process, called instrumental genesis, involving the construction of personal schemes or, more generally, the appropriation of social pre-existing schemes (p. 250).

This instrumental genesis can be explained in terms of constraints and potentialities of the artifact and their relation to the cognitive schemes that students develop as a result of progressively using the tool in problem solving activities:

> In order to understand and promote instrumental genesis for learners, it is necessary to identify the constraints induced by the instrument; ... "command constraints" and "organizational constraints" ... It is also necessary, of course, to identify the new potentials offered by instrumental work. (p. 250)

The students' process for transforming an artifact, in this case the *Derive* software, into a problem solving instrument shapes not only their ways of using the tool, but also their sense and conceptualization of mathematics and problem solving activities. That is, use of the tools influences directly students' ways of dealing with mathematical activities. The transformation process is linked to the tool characteristics (potentialities and constraints) and to the person's activity, including a learner's knowledge and ways to use it (Trouche, 2005). With the use of CAS, students build a conceptual system that guides their mathematical behavior. Ruthven (2002) has pointed out that "building a coherent conceptual system and an overarching concept of framing involves the progressive coordination of many other specific schemes" (p. 279).

From this perspective, in order to analyze how students performed during the study it was also important to document students' use of representations during the construction of concepts (Goldin, 1998). In this context, it was essential to analyze the type of basic resources and strategies that students utilized when dealing with a particular representation of the problem and also the extent to which students were able to transit, in terms of meaning, from one type to another type of representation (graphic, algebraic, and numeric representations).

The ideas embedded in this framework played a fundamental role not only in analyzing students' work but also influenced the design and structure of the study. In particular, the design of a Utility File was based on the idea that students could use the Utility File as an aid to calculating a set of definite integrals in which the primitives of the function to be integrated could not be expressed through elementary functions. Thus, the use of the Utility File could help students develop both an image of the integration processes and the relationship with the area concept.

Duval (1993) recognized that for students to achieve mathematical understanding, they need to distinguish between a mathematical object and its representation. Thus, different semiotic representations of a mathematical object appear important in problem solving processes that include students' recognition of different representations and treatments (within and between representations), and conversion of algebraic, graphic, and numerical representations. Indeed, Santos (2000) documented that an important aspect favoring the process of making connections between distinct representations of problems or mathematical objects is that students have the opportunity to reflect on the information that each representation provides in order to discuss and relate other systems of representation.

Design, Methods, and General Procedures

Thirty-one first-year university engineering students participated in the study. The study was carried out in a regular calculus class during one semester. Students met three times weekly in two-hour problem-solving sessions; in one of those sessions the students worked in the computer laboratory. The theme of definite integral is part of the official calculus program that includes the study of topics such as functions, limits, and derivative. As part of the course, students used *Derive* software to work on a series of problems. In particular, a special Utility File was designed to help students calculate approximations of areas. Appendix 1 includes the Utility File and two clearly differentiated parts can be seen: graphic aspects (Lines 1 to 25) and numerical aspects (Lines 26 to 39). The goal here was for students to develop a conceptual understanding of integration processes. Ideas like partition, refinements, limits, and area approximation appeared important during the students' use of the Utility File. Here it was also important for students to recognize linkages between numerical, algebraic, and graphic representations associated with the integral concept.

To collect data, a questionnaire was given to students at the end of the course. In addition, 6 of the 31 students took part in task-based interviews in which they had the opportunity to reflect on the use of different problem solving strategies and representations and the use of the software. In particular, the research team asked students to explain and elaborate their approaches to the problems.

The course was problem-solving oriented and students were regularly asked to respond and discuss questions concerning the concept of area and definite integral. In general, three related phases distinguished the instructional approach used in the course. In Phase 1, the instructor presented and discussed with the whole class ideas about fundamental concepts involved in the definite integral themes. In Phase 2, students worked in pairs on a series of problems in which they used the *Derive* software – students worked in the computer laboratory and each pair handed in a written report. In Phase 3, the instructor reviewed students' approaches and discussed what students did during the session with the whole class. These three instructional phases appeared consistently throughout the course for one semester.

Conceptual ideas that emerged during students' discussions included themes like the limit of sums (Riemann's approach), area of bounded regions, and their relation to the Fundamental Theorem of Calculus. In particular, students analyzed the relationship between a given function $f(x)$ and its integral function $F(x)$ (that is, $F'(x) = f(x)$). In addition, students used the software to represent functions graphically and evaluate definite integrals.

To analyze students' understanding of fundamental ideas related to definite integral concepts, students were asked at the end of the course to work on a questionnaire that included nine non-routine problems. In general, the problems were organized into three groups in accordance with the following characteristics:

Problem Type 1. These were problems in which there was a geometric representation and students were asked to determine, whenever possible, the area of some regions (students had to analyze features of the graph in order to identify corresponding integration limits). Otherwise, students needed to provide a mathematical argument to explain why it was not possible to calculate the area (see Appendix 2, Tables 2 and 3, questions Q1, Q2, Q8, and Q9).

Problem Type 2. These were problems that involved an algebraic expression and students were asked to find the integral. Here, it was important to represent graphically the given expression in order to solve the problem. In addition, it was important for students to identify the location of the region (above or below the x-axis) and in some cases recognize discontinuities of the function (see Appendix 2, Table 4, questions Q3, Q4, and Q7)

Problem Type 3. These were problems in which there was a statement and students needed to discuss whether it was true or false (see Appendix 2, Table 5, questions Q5 and Q6)

In addition to working on the questions both with and without the use of the software, 6 students were chosen from the group of 31 students to be interviewed later. Thus, students worked on the questionnaire in three different scenarios (see Appendix 2):

Scenario 1. In this scenario, the 31 students worked individually on the problems in the questionnaire using pencil and paper only (Q2, Q3, Q4, Q5, Q6, Q8, and Q9) and then reported their problem approaches or solutions in a written way.

Scenario 2. The second scenario corresponded to the work shown by the students when they answered the questionnaire using *Derive* software. Only 26 students took part in this second scenario. They worked in pairs to answer the questions. Here, they handed in an electronic copy of their work showing their approaches towards the problems and relevant comments. In Scenario 2 students addressed the same questions as in Scenario 1.

Scenario 3. In the third scenario, the 6 selected students participated in a semi-structured interview where they were directly questioned on their approaches toward solving the problems. Here they freely chose the type of tool they needed in order to explain and solve the problems. In this case, some of the questions used in the first two scenarios were eliminated while others were added. Specifically, in Scenario 3, questions Q1, Q2, Q3 and Q5, Q6, Q7 were used.

Students' responses to the questionnaire using the software provided relevant information regarding ways they employed the software to represent and think of the problems. Later, the students' interviews provided data to analyze in detail the extent to which they used the software as a problem solving tool to approach the tasks. Thus, the three scenarios in which students responded to the questionnaire provided useful information to trace the students' use of the software in terms of transforming the artifact into a problem-solving tool.

It is important to remark on the fact that not all the questions were answered by the students in the three scenarios. The questions used during the students' interviews were chosen based on analysis of how students answered in Scenarios 1 and 2. Tables 2, 3, 4 and 5 (Appendix 2) show the question sets, the mathematical qualities taken into account in the subsequent analysis, the scenarios where they were used, the theoretical objectives proposed for each question, and the performance expected from the students. All these components played an important role during the analysis of students' mathematical competences. The information gathered while students worked on the problems using paper and pencil and when they used the *Derive* software was analyzed and eventually used to group and categorize students' work (see Appendix 3, Tables 6 to 10). This categorization of the students helped in organizing and structuring an interview protocol and in selecting the six students to be interviewed. The students chosen are referred to in this report as: Samuel, George, Blenda, Javier, María and Mirvic.

Results in Scenarios 1 and 2

The problems that appear in the tables in Appendix 2 were discussed in detail by the research team before they were given to students. The idea was to choose problems that involved the use of concepts and resources that were relevant to comprehending and applying the concept of definite integral and relating it to the area concept. In particular, each question had a theoretical purpose and an associated description of mathematical problem properties and types of representations

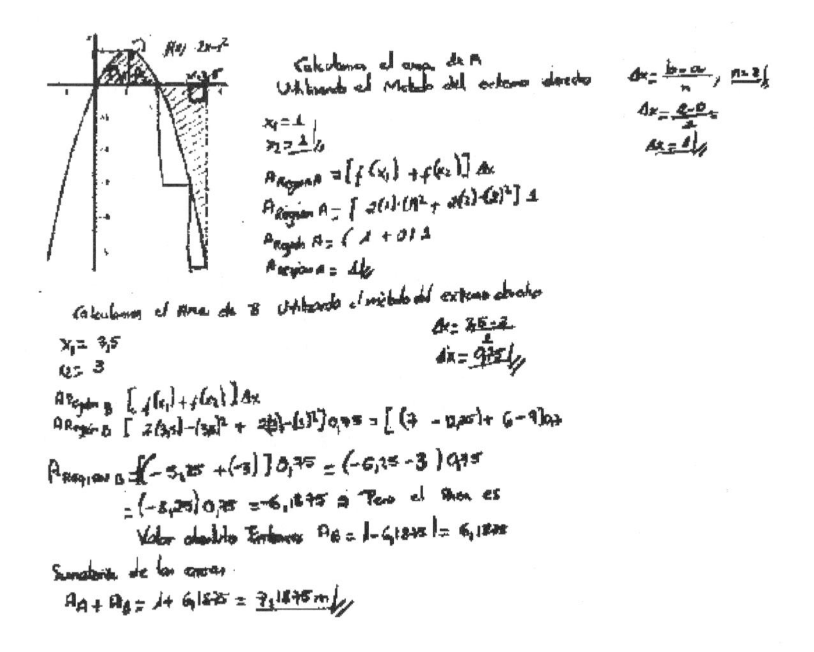

FIGURE 1. Approximating the area using rectangles and trapezoids.

relevant to dealing with the problem. Thus, the information that characterized each problem was useful to orient the analysis of students' work. To report on the information gathered throughout the study, we focus initially on examining the work shown by the students when working on a set of problems individually; later we characterize and contrast the students' approaches to the problems when they worked in pairs and during interviews. Based on these findings, we discuss the research questions in terms of analyzing, first, what students showed in each group of problems and later contrasting students' competences across all the groups of problems. Thus, the findings or main results that emerged from this research inform us about ways in which the students' initial approaches to the problems evolved as a result of discussing and reflecting on those problems while working in pairs and during the interviews. Each scenario offered students various conditions for working and reflecting on the problems and we documented the manner of reasoning demonstrated by students during the three scenarios. The use of the *Derive* software becomes relevant to representing and exploring the problems in Scenarios 2 and 3.

When students worked on the first group of problems individually it was observed that 17 students out of 31 used a numeric approach to determine the area of the limited region (see Appendix 2, Table 2 and Appendix 3, Table 6, Q2). They used rectangles and trapezoids to approximate the corresponding area. Some students experienced difficulties identifying a proper partition to draw for simple figures to approximate the area. For example, María divided the integration interval into equal parts to draw rectangles but she did not show the corresponding values associated with the base of the rectangles and the function value (height) to calculate the area. It was evident that she knew a mechanical procedure to approximate the area but she experienced serious difficulties in applying this procedure within a particular problem (Figure 1).

Other students relied on applying the procedure (Riemann sums) to determine the area. In this case, it was evident that students were aware of using the concept of limit to determine the area. Mirvic, for instance, determined the length of each interval partition algebraically and expressed the area of the region as the sum of the areas (Figure 2)

Six students approached the problem by solving directly the two integrals associated with the problem. Here, students did not consider the algebraic expression to determine the integration limits, instead, they visually recognized the intersection points between the function and the x-axis as those limits. For example, Blenda calculated the area of the region, in terms of finding the integral directly, by selecting two intervals $]0,2[$ and $]2,3.5[$. The result of calculating the integral was -2.04 and she did not explain what this number meant (Figure 3). That is, she did not elaborate on why the value was negative since her work showed that she solved directly the integrals involved. It seems that she was not clear about the type of relationship there is between the area concept and the definite integral. Students at this stage, in general, relied on the use of numeric, algebraic, and graphic approaches to represent and deal with the problems.

However, they experienced difficulties in using proper resources and strategies to examine mathematical properties or relations embedded in those representations. In particular, it was evident that when students worked on the problems individually they did not show monitoring strategies to evaluate what they were doing. Therefore, it was difficult for them to connect or transit, in terms of meaning, among those representations of the problem (Duval, 1993).

However, when students worked on this problem in pairs with the help of the software, in Scenario 2, they focused on calculating the two integrals and changed the sign of the result in the case in which the area was located below the axis. One pair of students, Javier (one of the interviewed students) and Valeria (not interviewed) used the software and found the value of the two integrals and later found the corresponding area (Figure 4)

Within this first group of questions, on the problem that involves a non-limited region (Appendix 2, Table 2, problem Q8), most of the students who worked individually in Scenario 1 responded that it was not possible to calculate the area while all of the students who worked in pairs in Scenario 2 responded that it could not be calculated (see and Appendix 3, Table 7). However, there were differences in the ways students supported their responses, some students (working individually) solved the integral, others only mentioned that the shaded area was infinite, while others mentioned that the function was not continuous and it had an asymptote. Some students working on this task in pairs argued that the graph was not bounded, "the value of the function goes to infinity when x approaches to zero and there was an asymptote" and others mentioned that the problem did not fulfill the conditions to calculate the integral. For example, George responded individually that "the area cannot be calculated because on $x1 = -1$ and $x = 1$ there is a vertical asymptote, that is, there is no exact value and neither a ceiling value of that region under the curve." Later, when George worked with Lorena (not interviewed) in Scenario 2, they argued that the function was discontinuous (without specifying where) and busco el área [0,2] $f(x_i) = 2x_i - x_i$

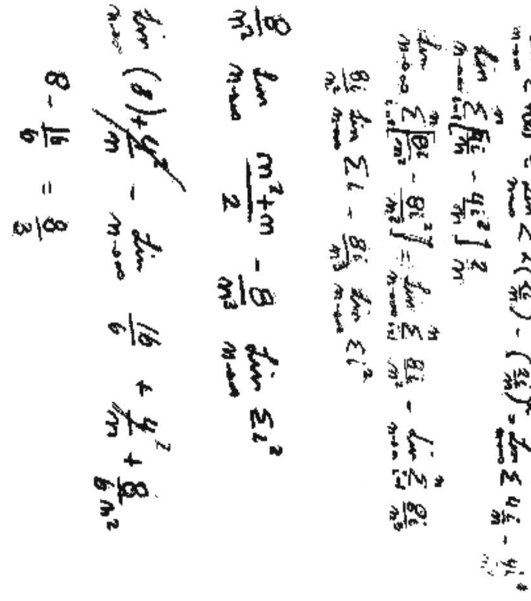

$\Delta x = \frac{b-a}{n} = \Delta x = \frac{2-0}{n} = \frac{2}{n}$

$x_i = a + i\Delta x$ $x_{(i-1)} = a + (i-1)\Delta x$

$x_i = 0 + i\Delta x$ $x(i-1) = 0 + (i-1)\frac{2}{n}$

$x_i = \frac{2i}{n}$ $x(i-1) = \frac{2}{n}i - \frac{2}{n}$

$\lim_{n\to\infty} \sum f(x_i)$

$\lim_{n\to\infty} \sum^n f(x_i) = \lim_{n\to\infty} \sum^n 2\left(\frac{2i}{n}\right) - \left(\frac{2i}{n}\right)^2 = \lim_{n\to\infty} \sum^n 4\frac{i}{n} - \frac{4i^2}{n^2}$

$\lim_{n\to\infty} \sum_{i=1}^n \left[\frac{4i}{n} - \frac{4i^2}{n^2}\right]\frac{2}{n}$

$\lim_{n\to\infty} \sum_{i=1}^n \left[\frac{8i}{n^2} - \frac{8i^2}{n^3}\right] = \lim_{n\to\infty} \sum_{i=1}^n \frac{8i}{n^2} - \lim_{n\to\infty} \sum_{i=1}^n \frac{8i}{n^3}$

$\frac{8i}{n^2} \lim_{n\to\infty} \sum i - \frac{8i}{n^3} \lim_{n\to\infty} \sum i^2$

$\frac{8}{n^2} \lim_{n\to\infty} \frac{n^2+n}{2} - \frac{8}{n^3} \lim_{n\to\infty} \sum i^2$

$\lim_{n\to\infty} (8) + \frac{4}{n} - \lim_{n\to\infty} \frac{16}{6} + \frac{4}{n}^2 + \frac{8}{6n^2}$

$8 - \frac{16}{6} = \frac{8}{3}$

FIGURE 2. Using the Riemann sum to determine the area.

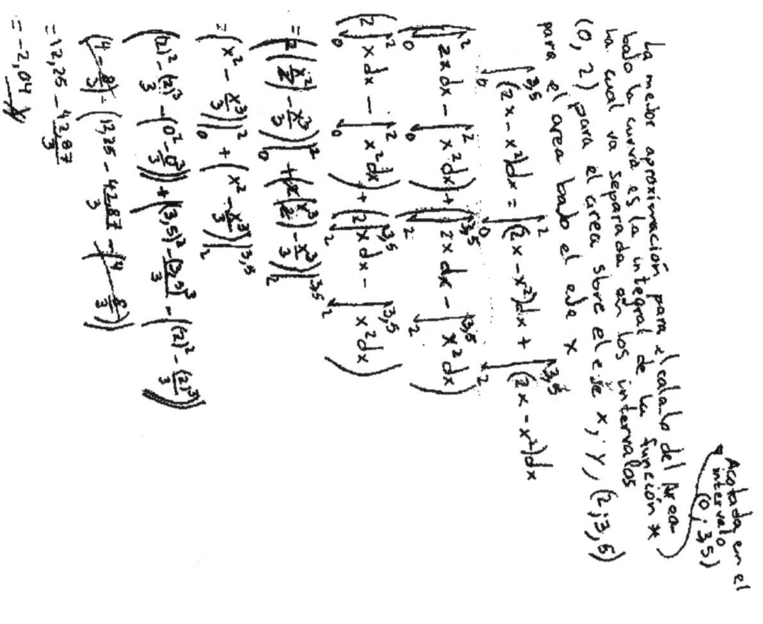

Acotada en el intervalo (0; 3,5)

La mejor aproximación para el calculo del Area bajo la curva es la integral de la función * la cual va separada en los intervalos (0, 2) para el area sobre el eje x, y, (2; 3,5) para el area bajo el eje x

$\int_0^{3,5} (2x - x^2)dx = \int_0^2 (2x - x^2)dx + \int_2^{3,5} (2x - x^2)dx$

$\left(\int_0^2 2x\,dx - \int_0^2 x^2dx\right) + \left(\int_2^{3,5} 2x\,dx - \int_2^{3,5} x^2dx\right)$

$\left(2\int_0^2 x\,dx - \int_0^2 x^2dx\right) + \left(2\int_2^{3,5} x\,dx - \int_2^{3,5} x^2dx\right)$

$= \left(2\left(\frac{x^2}{2}\right) - \frac{x^3}{3}\right)\Big|_0^2 + \left(2\left(\frac{x^2}{2}\right) - \frac{x^3}{3}\right)\Big|_2^{3,5}$

$= \left(x^2 - \frac{x^3}{3}\right)\Big|_0^2 + \left(x^2 - \frac{x^3}{3}\right)\Big|_2^{3,5}$

$\left((2)^2 - \frac{(2)^3}{3} - \left(0^2 - \frac{0^3}{3}\right)\right) + \left((3,5)^2 - \frac{(3,5)^3}{3} - \left((2)^2 - \frac{(2)^3}{3}\right)\right)$

$\left(4 - \frac{8}{3}\right) + \left(12,25 - \frac{42,87}{3} - \left(4 - \frac{8}{3}\right)\right)$

$= 12,25 - \frac{42,87}{3}$

$= -2,04$

FIGURE 3. Finding the definite integral of the function

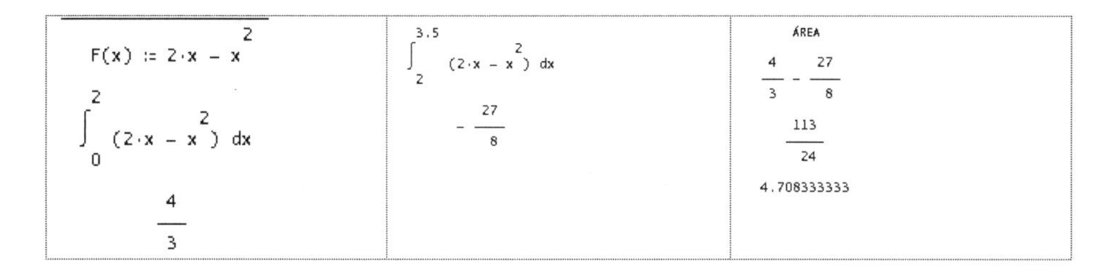

FIGURE 4. Changing the sign to the value of the integral to calculate the area.

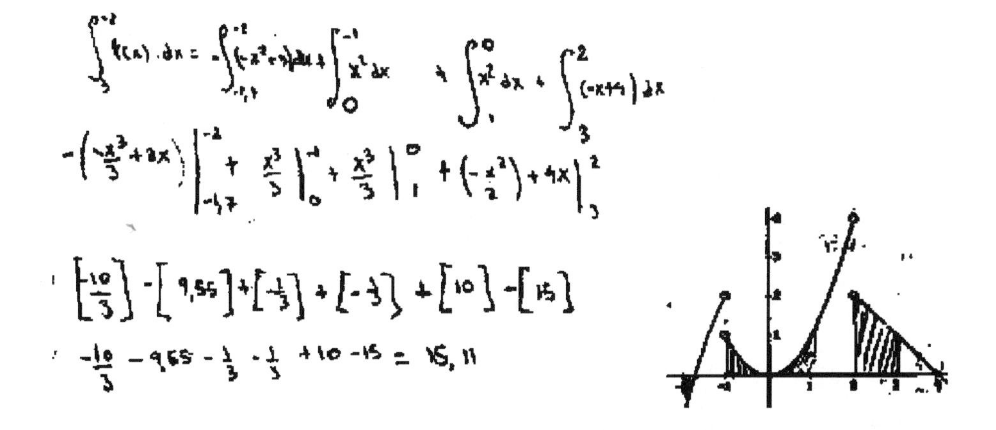

FIGURE 5. Finding the definite integral of the shaded regions.

affirmed that to calculate the shaded area the function had to be continuous on its domain. Thus, when students worked in pairs, they had opportunities to discuss their ideas and look for arguments or explanations to support their responses or solutions to the problems. Convincing or agreeing with each other to solve the task was an activity that led them to evaluate their responses (monitoring strategies) in terms of the arguments or explanations.

While working on the problem that involves a function defined by parts (see Appendix 2, Table 3, Q9 and Appendix 3, Table 7), nine students working individually in Scenario 1 solved the integral. Two of these students shaded the regions, George, however, avoided shading part of the regions and focused on finding the definite integral associated with the shaded regions (Figure 5).

Likewise, most students working in pairs in Scenario 2 thought that the integral could not be solved because the function was not continuous. Only two pairs of students calculated the integral by considering limit values associated with the discontinuity points. For example, Mirvic and Gabriela identified the points where the function is discontinuous as the integration limits and calculated the definite integrals associated with those intervals (Figure 6).

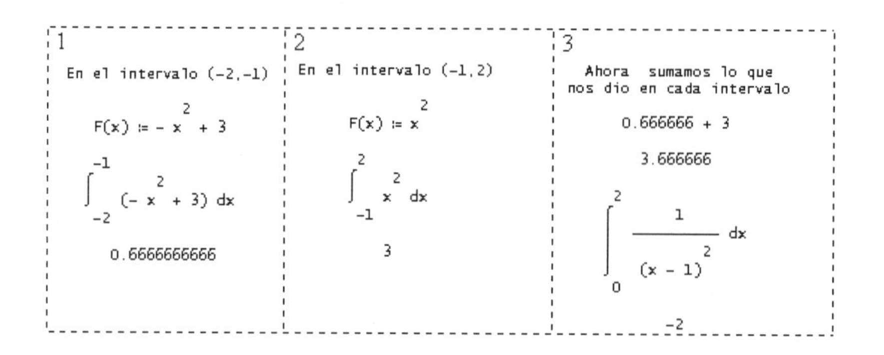

FIGURE 6. Calculating the definite integrals associated with two regions.

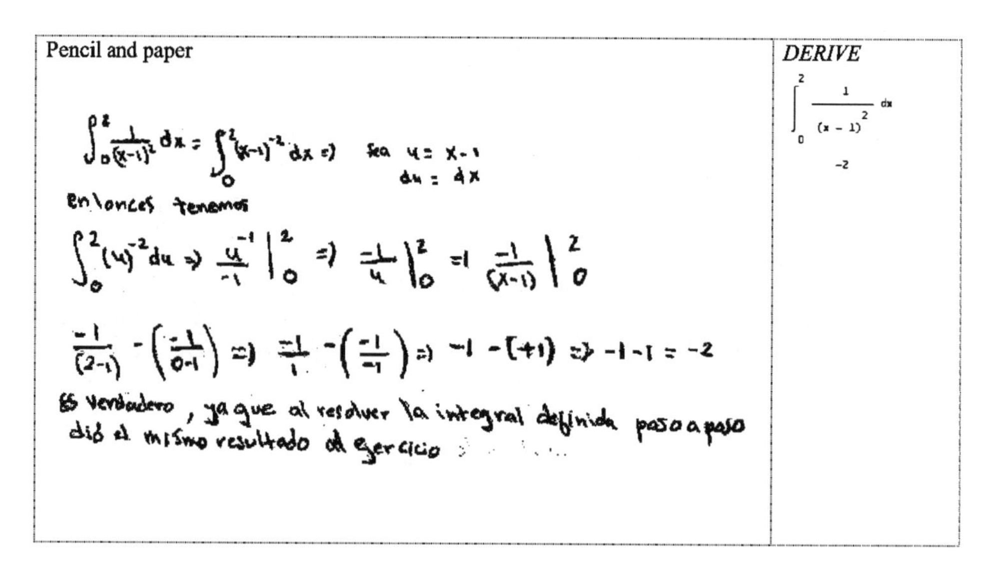

FIGURE 7. Justifying that the statement of the problem was true.

When students worked on the second group of problems, which basically involved algebraic expressions, they generally failed to represent the problem graphically (see Appendix 3, Table 8). Instead, they tended to identify the integration limits directly from the problem information and solve the corresponding definite integral. It was interesting to observe that student, while working on Q4 (Appendix 2, Table 4, Q4), in general affirmed that the statement of the problem was true and they justified their response by either solving the integral again or using the software (Figure 7).

Finally, when students worked on the third group of problems (Appendix 2, Table 5) it was observed that, in general, they based their responses on considering a particular example or sketching a geometric representation of the statement (see Appendix 3, Tables 9 and 10). Thus, George, for example, drew two functions and shaded the regions for which he calculated the definite integral and concluded that, in this case, the proposition was true. In addition, he provided a particular example

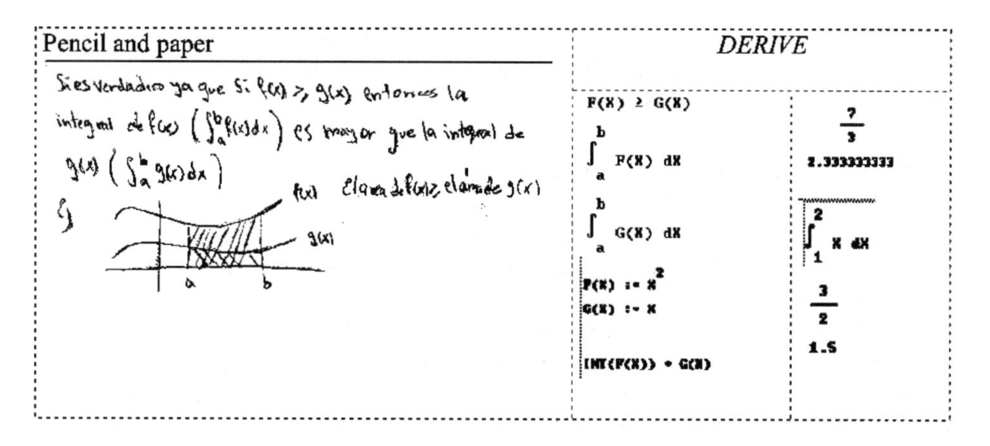

FIGURE 8. Supporting the proposition statement.

including the proposition conditions and calculating the two definite integrals to confirm the result. It is noticeable that he did not consider other options to verify other cases in order to discuss the generality of the statement (Figure 8).

Students' processes for understanding a complex domain – such as the concept of definite integral – demand that they develop resources, representations and strategies that help them unfold fundamental ideas associated with that concept and related problem solving activities. It was observed that students' ways of dealing with the activities generated useful information regarding initial difficulties they exhibited to approximate the area of particular regions in terms of selecting an interval partition and identifying relevant information to calculate the area of the selected figures. Providing students with ways to approximate the area, via the Utility File, seemed to help them to visualize the procedure; but it did not necessarily contribute to their conceptual understanding of the process of constructing and refining the interval partition. In addition, a lack of analysis of the algebraic representations to graph the corresponding expression was another difficulty that students faced while calculating areas or the value of the definite integral. The use of the software often became an instrument only for facilitating calculation of the definite integral but without really understanding the meaning of that result. Thus, working on the problems within different instructional scenarios became an opportunity for students to elicit their initial understanding of the definite integral concept and to continue refining their ideas. In particular, this information was used to orient the student tasks in interviews, where students were encouraged to think deeply about any inconsistencies in understanding and using the concepts. Based on the students' work during the interviews, three student profiles emerged that describe the students' use of the software to deal with the problems.

Recognition of Students' Problem-Solving Profiles

Relevant features of students' ways of approaching the problems began to appear in the analysis of students' work in the first two scenarios. These initial features led to identification of student problem-solving profiles that were also identified from analysis of students' work during the task-based interviews. We present next a characterization of the profiles associated with the work shown by the students and some of the evidences that has allowed us to categorize the profiles based on the behavior observed during the interviews.

Problem-Solving Profile 1. Students who were grouped in the first profile (George, Javier and Mirvic) showed a tendency to use the software as a tool to carry out algebraic operations or to find points of intersections of the curve with the x-axis. However, there is no evidence that they used the software as a problem solving tool since, in some cases, they moved the cursor on the curve to find roots values rather than identifying and solving corresponding equations. That is, their use of the software focused mainly on calculating algebraic or numeric operations involved in the problem or situation without including a graphical approach. In those problems that involved graphic representation, they generally had difficulties making sense of the situation and tried to use algebraic methods persistently. However, it was important to observe that this group of students tried to examine the validity of general relationships (problems of the third group, Appendix 2, Table 5) by analyzing particular examples graphically. This led us to consider that for this group of students the way the problems were stated influenced their use of the tool. Moreover, these students seemed to perceive the process of solving definite integrals as the application of some rules or procedures by taking into account the context of the problem. For example, when George was asked to explain the meaning of definite integral he answered:

George: If I have the integral

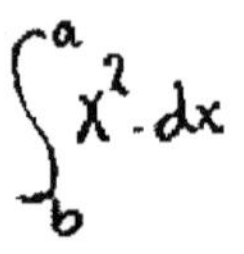

then to calculate it (pointing to the integrand), I would get the primitive, that is, I would obtain it; and this means that if I have function (he drew)

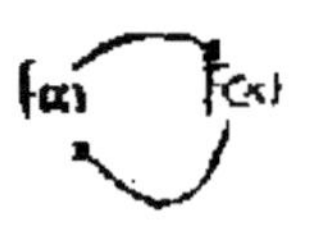

if I solve this (left side), I get this (right side), likewise if I get the derivative of this (right side), I get this function (left side). Thus, I get the derivative of, I would get

$$\frac{3x^2}{3} = x^2$$

George's comments involved the use of the Fundamental Theorem of Calculus. However, he did not rely on those methods widely used in laboratory practices. It might be that class instruction did not influence the students' way of understanding the concept of definite integral.

Problem-Solving Profile 2. A second group of students generally recognized the importance of finding areas of limited curves through the idea of approximation. They were aware of the need to get better and better approximation by the process of refining a partition within an interval. However, they had not developed a clear understanding of how the process of selecting a particular partition on the interval was done; particularly when they tried to do this task without using software. They also associated the definite integral concept with a set of procedures to calculate its value and failed to identify particular necessary conditions to apply those procedures. It seemed that students considered the use of the software as a tool that allowed them to facilitate those approaches that involved only the use of paper and pencil. Indeed, the software was used often as a means to support what they had done with paper and pencil.

Another important result is that when students worked on a problem embedded in a graphical representation, they were often able to identify limits of integration and ways of calculating areas of limited regions; however, when the problem was expressed in an algebraic form they seldom relied on graphic representations to solve the problem. Here the use of the software seemed to be sufficient to solve the problem. In one problem, which involved calculating the area of a region bounded by simple triangles (calculate $\int_{-3}^{4} |x+1| dx$), students immediately began to apply integral definite methods rather than calculating the area through the formula $A = \frac{b \cdot h}{2}$. In particular, during her interview María mentioned that since the topic was the definite integral, then they had to solve it using methods associated with that topic:

Researcher (R): Assume that you ask me to solve this problem and I solve it by calculating directly the triangles' area applying the formula. What would you say?

María (M): I'd say that...

R: Would I receive full marks, partial or none using an evaluation scale zero to ten points?

M: From zero to ten, if the problem were solved in that way, ...if I am assessing the content integral and I want the properties of the integral, I think, "here the student got this absolute value and asked: How can I get my integral and my intervals? For the student the property is not clear, but I would take into account that he understands what he is doing, ...he is conscious," he analyzed the problem, I would give him a nine.

R: If I don't mention anything about integrals, would I get nine? Why not ten?

M: Because what I'm saying, I'm evaluating the property and it may be that he got scared by this absolute value.

Problem-Solving Profile 3. A third student profile is associated with those students who successfully applied the idea of approximation to determine areas of bounded regions. They were fluent not only in deciding what type of partition to take on the interval but also in using algebraic tools to carry out the operations involved in calculation of the corresponding areas. This group of students showed a clear disposition to use the Utility File designed to approximate areas. In general, they identified and properly used important information connected with both algebraic and graphic representations in order to calculate definite integrals.

There is evidence that these students grasped the relationship between area and definite integral concepts. This was evident in the way they used the Utility File to approximate bounded areas. Here, students recognized that calculating integrals goes beyond applying a set of formulae or using a particular software command, it involves a process that they could visualize through the use of the Utility File. However, when they were asked to examine general statements about properties of functions and their relationships with the definite integral, they failed to provide a coherent argument to support their claims. In particular, they seemed to lack specific problem-solving strategies (analyzing particular cases, providing counterexamples, or using graphic representations) to make sense of or interpret this type of problem. For example, during the interview, when the student Blenda was asked to reflect on a possible relationship between the graphs of two functions and their integrals, she provided contradictory arguments:

Researcher (R): If you have these two functions (see picture) with $f(x) > g(x)$, is the integral of $f(x)$ grater than or equal to the integral of $g(x)$ in the interval $]a, b[$?

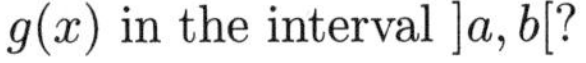

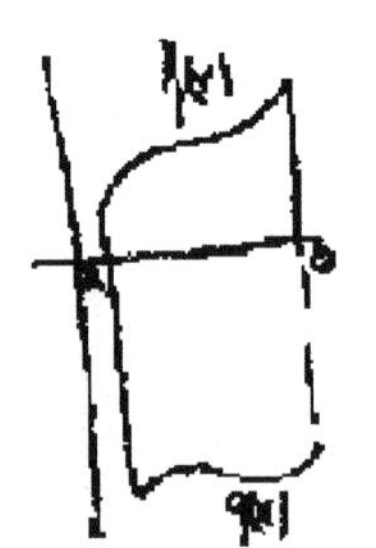

Blenda (B): This is greater (pointing to the g region), this area is greater than this one (pointing to the g region in comparison to f region) but it is negative (pauses and seems to be thinking),

R: Then what?

B: I am checking whether $f(x) \geq g(x)$?

R: We have assumed that $f(x) > g(x)$.

B: The values here (pointing at the segment between a and b) when evaluated under the function we get positive values (draws a line under the graph of f) These [values] will give you negative (draws a line above the graph of g). Here I can verify that this is true (pointing to the inequality of the functions). These values are positive and these are negative values (pointing to the graph of f and g).

However, it seems that the student is not convinced about his initial response. He confuses the use of the inequality sign between the functions and between the integrals seen as area.

The Profile 3 group showed a strong inclination to use the software to approach all the problems without considering the graphic representation which is often necessary to solve them. This group experienced serious difficulties in solving problems that involved proving or rejecting some propositions. In addition, they seemed to think that the software provided them not only with an efficient way of approaching the problem but it also became the only way of actually solving the problems correctly. Indeed, during the interview, when these students were asked to explain graphically the meaning of what they had obtained through the software, they were not able to explain on their own what happened when the value of the integral was negative. Again, these students, in general, seemed to use the software mechanically to make calculations and failed to identify and relate embedded representations in order to make the transition between and within them. To illustrate this type of behavior, we show part of an interview with a student (Samuel) who decided to use the Utility File to approach a problem that can be easily solved directly. During the interaction between the researcher and the student it became clear that he had fluency in the use of the software but experienced difficulties in interpreting his work.

To calculate the area limited by the function and the x-axis, the student chose the command `RECT_EXTREMO_DERECHO(a,b,n)` from the Utility File and substituted values of -1 for a, 3 for b and 10 for n, then calculated the matrix and represented the rectangles (see Appendix 1 and Camacho & Depool, 2003, for more details about the Utility File).

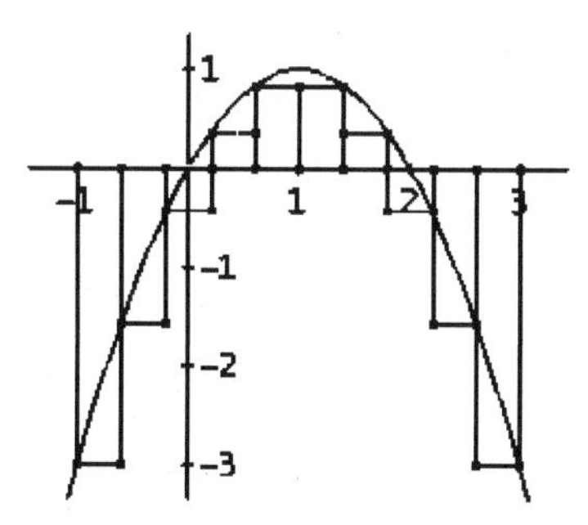

Samuel (S): This is the area taking the right side of this interval (the interval $]-1,3[$), here we can draw the rectangle, but we need to provide a number for x.

Researcher (R): Well, you may notice that you are considering from -1 to 3, but we want to find the area in the interval from 0 to 3. That is, I am asking you to find the area of this shaded region (pointing at the figure).

S: From 0 to 3?

R: Yes.

S: I can do that by a numerical method as well.

R: What are you doing?

S: Calculating numerically (selects the `MEDIDA_EXTREMO_DERECHO(a,b,n)` command). We are going to do it by drawing rectangles. I substitute a by 0, b by 3 and n by 10 and the result is -0.495. Why is the value negative?

R: I was going to ask you why it is negative.

S: (student tries to use the Utility File again but it does not work, that is, he does not get a response).

R: Why do you think you get a positive value on one side and negative value on the other side?

S: Why is that?

R: What do you think?

S: ... (silence).

TABLE 1. Summary of Problem-Solving Profiles.

Profile	Students	Characteristics
1	George Javier Mirvic	• Software is a means for algebraic calculations. • Scarcely uses graphic representations. • Perceives the calculation of the definite integral as the application of an algorithm procedure without context of the problem. • Does not follow the methods used in Laboratory Practices.
2	María	• Recognizes the importance of finding areas of limited curves by means of the idea of approximation. However, has problems when trying to refine the partition of an interval to optimize the calculation. • Associates the definite integral concept with the process of calculating its value, without checking the conditions to apply the procedure. • Software is used to support calculations made with pencil and paper. • Graphic representation is only used if it is provided.
3	Blenda Samuel	• Tends to use *Derive* and/or the Utility File while solving problems • Applies the idea of approximation to determine areas of regions. • Uses graphic representations when solving problems. • Perceives that the calculation of the definite integral is not only the application of a formula or the use of software and/or Utility File. • Faced with a general proposition, not able to give a coherent argument to support an answer.

This student's tendency to approach the problem through the use of the software is evident. At this stage, the interviewer directs the dialogue to understand the meaning of this process and eventually the student seems to comprehend what he is doing.

Table 1 represents the type of profile, the students, and main features associated with these types of student problem-solving behavior. We can not say that there is any relationship between the profiles we have established and the typologies of behavior defined by Guin and Trouche (1999) during the students' learning process. It could be feasible to compare both categorizations in the future.

Discussion of Results

It is important to reflect on relevant aspects that appear throughout the study in accordance with ideas set forth in the framework used to structure it. An overarching idea around the design of this study is the recognition that in order to develop students' understanding they need themselves to problematize the concept of definite integral in terms of posing and pursuing substantial questions associated with that concept. However, the process of formulating questions and looking for arguments to respond to or support a particular relation demands that students develop a mathematical disposition to access and examine constantly previously learned knowledge. There is evidence that the use of the tool became important for the students to identify and explore basic concepts involved in the study of definite integral. For example, when they used the Utility File, they had the opportunity to work on various examples to calculate approximately area values associated with bounded regions. Here, they focused on examining aspects that included the relationship between the domain of the function in question and feasible partitions to calculate approximately corresponding areas. In particular, the use of the Utility File provided students with basic elements to visualize the concept of limit involved in calculating definite integrals. In addition, students, in general, used the *Derive* software to represent functions graphically and to calculate integrals. In this context, the software became an important tool for students to identify intersections of the graph and the x-axis and the position of the region (above or below the x-axis). In some cases, they utilized the software as a means to validate results that they had obtained through paper and pencil procedures. Thus the use of the tool seemed to help students identify a set of basic referents (e.g., domain of a function, interval partitions, concept of limit, function graphics) that are important in understanding and solving problems associated with the concept of definite integral. Students' recognition of a set of referents associated with the concept is an important step to reflect on ways to discuss their relations and applications in problem solving activities (Thurston, 1994).

It was also noticeable that the type of problems that students were asked to solve demanded that they reflect on the meaning of fundamental ideas associated with the concept of definite integral and area. Thus, students needed to interpret graphical representation of a particular function in order to identify integration limits or conditions to calculate the corresponding integral. There is evidence that students, in general, experienced difficulties in making sense of graphical representations at particular points (e.g., discontinuous functions) in order to identify and calculate the corresponding area. Guin and Trouche (2002) suggest paying attention to the process by which students transform an artifact into a problem-solving tool. In this case, students initially seemed to read or interpret the software output superficially without relying on conceptual ideas. For example, some students tried to identify the roots of a function by moving the cursor close to the intersection point of the function graph (provided by the software) and the x-axis, instead of solving the corresponding equation. Thus, the transformation of the artifact into problem solving tool not only involves design and command use, but also the constraints induced by its use (Artigue, 2002).

An important issue that emerged from analysis of students' work is that in order to comprehend and explore connections and relationships between concepts connected with the study of definite integral, they need to make the transition, in terms of meaning, between the distinct representations of the concept (Duval, 1993; Santos, 2000). For example, when dealing with the graphical representation of a function it was important to explain the position of the graph and both its relation to the sign associated with the value of its integral and the process of approximation to the value of its area through a numerical approach (area of small rectangles). In addition, students need to develop problem-solving strategies that help them think of cases beyond those embedded in particular problems. That is, it was evident that when students were asked to work on problems that involved general statements, they experienced serious difficulties in constructing examples or counter-examples that could help them to understand and explore the situation in general terms. In this respect, it becomes important to design and implement instructional activities that include the use of the software in problem-solving contexts in which students have opportunities to develop basic problem solving strategies (including metacognitive ones). Finally, implementation of students' learning activities with the use of technology should value the ways in which students present and communicate their results. In this study, the task-based interviews functioned as tools for reflection by which students had opportunities to reveal their ideas and at the same time to explain and examine in detail connections between the various representations of the problem. At this stage, it was observed that students had not often thought of those connections and the very questions asked by the interviewer (researcher) became an opportunity for students to enhance their understanding. In this regard, it was important that the class itself should be seen as a community that demands constant reflection from each of its members.

Concluding Remarks

Learning or developing mathematical ideas is a continuous process in which students need to constantly examine and refine previous knowledge to expand and learn new knowledge and solve new mathematical problems. Results from this research indicate that the first year engineering students who participated in this study experienced serious difficulties in accessing and coordinating previously studied basic knowledge to approach problems in a flexible way. It seems that they might have acquired sets of isolated facts or mathematical results, but had not developed problem-solving approaches to move flexibly from one setting to another. The problem solving activities implemented throughout the course helped students be aware of the importance for them to share and discuss their ideas with other students and also to explore ideas and problems from various perspectives. In general, students also recognized the potential of using the *Derive* software to deal with the problems; also they became aware of the process involved in transforming their ways of thinking of the problems in order to make efficient use of the tool. This process demands that they explain the meaning and interpret in different ways (using various resources) representations and results obtained through the use of the tool.

Although the use of the software may provide an interesting means for students to free themselves from memorizing formulae or calculation procedures, it is also important to recognize that students need time to mature and develop a firm conceptual understanding of the definite integral. In particular, students need to pay attention to the process of transforming and connecting relationships among graphic, algebraic and numerical representations. This seems to be a crucial step in order for students to develop deep understanding of the definite integral concept. The initial difficulties that students showed during the interaction with the tasks offered important information to orient and consider relevant questions that students needed to focus their reflection. Thus, working on the three different scenarios (paper and pencil, use the software, and interviews) generated useful information to identify not only what conceptual aspects or ideas students needed to review, but also to recognize the importance of students expressing their ideas in writing and oral forms. In addition, students need to develop a set of problem-solving strategies that would help them decide when to use and monitor the work done through the use of the software. Students' task-based interviews not only provided important information regarding students' competences but also became a tool for reflection by which students had the opportunity to extend their knowledge. It was also observed that students can initially study themes related to the definite integral to develop meaning attached to fundamental concepts and later study topics that include indefinite integrals.

Acknowledgments

This work has been partially supported by Contract No SEJ2005-08499 from DGI of the Spanish Ministry of Education, CIMAC Foundation (Centro de Investigación Matemática de Canarias) and Conacyt, Mexico, reference # 47850. A short version of this paper appeared in *Proceedings of the 26th annual meeting of the North American Chapter of the International Group for the Psychology of Mathematics Education,* 2004, Vol. 2, pp. 447-454. Toronto: OISE/UT.

References

Artigue, M. (2002). Learning mathematics in a CAS environment: The genesis of a reflection about instrumentation and the dialectics between technical and conceptual work. *International Journal of Computers for Mathematical Learning*, *7*(3), 245–274.

Camacho, M., & Depool, R. (2003). Un estudio gráfico y numérico del cálculo de la Integral Definida utilizando el Programa de Cálculo Simbólico (PCS) DERIVE. *Educación Matemática*, *15*(3), 119–140.

Cuoco, A. (2002). Thoughts on reading Artigue's "Learning mathematics in a CAS environment". *International Journal of Computers for Mathematical Learning*, *7*(3), 293–299.

Drijvers, P. (2002). Learning mathematics in a computer algebra environment: Obstacles are opportunities. *ZDM*, *34*(5), 221–228.

Duval, R. (1993). Registres de représentation sémiotique et fonctionnement cognitif de la pensée. *Annales de Didactique et de Science Cognitives*, *5*, 37–65.

Goldin, G. A. (1998). Representations, learning, and problem solving in mathematics. *Journal of Mathematical Behavior*, *17*(2), 137-165.

Guin, D., & Trouche, L. (1999). The complex process of converting tools into mathematical instruments: The case of calculators. *International Journal of Computers for Mathematical Learning*, *3*(3), 195–227.

Guin, D., & Trouche, L. (2002). Mastering by the teacher of the instrumental genesis in CAS environments: Necessity of instrumental orchestrations. *ZDM*, *34*(5), 204–211.

Heid, K. M. (2002). How theories about the learning and knowing of mathematics can inform the use of CAS in school mathematics: One perspective. *International Journal of Computer Algebra in Mathematics Education*, *9*(2), 95–112.

Ruthven, K. (2002). Instrumenting mathematical activity: Reflections on key studies of the educational use of computer algebra systems. *International Journal of Computers for Mathematical Learning*, *7*(3), 275–291.

Santos, M. (2000). The use of representations as a vehicle to promote students' mathematical thinking in problem solving. *International Journal of Computer Algebra in Mathematics Education*, *7*(3), 193–212.

Thompson, P. W., & Saldanha, L. A. (2003). Fractions and multiplicative reasoning. In J. Kilpatrick, W. G. Martin, & D. Schifter (Eds.), *A research companion to the Principles and Standards for School Mathematics* (pp. 95–113). Reston, VA: National Council of Teachers of Mathematics.

Thurston, W. P. (1994). On proof and progress in mathematics. *Bulletin of the American Mathematical Society*, *30*(2), 161-177.

Trouche, L. (2005). An instrumental approach to mathematics learning in symbolic calculator environment. In D. Guin, K. Ruthven, & L. Trouche (Eds.), *The didactical challenge of symbolic calculators. Turning a computational device into a mathematical instrument* (pp. 137–162). New York: Springer.

Appendix 1: Utility File

```
#1:  F(x) :=

#2:  RECTANGULO(a, b, h) := [[a, 0], [a, h], [b, h], [b, 0]]

#3:  TRAPECIO(a, α, b, β) := [[a, 0], [a, α], [b, β], [b, 0]]

#4:  MEDIDA_RECT(c, d) := c·d

#5:  MEDIDA_TRAP(c, d, h) := (c + d)·h / 2

#6:  H(a, b, n) := (b - a) / n

#7:  FMIN(a, b, n) := MIN(F(a + i·H(a, b, n)), F(a + (i + 1)·H(a, b, n)))

#8:  FMAX(a, b, n) := MAX(F(a + i·H(a, b, n)), F(a + (i + 1)·H(a, b, n)))

#9:  XL(a, b, n) := a + (i - 1)·H(a, b, n)

#10: XR(a, b, n) := a + i·H(a, b, n)

#11: XM(a, b, n) := a + (i - 0.5)·H(a, b, n)

#12: XLP(a, b, n) := a + 2·(i - 1)·H(a, b, n)

#13: XRP(a, b, n) := a + 2·i·H(a, b, n)

#14: XMP(a, b, n) := a + (2·i - 1)·H(a, b, n)

#15: RECT_INF_SOBRE_EL_EJE_X(a, b, n) := VECTOR(RECTANGULO(a + i·H(a, b, n), a + (i + 1)·H(a,
       b, n), FMIN(a, b, n)), i, 0, n - 1)

#16: RECT_INF_BAJO_EL_EJE_X(a, b, n) := VECTOR(RECTANGULO(a + i·H(a, b, n), a + (i + 1)·H(a,
       b, n), FMAX(a, b, n)), i, 0, n - 1)

#17: RECT_SUP_SOBRE_EL_EJE_X(a, b, n) := VECTOR(RECTANGULO(a + i·H(a, b, n), a + (i + 1)·H(a,
       b, n), FMAX(a, b, n)), i, 0, n - 1)

#18: RECT_SUP_BAJO_EL_EJE_X(a, b, n) := VECTOR(RECTANGULO(a + i·H(a, b, n), a + (i + 1)·H(a,
       b, n), FMIN(a, b, n)), i, 0, n - 1)

#19: RECT_EXTREMO_IZQUIERDO(a, b, n) := VECTOR(RECTANGULO(a + i·H(a, b, n), a + (i + 1)·H(a,
       b, n), F(a + i·H(a, b, n))), i, 0, n - 1)

#20: RECT_EXTREMO_DERECHO(a, b, n) := VECTOR(RECTANGULO(a + i·H(a, b, n), a + (i + 1)·H(a, b,
       n), F(a + (i + 1)·H(a, b, n))), i, 0, n - 1)

#21: RECT_PTO_MEDIO(a, b, n) := VECTOR(RECTANGULO(a + i·H(a, b, n), a + (i + 1)·H(a, b, n),
       F(a + (2·i + 1)·H(a, b, n)/2)), i, 0, n - 1)

#22: TRAPECIOS(a, b, n) := VECTOR(TRAPECIO(a + i·H(a, b, n), F(a + i·H(a, b, n)), a + (i +
       1)·H(a, b, n), F(a + (i + 1)·H(a, b, n))), i, 0, n - 1)

#23: PARAB_SIMP(a, b, n) := VECTOR(FIT([[x, p·x^2 + q·x + r],
                                         [XLP(a, b, n), F(XLP(a, b, n))],
                                         [XMP(a, b, n), F(XMP(a, b, n))],
                                         [XRP(a, b, n), F(XRP(a, b, n))]]), i, 1, n/2)

#24: CURVA_SIM(a, b, n) := VECTOR(CHI(XLP(a, b, n), x, XRP(a, b, n))·ELEMENT(PARAB_SIMP2(a,
       b, n), i), i, 1, n/2)

#25: SEGMENTOS_SIM(a, b, n) := VECTOR([[XLP(a, b, n), F(XLP(a, b, n))],
                                       [XLP(a, b, n), 0],
                                       [XRP(a, b, n), 0],
                                       [XRP(a, b, n), F(XRP(a, b, n))]], i, 1, n/2)
```

```
                                                   n - 1
#26: MEDIDA_RECT_INF_SOBRE_EL_EJE_X(a, b, n) :=     Σ    MEDIDA_RECT(FMIN(a, b, n), H(a, b, n))
                                                   i=0

                                                 n - 1
#27: MEDIDA_RECT_INF_BAJO_EL_EJE_X(a, b, n) :=    Σ    MEDIDA_RECT(FMAX(a, b, n), H(a, b, n))
                                                 i=0

                                                   n - 1
#28: MEDIDA_RECT_SUP_SOBRE_EL_EJE_X(a, b, n) :=     Σ    MEDIDA_RECT(FMAX(a, b, n), H(a, b, n))
                                                   i=0

                                                 n - 1
#29: MEDIDA_RECT_SUP_BAJO_EL_EJE_X(a, b, n) :=    Σ    MEDIDA_RECT(FMIN(a, b, n), H(a, b, n))
                                                 i=0

                                           n - 1
#30: MEDIDA_EXTREMO_IZQUIERDO(a, b, n) :=    Σ    MEDIDA_RECT(F(a + i·H(a, b, n)), H(a, b, n))
                                           i=0

                                         n - 1
#31: MEDIDA_EXTREMO_DERECHO(a, b, n) :=    Σ    MEDIDA_RECT(F(a + (i + 1)·H(a, b, n)), H(a, b,
                                         i=0

     n))

                                    n - 1                ⎛ ⎛                   H(a, b, n) ⎞
#32: MEDIDA_PUNTO_MEDIO(a, b, n) :=   Σ    MEDIDA_RECT   ⎜F⎜a + (2·i + 1)·──────────⎟, H(a, b,
                                    i=0                  ⎝ ⎝                       2      ⎠

     n)⎞
       ⎠

                                  n - 1
#33: MEDIDA_TRAPECIO(a, b, n) :=    Σ    MEDIDA_TRAP(F(a + i·H(a, b, n)), F(a + (i + 1)·H(a, b,
                                  i=0

     n)), H(a, b, n))

                              n/2
#34: MEDIDA_SIMP(a, b, n) :=   Σ
                              i=1

      H(a, b, n)·(F(XLP(a, b, n)) + 4·F(XMP(a, b, n)) + F(XRP(a, b, n)))
      ──────────────────────────────────────────────────────────────────
                                      6

                                         ⎡                    REC.INF
#35: MEDIDA_APROX_SOBRE_X(a, b, n) :=    ⎢
                                         ⎣ MEDIDA_RECT_INF_SOBRE_EL_EJE_X(a, b, n)

          PUNTO MEDIO                    TRAPECIOS                  SIMPSON
     MEDIDA_PUNTO_MEDIO(a, b, n)  MEDIDA_TRAPECIO(a, b, n)  MEDIDA_SIMP(a, b, n)
                     RECT.SUP                   ⎤
     MEDIDA_RECT_SUP_SOBRE_EL_EJE_X(a, b, n)    ⎦

                                         ⎡                    REC.INF
#35: MEDIDA_APROX_SOBRE_X(a, b, n) :=    ⎢
                                         ⎣ MEDIDA_RECT_INF_SOBRE_EL_EJE_X(a, b, n)

          PUNTO MEDIO                    TRAPECIOS                  SIMPSON
     MEDIDA_PUNTO_MEDIO(a, b, n)  MEDIDA_TRAPECIO(a, b, n)  MEDIDA_SIMP(a, b, n)
                     RECT.SUP                   ⎤
     MEDIDA_RECT_SUP_SOBRE_EL_EJE_X(a, b, n)    ⎦

                                        ⎡                    REC.INF
#36: MEDIDA_APROX_BAJO_X(a, b, n) :=    ⎢
                                        ⎣ MEDIDA_RECT_INF_BAJO_EL_EJE_X(a, b, n)

          PUNTO MEDIO                    TRAPECIOS                  SIMPSON
     MEDIDA_PUNTO_MEDIO(a, b, n)  MEDIDA_TRAPECIO(a, b, n)  MEDIDA_SIMP(a, b, n)
                     RECT.SUP                   ⎤
     MEDIDA_RECT_SUP_BAJO_EL_EJE_X(a, b, n)     ⎦

#37: MEDIDA_MATRIZ_APROX__SOBRE_X(a, b, j, k, m) := VECTOR([n,
     MEDIDA_RECT_INF_SOBRE_EL_EJE_X(a, b, n), MEDIDA_PUNTO_MEDIO(a, b, n),
     MEDIDA_TRAPECIO(a, b, n), MEDIDA_SIMP(a, b, n), MEDIDA_RECT_SUP_SOBRE_EL_EJE_X(a, b,
     n)], n, j, k, m)

#38: MEDIDA_MATRIZ_APROX__BAJO_X(a, b, j, k, m) := VECTOR([n,
     MEDIDA_RECT_INF_BAJO_EL_EJE_X(a, b, n), MEDIDA_PUNTO_MEDIO(a, b, n),
     MEDIDA_TRAPECIO(a, b, n), MEDIDA_SIMP(a, b, n), MEDIDA_RECT_SUP_BAJO_EL_EJE_X(a, b,
     n)], n, j, k, m)

                                             n - 1
#39: LIMITE_SUMA_DE_RIEMANN(a, b) := lim      Σ    REGION_RECT(F(a + i·H(a, b, n)), H(a, b, n))
                                     n→∞     i=0
```

Appendix 2: The Problems

TABLE 2. First Group of Questions (Problems Type 1)

In these problems it is expected that students were able to:

- Recognize algebraic and graphic representations of each problem, and make corresponding transformations and interpretations to explore properties associated with each representation.
- Construct algebraic and or numeric representations of the problem, and make transformations to analyze the problem.
- Make a coordinated conversion among the representations of the problem.

Question	*What does the question involve?*	*Scenario*	*Purposes (Theoretical)*
(Q2) Given the graph, calculate the area of the shaded region. $f(x) = 2x - x^2$	2.1. Students are asked to calculate the area. 2.2. The region is graphically identified. 2.3. The word integral does not appear. 2.4. The intervals are not given though identified in the graph. 2.5. There are two regions (over and under). 2.6. There is an algebraic expression associated with the curve. 2.7. They are not asked to approximate the area. 2.8. The function is continuous.	1. Pencil & paper 2. CAS 3. Interview	Determine if the student understands the way to get the area of the regions that are under the x-axis in terms of definite integral, as well as the relationships established with the regions over the x-axis.
(Q8) Given the graph of the function, if possible, calculate the area of the shaded region. If not, justify your answer. $f(x) = \frac{1}{x^2}$	8.1. Students are asked to calculate the area. 8.2. The region is graphically identified. 8.3. The word integral does not appear. 8.4. The intervals are not given though identified in the graph. 8.5. There is an algebraic expression associated with the curve. 8.6. There is an associated graph. 8.7. There is a region with a lack of continuity at one point. 8.8. A justification is asked for in case a calculation cannot be made.	1. Pencil & paper 2. CAS	Analysis of how students deal with the area of the region in question when this is infinite.

TABLE 3. First Group of Questions (Continuation, Problems Type 1)

Question	*What does the question involve?*	*Scenario*	*Purposes (Theoretical)*
Q9) Given the function defined $f(x) = \begin{cases} -x^2+3 & \text{if} \quad x < -1 \\ 1.36 & \text{if} \quad x = -1 \\ x^2 & \text{if} \quad -1 < x < 2 \\ -x+4 & \text{if} \quad x > 2 \end{cases}$ 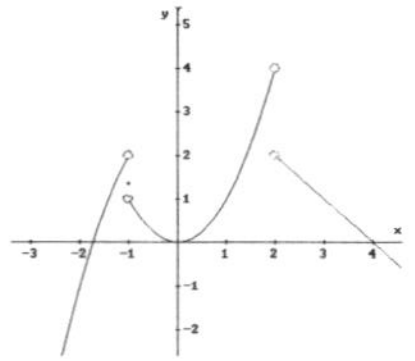 • Calculate, if possible, the area of the region bounded by the curve in the interval $[-2, 3]$ • If possible, estimate the value of the definite integral in that interval. • If it is not possible to calculate the whole area, calculate the portion or portions that can be calculated in the interval $[-2, 3]$. In case it is not possible to calculate any portion, explain why.	9.1. The whole domain is defined. 9.2. It is set with both the algebraic and graphic registers. 9.3. Students are asked to calculate the area. 9.4. The function has two points of discontinuity. 9.5. The word integral does not appear. 9.6. There is an algebraic expression associated with the curve. 9.7. There is an associated graph.	1. Pencil & paper 2. CAS	Determine if the student is able to apply the approximation method to find the area or to use the Fundamental Theorem of Calculus in spite of the lack of continuity.
Q1) How would you explain to someone the meaning of $\int_a^b f(x)dx$?	1.1. The term definite integral does not appear. 1.2. The explicit or particular expression of the function is not given. 1.3. The term area does not appear. 1.4. There is no associated graph. 1.5. The values of the integration limits are not given. 1.6. The expression is given without referring to the definite integral or the area.	1. Pencil & paper 3. Interview	Investigate the elements students use to translate into their own vocabulary what each understands by definite integral.

TABLE 4. Second Group of Questions (Problems Type 2)

In this group of problems it was expected that students were able to:
- Construct graphic and/or numeric representations, and make corresponding transformations (treatments-processing) within each.
- Recognize the algebraic representations and carry out corresponding transformations (treatments-processing).
- Make coordinated conversions among the representations.

Question	*What does the question involve?*	*Scenario*	*Purposes (Theoretical)*
Q3) Calculate the Definite Integral $\int_{-3}^{4} \lvert x+1\rvert dx$, in the simplest way.	3.1. It is presented in the algebraic register – a lot of implicit information. 3.2. The word area is not mentioned. 3.3. The integration interval is given. 3.4. There is no associated graph. 3.5. The solution can be thought of in a simple way since it involves the area of right triangles.	1. Pencil & paper 2. CAS 3. Interview	Analyze: • Whether students transfer basic knowledge from using area formulae to solve definite integrals. • Whether the student knows and correctly works with some properties of the definite integral.
Q4) Indicate if the next statement is true or false. Justify your answer. $\int_0^2 \frac{1}{(x-1)^2} dx = \int_0^2 (x-1)^{-2} dx$ $= \frac{(x-1)^{-2+1}}{-2+1}\Big\rvert_0^2$ $= -\frac{1}{(x-1)}\Big\rvert_0^2$ $= -\frac{1}{1} - \left(-\frac{1}{-1}\right)$ $= -2$	4.1. It is presented in the algebraic register with implicit information. 4.2. The word area does not appear in "calculate the integral." 4.3. The software directly solves it (in a wrong way). 4.4. The integration interval is given. 4.5. There is no associated graph. 4.6. Integration techniques are directly applied.	1. Pencil & paper 2. CAS	Determine if the student: • Uses the graphical representation to justify an answer. • Coherently interprets the hypothesis that functions must fulfill so that Barrow's rule can be applied.
Q7) Calculate the area formed with the x-axis and the function $f(x) = 2x^4 - 2x^3 - 14x^2 + 2x + 12$.	7.1. Algebraic representation is used (4th order polynomial). 7.2. The interval is not given. 7.3. The word integral is not mentioned. 7.4. Students are asked to calculate an area. 7.5. There is no associated graph.	3. Interview	Determine if the student: • Uses software or pencil-paper techniques to graph the function. . • Identifies the regions for integration through the intersection of the function with the x-axis. • Presents and calculates the integrals with Barrow's rule. Student might use the Utility File when calculating the approximations.

TABLE 5. Third Group of Questions (Problems Type 3)

In this group of problems it was expected that students were able to:
- Construct graphic and/or numeric representations, and make corresponding transformations (treatments-processing) within each.
- Recognize the algebraic representations and carry out corresponding transformations (treatments-processing).
- Make coordinated conversions among the representations.

Question	*What does the question involve?*	*Scenario*	*Purposes (Theoretical)*
Q5) Indicate whether the following proposition is true or false. Justify your answer. If $f(x) \geq g(x)$, then $\int_a^b f(x)dx \geq \int_a^b g(x)dx$ for all x that belong to $[a, b]$.	5.1. Explicit expressions of the intervening functions are not given. 5.2. The term area does not appear. 5.3. The word integral does not appear. 5.4. There is no associated graph. 5.5. Values for integration limits not given. 5.6. The integration interval in the statement is not mentioned. 5.7. No direct reference to the Definite Integral but area is mentioned. 5.8. The expression is complex: it is a proposition which connects two situations. 5.9. In general, the proposition is true.	1. Pencil & paper 2. CAS 3. Interview	Determine if the student is able to understand the general terms presented in the statement and if he/she establishes relationships between the area and the Definite Integral.
Q6) Indicate whether the following proposition is true or false. Justify your answer. If $\int_a^b f(x)dx \geq \int_a^b g(x)dx)$, then $f(x) \geq g(x)$ for all x that belong to $[a, b]$.	6.1. Explicit expressions of the intervening functions are not given. 6.2. The term area does not appear. 6.3. The word integral does not appear. 6.4. There is no associated graph. 6.5. Values for integration limits not given. 6.6. The integration interval in the statement is not mentioned. 6.7. No direct reference to the Definite Integral; area is not mentioned. 6.8. The presentation is complex: it is a proposition connecting two situations. 6.9. In general, the proposition is false whether it is integral or area.	1. Pencil & paper 2. CAS 3. Interview	Determine if the student is able to understand the general terms presented and if he/she establishes relationships between the area and the Definite Integral. Also if he/she uses counterexamples in his/her justification

Appendix 3: Categorization of Student Work in Scenarios 1 and 2

TABLE 6. First Group of Questions

<table>
<tr><td></td><td colspan="2">Participants: 31 students answered the questionnaire individually using paper and pencil</td><td>Target students</td></tr>
<tr><td rowspan="9">Question 2</td><td colspan="2">6 students recognized and solved two integrals associated with the problem. 4 of these 6 students identified in the graphic representation the area regions explicitly (domains) and 2 directly wrote the two integrals to be solved.</td><td>Javier</td></tr>
<tr><td rowspan="2">17 students used a numeric approach.</td><td>11 drew elementary figures (rectangles and/or trapezoids) on the graphic representation; 8 of these calculated the heights using the algebraic expression for the function and the longitude of the base for $\Delta x = \frac{b-a}{x}$ and 3 showed only the elementary figures on the graph</td><td>Samuel
María</td></tr>
<tr><td>6 applied limit of Riemann Sums.</td><td>Blenda
George
Mirvic</td></tr>
<tr><td colspan="2">8 students did not respond.</td><td></td></tr>
<tr><td colspan="2">Participants: 26 students working in pairs answered the questionnaire using Derive software.</td><td>Target students</td></tr>
<tr><td rowspan="2">11 pairs did not identify the function and solved two integrals. No calculations shown identifying intersection with the x-axis.</td><td>9 pairs solved absolute value of the integrals or by changing the sign of the integral of the portion under the x-axis and summed up their values.</td><td>Blenda
George
María
Mirvic</td></tr>
<tr><td>2 pairs calculated the integrals, without changing the sign of the value of the integral that represents the portion under the x-axis and added up both values.</td><td></td></tr>
<tr><td colspan="2">1 pair represented the function and calculated the values of the absolute value of the integral and added them up.</td><td>Javier</td></tr>
<tr><td colspan="2">1 pair did not respond.</td><td>Samuel</td></tr>
</table>

TABLE 7. First Group of Questions

<table>
<tr><td></td><td colspan="3">Participants: 31 students answered the questionnaire individually using paper and pencil</td><td>Target students</td></tr>
<tr><td rowspan="9">Question 8</td><td rowspan="3">29 affirmed that it was not possible to calculate the area.</td><td colspan="2">5 solved the integral.</td><td>Blenda
George</td></tr>
<tr><td colspan="2">16 stated that the shaded region was infinite.</td><td>María
Mirvic
Samuel</td></tr>
<tr><td colspan="2">8 students mentioned that the function was not continuous or that it had an asymptote.</td><td>Javier</td></tr>
<tr><td colspan="3">1 student stated that it was possible to calculate the area, and solved the integral.</td><td></td></tr>
<tr><td colspan="3">1 student did not respond.</td><td></td></tr>
<tr><td colspan="3">Participants: 26 students, working in pairs, answered the questionnaire using Derive software.</td><td></td></tr>
<tr><td rowspan="3">All 13 pairs mentioned that it was not possible to calculate the area.</td><td colspan="2">3 pairs represented the function and said the region was infinite or that the graph was not bound.</td><td>Samuel</td></tr>
<tr><td rowspan="2">10 pairs did not represent the function.</td><td>7 pairs mentioned that the function was not continuous or that the function spreads to infinite when x approaches zero or that there is an asymptote at $x = 0$.</td><td>Blenda
George
Javier
María
Mirvic</td></tr>
<tr><td>3 pairs mentioned the region was closed and conditions for calculating the integral were not fulfilled.</td><td></td></tr>
</table>

<table>
<tr><td></td><td colspan="2">Participants: 31 students answered the questionnaire individually using paper and pencil</td><td>Target students</td></tr>
<tr><td rowspan="10">Question 9</td><td colspan="2">9 students solved the integral. 2 of these students identified the regions to calculate the area.</td><td>Blenda
George
Javier
Samuel</td></tr>
<tr><td colspan="2">9 affirmed that it was not possible to calculate the area since the function was not continuous.</td><td></td></tr>
<tr><td colspan="2">1 student calculated the area for numeric approach and drew rectangles on the region to calculate it.</td><td></td></tr>
<tr><td colspan="2">12 students did not respond.</td><td>Mirvic</td></tr>
<tr><td colspan="2">Participants: 26 students, working in pairs, answered the questionnaire using Derive software.</td><td></td></tr>
<tr><td rowspan="3">10 pairs mentioned that the integral could not be solved since the function was not continuous.</td><td>4 pairs said they would calculate the area approximating values near the discontinuity points, and solve the corresponding integral.</td><td>George
Javier</td></tr>
<tr><td>3 pairs solved the integral by considering the end points (x-axis) as the integration limits.</td><td>Blenda
Mirvic</td></tr>
<tr><td>3 pairs did not show calculations.</td><td></td></tr>
<tr><td colspan="2">2 pairs mentioned that it was possible to calculate the integral by considering values near the discontinuity points.</td><td>María
Samuel</td></tr>
<tr><td colspan="2">1 pair did not respond.</td><td>Samuel</td></tr>
</table>

TABLE 8. Second Group of Questions

	Participants: 31 students answered the questionnaire individually using paper and pencil		*Target students*
Question 3	28 students did not graphically represent the function absolute value and solve the integral directly.	14 students completed two integrals respectively taking the intervals $[-3,-1]$ and $[-1,-4]$.	George Javier María Mirvic
		11 students considered zero as one of the limits of integration.	Blenda
		3 students solved two integrals in $[-3,4]$.	Samuel
	3 students did not respond.		
	Participants: 26 students, working in pairs, answered the questionnaire using Derive *software.*		
	All 13 pairs calculated the integral using the command "Calculation" in *Derive.*		All six

	Participants: 31 students answered the questionnaire individually using paper and pencil		*Target students*
Question 4	23 students affirmed that the proposition was true and they justified their answers redoing the calculations.		All six
	8 answered that the proposition was false and their arguments were that (a) the denominator could be expanded and (b) others indicated that the result was zero.		
	Participants: 26 students, working in pairs, answered the questionnaire using Derive *software.*		
	All 13 pairs affirmed that the proposition was true.	5 pairs justified their answer redoing the calculations.	María
		5 pairs justified their answers through the application of the Fundamental Theorem of Calculus.	Javier Mirvic
		3 pairs calculated the integral with *Derive* and used the result to support their answer.	Blenda George Samuel

TABLE 9. Third Group of Questions

<table>
<tr><td></td><td colspan="2">Participants: 31 students answered the questionnaire individually using paper and pencil</td><td>Target students</td></tr>
<tr><td rowspan="11">Question 5</td><td rowspan="3">9 students answered that the proposition was true.</td><td>1 student justified the statement by calculating the integral of particular functions.</td><td>Samuel</td></tr>
<tr><td>2 students relied on a graphical representation to justify their responses.</td><td>George</td></tr>
<tr><td>6 students mentioned that it was a property of the definite integral.</td><td>Blenda</td></tr>
<tr><td colspan="2">8 students stated that the proposition was false because the limits of the integral were not given and the function was not continuous.</td><td>Javier
María
Mirvic</td></tr>
<tr><td colspan="2">14 students did not respond.</td><td></td></tr>
<tr><td colspan="2">Participants: 26 students, working in pairs, answered the questionnaire using Derive software.</td><td></td></tr>
<tr><td rowspan="3">6 pairs mentioned that the proposition was true.</td><td>3 pairs solved particular examples to prove that statement was true.</td><td>George</td></tr>
<tr><td>2 pairs mentioned that it was a property of the definite integral.</td><td></td></tr>
<tr><td>1 pair wrote the proposition in words.</td><td>Blenda</td></tr>
<tr><td colspan="2">7 pairs affirmed that the proposition was false, indicating that the function was not continuous.</td><td>Javier
María
Mirvic
Samuel</td></tr>
</table>

TABLE 10. Third Group of Questions (continued)

	Participants: 31 students answered the questionnaire individually using paper and pencil		*Target students*
Question 6	18 affirmed that the proposition was true.	1 student justified the proposal by writing the integral of particular functions.	Samuel
		1 student sketched a graph to explain his answer.	
		16 students provided theoretical justifications, such as that the proposition was the reciprocal of Proposition 5, it was a consequence of the Fundamental Theorem of Calculus, etc.	Blenda George Javier María Mirvic
	2 students affirmed that the proposition was false and argued that the functions were not defined.		
	11 students did not respond.		
	Participants: 26 students, working in pairs, answered the questionnaire using Derive *software.*		
	All 13 pairs affirmed that the proposition was true.	8 pairs used particular cases to explain their response.	Blenda George María Samuel
		3 pairs mentioned that it was a property of the definite integral.	Javier Mirvic
		1 pair mentioned that it was the reciprocal of Proposition 5.	
		Only 1 pair stated the proposition in their own words.	

DEPARTMENT OF MATHEMATICAL ANALYSIS, UNIVERSITY OF LA LAGUNA, AVDA. FRANCISCO SÁNCHEZ S/N 38271, CANARY ISLANDS. SPAIN.
E-mail address: mcamacho@ull.es

UNEXPO. BARQUISIMETO. MATHEMATICS DEPARTMENT, AVENIDA CORPAHUAICO, 3001, BARQUISIMETO, LARA STATE. VENEZUELA
E-mail address: rdepool@bqto.unexpo.es

CINVESTAV. MATHEMATICS EDUCATION DEPARTMENT, AV. IPN 2508; SN PEDRO ZACATENCO 07360, D.F. MÉXICO
E-mail address: msantos@cinvestav.mx

CBMS Issues in Mathematics Education
Volume **16**, 2010

Mathematicians' Perspectives on the Teaching and Learning of Proof

Lara Alcock

Abstract. This paper reports on an exploratory study of mathematicians' views on the teaching and learning that occurs in a course designed to introduce students to mathematical reasoning and proof. Based on a sequence of interviews with five mathematicians experienced in teaching the course, I identify four modes of thinking that these professors indicate are used by successful provers. I term these instantiation, structural thinking, creative thinking and critical thinking. Through the mathematicians' comments, I explain these modes and highlight ways in which students sometimes fail to use them effectively. I then discuss teaching strategies described by the participants, relating these to the four modes of thinking. I argue that teaching aimed at improving structural thinking tends to dominate, and that courses which introduce proof, regardless of classroom organization, should address all four modes in a balanced and integrated way.

1. Introduction

The research reported here addresses the teaching and learning of mathematical proof through the perspectives of mathematicians who teach a course entitled Introduction to Mathematical Reasoning. This is a "transition course," designed to facilitate the move from calculation-oriented mathematics courses to proof-oriented courses such as real analysis and abstract algebra. Like other such courses, it involves elementary work on logic, set theory, relations, functions, and various types of proof (contradiction, mathematical induction, etc.), with some of this done within mathematical topics such as number theory and combinatorics. One goal is to familiarize students with the mechanics of proof, so that when they enter upper level courses they will be able to concentrate on the content.

Research on transition courses has focused largely on two issues. One is student learning in such contexts, with a frequent finding that this is less than we might hope (Moore, 1994). The second is alternative pedagogical approaches to introducing proof, with researchers investigating collaborative classes (Blanton & Stylianou, 2003). The study reported here took a different approach, investigating the teaching and learning that occurs in this particular transition course from the

The author wishes to thank the mathematicians who took part in this research, along with Keith Weber and Adrian Simpson for their valuable comments on earlier drafts.

©2010 American Mathematical Society

perspective of the mathematicians for whom it is part of their regular teaching. It involved in-depth interviews with five such mathematicians, and I aimed to investigate the participants' experiences *as they reported them*, with no attempt to establish whether their comments constituted an accurate report of their own or their students' behavior in the classroom. This was a deliberate decision, taken for three reasons. First, because I wanted to establish a relationship with the professors based on good faith and respect, which would have been difficult if they felt that their comments and their teaching stood to be negatively evaluated. Second, because I believe that if the mathematics education community wishes to engage mathematicians in discussions about their teaching, it needs to understand what they perceive as their most pressing concerns and to take these seriously. Third, because the participants have an enormous amount of collective experience that constitutes a rich source of information about the learning and teaching of proof in its own right.

This paper begins with a review of the mathematics education literature on the teaching and learning of proof and a detailed description of the methods used to collect and analyze the interview data. Following this, the results are reported in two distinct sections. The first of these, Section 5, describes the four modes of thinking that I discerned in the professors' discussions of proving, together with their comments on their students' use of these modes. The other, Section 6, describes the teaching strategies the professors use, relates these to the four modes, and concludes with the main points of an ongoing debate among the mathematician participants about what should be the nature of the course. In Section 7, I discuss ways in which I believe these results should influence our thinking about how we teach transition-to-proof courses and other advanced mathematics courses. The paper concludes with a short discussion of some remaining issues and directions of ongoing research.

2. Previous Work on the Teaching and Learning of Proof

There has been much research on university students' difficulties with both constructing and evaluating proofs. Students' standards of justification may differ from those of mathematical proof, so that they offer empirical arguments where deductive ones are required (Harel & Sowder, 1998; Recio & Godino, 2001). Also, students may be unaware of the importance of definitions in constructing proofs (Edwards & Ward, 2004; Vinner, 1991), and they often find it difficult to understand or work with many of the definitions of advanced mathematics (Dubinsky, Elterman, & Gong, 1988; Moore, 1994). This may be partly attributed to the fact that they have less familiarity with examples of concepts than do their teachers (Moore, 1994) and less inclination to generate examples in order to facilitate understanding of a new definition (Dahlberg & Housman, 1997). When evaluating a proof given by others, students may base their judgment on its surface resemblance to formal arguments rather than on its content (Segal, 2000). They may not have the skills necessary to identify the logical structure of a mathematical statement or proof, particularly when this involves interpreting quantifiers and conditional statements (Dubinsky & Yiparaki, 2000; Durand-Guerrier, 2003; Weber & Alcock, 2005). As a result, students cannot reliably judge whether a proof establishes a given result (Selden & Selden, 2003). Overall, they may come to view proofs and proving as unrelated to their own ways of thinking about mathematical concepts and relationships (Moore, 1994; Raman, 2003).

There has been rather less research on how students can come to understand the nature of proof and master the art of proving. Much of the extant work involves theoretical arguments (e.g., Simpson, 1995) and case studies (e.g., Pinto & Tall, 2002), both of which suggest that there may be at least two "routes" to proof. One of these is a "formal" route in which formal routines and manipulations are first established and later come to be invested with meaning, and the other a "natural" route in which informal ideas and images are progressively refined and linked to formal definitions (Pinto & Tall, 2002; Simpson, 1995; Weber, 2001). Suggestions for ways to improve the teaching of proof tend to reflect one or other of these routes. Some suggest a systematic approach based on a solid grounding in logic and its associated linguistic expressions (Epp, 1998, 2003; Selden & Selden, 1999), or on a presentation or sequence of problems that can lead students to more easily see the structure of certain proof types (Harel, 2002; Leron, 1985). With a similar focus on structure, empirical studies have specified skills students need to produce simple proofs in a given area (Gholamazad, Lijedahl, & Zazkis, 2003; Weber, 2006), and have investigated technological environments that can focus students' attention on the axiomatic development of a mathematical theory (Cerulli & Mariotti, 2003). Suggestions and research in keeping with the "natural" route to proof tend to focus on classroom environments in which students and teacher work collaboratively, investigating problems, formulating conjectures, debating the validity of arguments and so forth (Alibert & Thomas, 1991; Blanton & Stylianou, 2003; Rasmussen, Zandieh, King, & Teppo, 2005; Yackel, Rasmussen, & King, 2000).

While we have these suggestions for how to improve the teaching of proof-oriented mathematics, there have been few empirical studies on how transition courses are usually taught. Both Moore (1994) and Weber (2004) have reported on the teaching of individual professors, and on the articulated rationales these professors give for presenting proof-based material in a certain way. However, I am not aware of any research that seeks to synthesize the knowledge, views and methods developed by mathematicians who teach introductory proof courses. This study takes a step toward addressing this.

3. Research Context

The Introduction to Mathematical Reasoning course described here is offered in a primarily lecture-based format at a large state university in the United States and is a prerequisite for the majority of the upper level mathematics courses. Around 100 to 120 students per semester take it, approximately 40 of whom will go on to be mathematics majors. The course is taught in classes of 20 to 25 students that meet for two 80-minute periods per week during one 14-week semester. The small class size allows professors to have relatively close contact with individuals and to become familiar with their work. The five mathematicians who took part in this exploratory research were invited to participate because they were teaching the course during the 2002–2003 academic year, when the interviews took place. Four of the five had taught the course several times and had also participated in more and less formal discussions with each other about such things as its design and the choice of textbook. All had also taught content-based upper level courses at either the research site or other institutions and they were familiar with the problems that Introduction to Mathematical Reasoning was introduced to alleviate. Their comments reflect direct experience in teaching the course as well as their knowledge of each others' views on the course.

4. Methods

4.1. Data Collection. The five participants were interviewed individually, and the interviews were audio recorded and transcribed. In an initial interview lasting an hour or more, each was asked first to speak in general about their experience with the course. They were then asked more specifically about their thoughts on:

(1) The most important things students should learn in the course.
(2) Common student mistakes or misunderstandings, and why these arise.
(3) Strategies they used in trying to teach this material, why they used these, and whether they found them to be successful.
(4) What they would like to see come out of this research.

It was clear that all of the participants had given considerable thought to the teaching and learning of proof. They offered many well-articulated conceptualizations of the problems they had encountered, and reflections upon their attempts to address these. However, these conceptualizations naturally varied from participant to participant, and the pedagogical approaches they reported were designed in response to individual interpretations of students' behavior. In addition, there appeared to be disagreement among the participants on some points. My goal was to synthesize the issues raised, so that each participant could recognize his or her own experience while relating this to a coherent overall framework. In order to accomplish this I analyzed the transcripts of these first interviews in the manner described by Glaser (1992), as detailed below.

4.2. Data Analysis. First, I added *conceptual descriptions* to each transcript. These were short phrases written to capture the primary conceptual content of each of the participant's comments. Concurrently, for each eight paragraphs of the participant's speech, I wrote a *summary memo.* These memos summarized the content of my conceptual descriptions, thereby providing a condensed version of the conceptual content of each interview. These descriptions and memos were written on a rotating basis, switching from one participant to the next after each summary memo, in order to facilitate synthesis of the ideas raised and to avoid becoming focused on the opinions of a single participant.

During this process, I also wrote *theoretical memos* about questions arising from the data. These typically contained my thoughts on points that needed clarifying, on possible theoretical relationships between issues raised, and on links to extant theoretical concepts and empirical studies. All memos were written in an informal style, the aim being to exhaust ideas rather than to produce a polished product (Glaser, 1992). However, I took care to indicate, through the language used, which ideas were grounded in the data and which were my own conjectures, questions or speculations. At this stage, all memos were kept in one file in the order in which they were written, and each was headed with a reference to the interview and line(s) that prompted its writing. An example of a section of transcript and my accompanying conceptual descriptions and summary and theoretical memos can be seen in the appendix.

Next, I sorted the file of memos. I identified the main substantive and/or theoretical content of each, and placed it in a new file under one or more short headings describing this content. Each memo added was assigned a new heading, or added under one or more existing headings, possibly prompting some reconsideration of the wording used for these. As this file became longer, these headings and their associated memos were progressively subsumed under higher-level headings. This resulted in a flexible system of categories that could be viewed at different levels, and in which lower level headings, together with their memos, could be moved around as necessary. A separate file was also kept of specific questions that I wanted to ask in subsequent interviews.

I found this to be a more satisfactory way of managing this exploratory data than straightforward "coding" of the transcripts. Such coding usually involves classifying data as it is read, and can tempt the researcher to assign short codes at an early stage, so that later data becomes skewed into prematurely-specified categories and the focus is upon labeling rather than upon the relationships between the concepts that the codes describe. The method described here separates memoing and sorting into two distinct stages, so that at the time of writing the initial summary and theoretical memos I was free to express each conceptual description and theoretical question as accurately as possible. The sorting then focuses on the conceptual and theoretical content of the memos, with the eventual category headings playing the role of short codes. Note that by the time these headings are fixed, they are automatically attached to references to the relevant data. Naturally, this does not mean that another researcher would use the same descriptions or ask the same theoretical questions. But it does avoid an early decision either to focus on some parts of the data at the expense of others or to prematurely specify terminology. It thus allows the data as a whole to drive the construction of the category system.

After this process was completed for the initial interviews, I designed second and third interviews based on the question file, and on reading the main file for any areas that seemed under-represented given their importance in the emerging category system. With one exception (Professor 4, who had retired), all of the professors took part in these follow-up interviews, which again lasted approximately one hour. I analyzed each new interview in a similar way to that described above: a file was created of summary and theoretical memos, and once it was completed these were sorted into the existing system, modifying this system where appropriate.

Finally, a two-hour seminar was arranged in which I presented my main findings to the four remaining professors and a colleague from mathematics education. This was an opportunity for all of the participants to respond both to each other's views and to my overall formulation of a theoretical framework on the basis of the data. The seminar was also audio recorded and the participants' comments analyzed.

4.3. Form of the Results. The participating professors spoke about many practical issues including ongoing feedback and assessment, the difficulty of finding an appropriate textbook, the absence of a coherent body of material in the course, and the need to maintain a certain level of emotional comfort on the part of the students. However, the discussions were primarily centered around how to account for the fact that students would frequently make errors in surprisingly simple logical situations, and how to help them to avoid such errors and make progress in writing correct proofs. As the analysis progressed, therefore, the category that emerged as core was "reasoning." This was subdivided into four modes of thinking

TABLE 1. Four Modes of Thinking

Mode	Class	Purpose
Instantiation	Semantic	To meaningfully understand a mathematical statement by thinking about the objects (particular or generic) to which it applies.
Structural Thinking	Syntactic	To generate a proof for a statement by using its formal structure (making formal deductions based on the statement and/or associated definitions and known results).
Creative Thinking	Semantic	To examine instantiations of mathematical objects in order to identify a property or set of manipulations that can form the crux of a proof
Critical Thinking	Semantic	To check the correctness of assertions (by checking for counterexamples and for properties that are implied or should be preserved)

that the professors, as a group, expected a person to engage in while proving, but with respect to which they often found their students to be lacking. Following language used by the professors, I term these modes *instantiation*, *structural thinking*, *creative thinking*, and *critical thinking*. This is a coarse breakdown, and I do not claim that the list is exhaustive, based as it is on self-reports of only five (albeit experienced) participants. Indeed, it was not the case that each professor placed equal emphasis on all four modes; each had developed a range of different emphases in thinking about these issues. However, each recognized all four in the seminar discussion, and none remarked that any major issue had been neglected. I therefore put the classification forth as appropriate for this paper's aims of describing these professors' experiences (their goals for student learning, their teaching strategies, and their concerns about these) and providing a framework for further discussion.

5. Results: Four Modes of Thinking

This section describes the four modes of thinking. Each is defined, as in Table 1, by its purpose for the prover, that is, by the interim goal during a proof attempt that its use addresses. To facilitate later discussion, each is also classified as semantic or syntactic (for more on this classification, see Weber and Alcock, 2004):

Semantic modes are those that involve thinking about the mathematical objects to which a statement refers.

Syntactic modes are those that involve thinking about and manipulating a statement based on its form.

In Sections 5.1 to 5.4, the four modes are described in detail through professors' comments, which also serve to highlight what they would like students to learn in the course. As mentioned above, the classification into modes is coarse, so there are cases in which several different skills or strategies would contribute to the goal of one mode. For the sake of clarity, the content of Sections 5.1 to 5.4 is restricted to presentation of interview excerpts. Commentary upon links with other research is postponed until Section 5.5 and teaching strategies are discussed separately in Section 6. Throughout the interview excerpts, "..." indicates a pause or hesitation and "[...]" indicates that one or more short phrases has been omitted.

5.1. Instantiation. Instantiation is a semantic mode, the goal of which is to meaningfully understand a mathematical statement by thinking about its referent objects. I use the term broadly, to include thinking about generic examples or images as well as particular objects. The professors talked about instantiation as:

- a natural response to a new definition,
- an activity that is necessary for making sense of a new definition,
- a way to understand the implications of choosing a certain property as a definition, and
- a way of making the material seem less abstract and more accessible.

Professors noted that students often do not spontaneously instantiate for such purposes and may resist inducements to do so.

The majority of the quotations in this section come from Professor 1, who had given the most thought to this mode. Both creative and critical thinking build on the use of instantiations, and their respective sections include more comments from other professors.

Professor 1 described instantiation as a natural activity that is necessary for making sense of a new definition.

> *P1:* So one of the things, again, that's second nature to me but it's not to them, is that if I see a definition, I immediately instantiate it. You know, just try some examples of this definition, and try to fit it in.
>
> *P1:* What happens is, you know, that you describe a new definition, you say "let f be a function, let x be a real number, we say that..." and then "some relationship between f and x holds if ... blah blah blah." So then what they have to do, they have to realize that this definition only makes sense in the context of, I have to have a function in mind and I have to have a [number] in mind.

Professor 5 described the way in which instantiations (in this case, imagined sets of integers and apples) can be used to understand the implications of using a certain property as a definition.

> *P5:* We started thinking through, um, you know, what are the implications of defining a set to be simply a collection of unordered objects? [...] So we think through, what happens with integers, and what happens with apples and different collections of things, and, um, you know if you have five apples in a set, is that the same as having one apple, for example?

Professor 4 described a way in which an instantiation can make the abstract concept of function seem more accessible.

> *P4:* My guess is that the notion of functions between arbitrary sets is too abstract for them. I think they can't... picture it. I mean I keep telling them to try picturing a function as like shooting the elements of one set to the other set. Make it very physical. Because, if they can see these things, these things are not hard.

Regarding students' behavior, Professor 1 noted that his students often do not instantiate mathematical statements at times when this would be appropriate, and that he encounters resistance when he encourages them to do so.

P1: And what they'll do is typically if you have a sequence, you know, if I have a sequence definition to use in the rest of the problem, and they don't understand the definition, they'll just skip that sentence and go on. I will... they will come in for help on a problem, and five or ten minutes into the discussion I'll realize that, that they never bothered to process this particular definition. They have no idea what this means.

P1: I also tell them, things like, before you try to prove a statement, let's say it's a "for all, something," then, you should look – illustrate. Write down a few illustrations of this theorem. So pick some examples, just so you can see how, how this theorem works. So maybe you don't see exactly how it's going to help you do what you have to do, but trust me, it will. [...] And they just refuse to do this. They just, they just won't do it.

5.2. Structural Thinking. Structural thinking is a syntactic mode, the goal of which is to generate a proof for a statement by using its form; that is, by introducing appropriate definitions and making deductions from either these or the statement itself according to the rules of logic. The professors talked about structural thinking as:

- allowing the logic to drive the steps of a proof,
- a way to approach proving systematically,
- a way to reduce the complexity of the proving process,
- a sensible first approach when trying to construct a proof, and
- a way to handle tasks such as negating statements involving multiple quantifiers.

With regard to students' work, they noted that students often lack skills that are needed for effective structural thinking. These include knowledge of, and inclination to introduce, relevant definitions, ability to identify the logical structure of a sentence, ability and inclination to make mathematically correct interpretations of connectives such as "and" and "or," care in the precise use of symbols and rules, and ability to write one's own statements clearly. In this section we hear from all five professors, with particular focus from Professor 3 on the importance and advantages of systematicity and precision, and from Professor 2 on the errors students make in the use of mathematical language.

Professor 1 remarked that it is often possible to make progress on a proof using structural thinking.

P1: Particularly early on when you're doing proofs, there's a certain amount of steps of the proofs which are just sort of... driven by the logic. You know if you just, carefully write down the definitions of what you're doing, and you look at what you're trying to show, then you almost have to do certain things, as long as you've mastered the logic.

Professor 3 aims to teach students to systematically make use of this logical structure. He sees this as a way to make the task of proving easier by reducing the number of decisions that need to be made.

P3: So what I try is to have things very systematic. [...] We should really get to the point, where if I give you, if I give the students the definition of limit, "for all epsilon blah blah blah...," just mechanically we want to prove, for all epsilon, blah blah blah. There is only one rule that enables us to "prove for all epsilon, blah blah blah." Therefore there is only one way to start the proof. "Let epsilon be arbitrary, we will prove blah blah blah."

P3: My dream is always – I mean it seems to me that when one makes a thing very minimal in this way, this has a very positive side, which is that... you know when there is only one thing you can do, then it's much easier to know what you can, what you should do.

With regard to the students' work, Professor 3 noted that students may not state definitions correctly and Professor 4 commented on the fact that they may not introduce definitions when this would be appropriate.

P3: For instance, Question: "Define even." Answer: "Even is when it's multiplied by 2." Things like that.

P4: I mean like for example... ah... show that two sets are equal. The only way you can really show that the – you give a definition, sets are equal if they have the same elements, okay? And... so that's what they have to show. And they sort of don't get that. They start with some formulas, they start some manipulations and they don't get the fact that... there's a definition.

Professor 2 commented on students' lack of experience in parsing sentences, and offered a variety of instances in which students' uses of connectives (in this case, "and" and "so") are inconsistent with their use in formal mathematics.

P2: There is no formal grammar taught. I – next to none of my students had ever heard the word "parse" used about English language. They'd heard it, some of them knew it from computer programming, but almost none of them had ever... they don't, they're unfamiliar with the names for parts of a sentence.

P2: I was once again surprised at how many kids don't hear the difference between... "the solutions to this equation are 3 and 4," and "x is a solution to this equation if x is 3 and x is 4."

P2: But students who would say, "we want to prove this, so that..." - it was never clear to me whether they meant "so we can conclude that," or "so we need to verify that." And there were a lot of proofs which would be correct, if I could only infer that they meant "so we need to verify... so we need to verify... so we need to verify... and then this last one is clearly true so everything stacks up."

Professor 3 commented on students' laxity regarding mathematical precision, in particular their tendency to write symbols incorrectly and to misquote given rules.

> *P3:* When I tell you that the symbol is this, it's exactly that. Not the symbol turned around, or upside down, or– okay. If I tell you that the rule says this, like there is a rule that says that if a is less than b and c is less than d, then a plus c is less than b plus d, that's it. You cannot change the b plus d into b times d. [...] Whereas the students seem to think that there is an enormous amount of flexibility.

Finally, both Professors 1 and 3 commented on the way in which an absence of precision in students' writing impedes their own efforts to evaluate students' proof attempts.

> *P1:* [I say] "So now I want you to read your argument, and I want you to identify – so there is a place in your argument where you lied to me. And there must be a critical place." [...] So of course at this point they're not writing proofs. They're not even writing arguments that you can apply this kind of analysis to. I mean it's sort of so vague or so diffuse.

> *P3:* Most of the proofs, ideas in the homework...I cannot even tell what's wrong with them. I cannot even pinpoint the step where it goes wrong.

5.3. Creative Thinking. Creative thinking is a semantic mode, the goal of which is to examine instantiations of mathematical objects in order to identify a property or set of manipulations that can form the crux of a proof. The professors spoke of two ways to do this:

- a direct method in which one experiments with an example in the hope of finding an argument or sequence of manipulations that will generalize, and
- an indirect method in which one attempts to build a counterexample and tries to identify a reason why this cannot be done (an informal analogue of proof by contradiction).

Once again, the professors' comments indicate that they consider such thinking natural, but that their students do not use this mode effectively. Reasons include a lack of inclination to instantiate (described in Section 5.1), incorrect understanding of the kinds of objects for which certain properties might hold, and the tendency to rely on empirical rather than property-based arguments. Professors 1, 2, 3, and 5 made comments of this type.

The professors often cited the direct method as a strategy for constructing an induction argument, as in this comment from Professor 5.

> *P5:* See if you can get from 2 to 3, if you can't get from n to n plus 1.

The indirect method was described in the context of universal statements, by Professors 1 and 2.

P1: The way I often think about a proof is that, you know you imagine this as, try to beat this. Meaning, try to find a counterexample. [...] If you think about the reason why you were failing to find a counterexample, okay, then, that sometimes gives you a clue, to why the thing is true.

P2: Then if it's an existential statement I look to see whether I can produce an example. And if it's a universal statement I probably try to show that I can't find a counterexample.

When the interviewer commented that it seems non-obvious that one would try to prove a statement by thinking about how it could never not hold, Professor 1 explained why he thinks this a natural way to proceed.

P1: So why do I think that this is a natural way to go about it? Um... well I, I'll make a statement which I haven't thought about before. The natural, the sort of natural thing that our brains can do, is sort of build examples and check them. Okay, and... all you, you know if one thinks of universal statements as saying that it's really a statement of impossibility, it's the negation, right? It's a statement that you can't... do something. And the way you understand that you can't do it, is by thinking about doing it.

Regarding students' work, both Professors 3 and 5 cited instances in which a student did at least cursorily examine instantiations of appropriate objects (numbers, in these cases), but failed to make an argument in terms of properties. In Professor 3's case, the student had been asked to establish that 1997 is a prime number. In Professor 5's case, a student was addressing an examination question requiring a proof that if the sum of two primes is odd, then one of the primes must be 2.

P3: So then I asked her in class, "How do you know that 1997 is prime?" The idea being that she would say "Well I checked," in which case I was going to answer, I was going to say "Okay, you tell me that you have checked. But why should I accept that?" Okay? That was going to be the second part, but then I didn't even get there because then she said, "It looks like a prime to me." That's what she said.

P5: I mean there was one student who just added... thirty things together... No, excuse me, she added 2 – she added 2 to every other integer, and got something prime. Got something odd, or prime or something like that.

5.4. Critical Thinking. The goal of critical thinking is to check the correctness of assertions in a proof. In theory this may be achieved syntactically, by checking the form of the deductions and perhaps providing subproofs (see, for example, Selden and Selden, 2003). However, the professors in this study rarely made comments that related to such a process. Rather, critical thinking emerged as a semantic mode involving:

- searching for possible counterexamples,
- checking for implied properties that are false, and/or
- checking for properties that should be preserved.

Regarding students' work, the professors' comments indicate that students may not perform checks using any of these methods, and they express some exasperation about this. This section includes comments from Professors 1, 2, 3 and 5.

Professor 5 described the process of checking for potential counterexamples as she tries to teach it to her students.

> *P5:* I taught them to, you know, [make a] mathematical claim, [then] stop. [...] Think through examples. Have a couple of examples, and then try to see if there's something fishy about them, that might not extend to all possible integers. Ah... look for a counterexample.

Professor 1's comment indicates that in his own work, the choice of what to check is based on a somewhat sophisticated classification within the domain of referents.

> *P1:* I guess I classify numbers into numbers which are prime, let's say, prime powers, and then there are sort of, "far from [prime]," you know so they have a few factors. But then they should have... but then there are also square-free numbers. [...] So I have a sort of classification in mind of different kinds of integers, and I just make sure that whatever I'm trying, I'm trying with these different kinds.

Professor 2 described the way in which her sensitivity to implied properties that cannot be true manifests itself.

> *P2:* [If] I've said something or written something that is wrong, I will get a sense of disquiet. [...] Part of it may be that at a semi-conscious or unconscious level, I'm in the habit of looking for implications of what I've written down. And if those implications are sufficiently absurd, then it gets up into the conscious level.

Professor 3 commented upon checking for preservation of properties in the following remarks about a student's mis-writing of the binomial formula as

$$(a+b)^n = \sum_{k=0}^{n} \binom{n}{k} a^k b^k.$$

He first noted that there is no doubt that the student had worked hard to memorize the formula, before continuing:

> *P3:* Of course, anybody who has a moderate amount of understanding of the question, should ah... realize for example that if you multiply a and b by 2, the right side should be multiplied by 2 to the n, and that – doesn't quite happen. In other words, there is something very wrong about those two powers k in there. It just cannot be.

This last is arguably a rather sophisticated way of checking the correctness of such a formula, and perhaps we would not expect students to use it often. However, the professors regularly noted that students often appear not to perform checks using any of these methods. Professor 3 gave the following instance in which checking for counterexamples would reveal a flaw in the argument.

P3: For instance, problem: "Express the number 30 as the difference of two squares, or show that it cannot be done." Answer: "It cannot be done because 30 is divisible by 6 and a number that is divisible by 6 cannot be written as the difference of two squares." Well, 12 is 16 minus 4. Ah... take any number that's a difference of two squares, multiply it by 36, you'll get a number that's the difference of two squares and is divisible by 6, so I mean... again I could give you loads of examples of the same kind.

Professor 3 expressed exasperation at such responses, considering them "obviously false" on the basis that counterexamples are readily accessible.

P3: I don't have a clue as to, what gets them to, to say things like that. In other words I would say, things that are obviously false. To a normal person with a little bit of mathematical education it would seem obvious that you could never say such a thing because it's so obvious that it's false. Take any example that you want, you see clearly that it's false.

5.5. Relationship to Other Research and Theories. We have now seen many comments on skills that the professors think students need to develop in order to produce mathematical proofs effectively. Many of their concerns are reflected in the mathematics education research literature, and in this section some specific links are discussed.

Others have observed that instantiation is used by mathematically successful individuals. Dahlberg and Housman (1997) note that more successful students spontaneously generate examples in response to a new definition, and in Weber and Alcock (2004) we observe that mathematicians are able to instantiate concepts like group and isomorphism in many ways. Pedagogically, Watson and Mason (2002) argue that student-generated examples can allow students to experience the range of variation in a concept and explore its boundaries, and Winicki-Landman and Leikin (2000a, 2000b) and van Dormolen and Zaslavsky (2003) use examples to explore the implications of basing instruction on different choices of definition. The observation that students often do not instantiate is also consistent with the considerable body of research on students learning mathematics as meaningless manipulations. It is well recognized that a student who knows "what to do" (Skemp, 1976) can often correctly complete a wide variety of algebraic calculations without considering their referent objects (Cerulli & Mariotti, 2003; Sfard & Linchevsky, 1994).

The component skills for structural thinking are widely discussed. In Alcock and Simpson (2002) we observe that students may not readily interpret "show that x is an X" to mean "show that x satisfies the definition of X," and it is recognized that students often reason instead in terms of their concept images (Edwards & Ward, 2004; Vinner, 1991). It is also recognized that in everyday situations, context can play a large role in allowing an individual to interpret a statement containing constructions such as "if... then..." or "for all... there exists...," (Dubinsky & Yiparaki, 2000; Epp, 2003; Zepp, Monin, & Lei, 1987) and that this can interfere with students' abilities to make use of logical structures in constructing and validating proofs (Selden & Selden, 1995, 1999). It is worth noting that effective structural thinking is "syntactic" in the sense in which we use the term in Weber and Alcock (2004) – to refer to correct reasoning based on form and logic – and not as this term is sometimes used to refer to the meaningless manipulation of symbols or ritual use of a certain format.

The skills associated with creative thinking are not so widely discussed in the literature. There are many reports about students who accept or offer perceptual or empirical evidence without going on to abstract some property and construct a general argument (Chazan, 1993; Harel & Sowder, 1998), but there is less discussion of how they might make this next step. Certainly the issue is considered important; Raman (2003) observes that mathematicians sometimes think about proofs in terms of "key ideas" that serve to link their formal and informal understanding, and Rowland (2001) suggests specific principles for using generic examples to help students understand and construct general proofs. However, I would contend that in general we who teach mathematics tend to believe that the creative aspect of proof construction, the "having a good idea," is to some degree a matter of talent and is less open to teaching than other modes. Rowland's work and the indirect strategy suggested by the mathematicians in this study suggest that this need not be the case, and I believe that this might be a fruitful avenue for further investigation.

As noted in Section 5.4, although the goal of critical thinking might in theory be achieved syntactically, in this study it appeared as a semantic mode, relying heavily on consideration of an assertion's referent objects and their properties. Checks based on preserved or implied properties are not widely discussed, though I would speculate that such an approach requires a more sophisticated and organized knowledge base than checking particular example objects. Counterexample checking is discussed in the psychological literature; for example, Johnson-Laird and Hasson (2003) report that people commonly use it to evaluate inferences based on statements about everyday situations. Research in mathematics education indicates that students do not necessarily make such checks when validating mathematical proofs, although relatively straightforward questions can prompt an improvement (Alcock & Weber, 2005; Selden & Selden, 2003). Part of the problem here may be that of deciding which objects might potentially serve as counterexamples. As we argue in Weber and Alcock (2005), evaluating a statement like "1007 is prime because 7 is prime," requires one to infer a warrant (in the sense of Toulmin, 1969) such as the general statement "If x is prime then $1000+x$ is prime," and evaluate it across the range of possible prime numbers. This highlights the complexity of counterexample checking, especially given the prior structural question of identifying a conditional statement if not expressed in the standard "if... then..." way.

6. Teaching Emphasis

This section discusses teaching strategies used by the participants and their reflections on these. It is organized so as to highlight which strategies may be thought of as addressing which mode, although the participants did not divide up their discussions in this way. The principal observation is that the majority of strategies are directed at developing the skills needed for effective structural thinking (the other three modes are therefore discussed first, departing from the previous order). The participants were aware of this emphasis and expressed mixed feelings about it. Their opinions are discussed in Section 6.5.

6.1. Teaching Strategies: Instantiation. Of the teaching strategies discussed, few seem to pertain directly to instantiation. Those that do, involve informally encouraging students to examine particular instantiations, and in one case (Professor 1), use more formal tasks that require instantiating a range of objects with different properties.

P4: I mean I keep telling them to try picturing a function as like shooting the elements of one set to the other set. Make it very physical.

P1: So, what I've been trying to do is to have these exercises where the whole purpose of the exercise is just for them to process a mathematical definition. [...] I have one where I, where I just define what [...] a partition of a set means. I define it formally [as] a collection of subsets, such that [1.] the empty set is not one of the subsets, [2.] for every element of the underlying set there is a subset that contains it, [3.] for any two sets in the partition the intersection is empty. [...] And then I just ask okay, construct three examples of a partition on the set $\{1, 2, 3, 4, 5\}$. And then, okay, construct an example of a collection of sets on $\{1, 2, 3, 4, 5\}$ which satisfies the first two properties but not the third. The first and the third properties but not the second, the second and the third properties but not the first.

6.2. Teaching Strategies: Creative Thinking. The five professors cited few pedagogical strategies that seem to pertain directly to creative thinking. One device that might provide an appropriate context is an "exploration" phase preceding a proof attempt, as used by Professor 2.

P2: Getting them to write down that thing following the caption "analysis" before they have to write down something following the caption "proof" seems to help. Giving them permission to just "think out loud," without having to jump immediately to a properly structured proof seems to help.

However, when asked about the specifics of what a student would do in such a phase, the professors found it difficult to articulate any general principles. Even Professor 1, who described the heuristic of trying to build a counterexample and attending to why this is impossible, noted that he had not systematized this in the classroom.

P1: I talk to the students about this somewhat, although I haven't quite figured out how to make it more systematic. But, the way I often think about a proof is that, you know you imagine this as, try to beat this. Meaning, try to find a counterexample...

6.3. Teaching Strategies: Critical Thinking. Teaching that pertains to critical thinking was described more regularly and systematically, at least as regards searching for possible counterexamples. Most of the professors commented that they model this search in class, and in one case this is institutionalized by designating a student whose job it is to try to give such counterexamples (Professor 3 reported that he assigns this role to a specific student, although he does not require the writing that he described):

P3: There is supposed to be this person over there that's designated "cat" - "creator of arbitrary things," for the semester. The cat's job, is that every time there is an arbitrary thing, he picks one, and writes it down somewhere, and puts it in a sealed envelope and we don't know what it is. [...] The cat's job, this is made very clear, the cat's job is to try to prove me wrong.

Another strategy that pertains to critical thinking is having students consider false statements and state what it would take to exclude counterexamples. Again Professor 3 goes further, deliberately including such statements on assignments with the injunction that students should always be alert to the possibility that something they are asked to prove could be false.

P1: And so, on the first homework assignment one of the questions gives six or seven universal propositions, and these are all false. So find a counterexample, and if possible, find a simple modification so, that makes it true.

P3: [I] give them lots of problems which they call trick questions. [...] We have a general principle that ah... every time you are asked to do something, either you do it, or you say that it cannot be done and you give a reason why.

The professors were also concerned that students should be alert to implied properties, although the more hesitant nature of these comments suggests that this is addressed less systematically.

P2: I guess sometimes when we're trying to construct a proof, and they suggest something which I think goes in the wrong direction, my reaction is that "Wouldn't that imply something that... something seriously wrong?"

6.4. Teaching Strategies: Structural Thinking. The professors' teaching strategies aimed at structural thinking were considerably more numerous and well-articulated, suggesting that the participants use them more systematically. Some involve rather informal modeling of the introduction of appropriate definitions and suggesting that this be done in an "exploration" phase.

P4: If I were to do those in class, the idea would be, how do we approach a problem like that? Well if f is injective, how do we show f is injective? Well we say, let's go back to the definition. If f of a is equal to f of a-prime, then we have to show that a is equal to a-prime.

P2: They are encouraged to write in rather free-form all the relevant definitions and all the apparently relevant theorems, in whatever order they come. And then when they see... think out from that maybe they can see how to put a proof together.

However, strategies pertaining to structural thinking are predominately based on a variety of writing guidelines or "rules." These guidelines pertain to correctly formulating definitions, to correctly introducing mathematical objects, to laying out proofs according to certain templates and/or clear general formats, and to restricting the use of certain terms. Examples of comments on each are given below.

On correctly formulating definitions:

P3: Every time you're asked to define [...] something... the first thing you need to ask yourself, because it's going to matter for what you write, is what kind of a predicate, that is how many arguments it has. For example, if you want to define prime, prime is about one number, it's something that can be true or false for one integer, and therefore you should start by saying "Let x be an integer. We say that x is prime if...." And then comes the condition, right? Whereas ah... divisible is a two-variable predicate. So you would start with "Let x and y be integers. We say that x is divisible by y if...," and then comes the mathematical content.

On correctly introducing objects that will be used in a proof:

P1: If you want to talk about a letter in more than one sentence, then the first sentence in which that letter appears must be "Let x be a blah." And I give them a bunch of rules, which are that they have to introduce them, they have to give everything a different name, to introduce it you must list all the properties that that thing has, you can't later on sneak in something.

On "templates" for writing certain types of proof:

P5: [I] put out a template for writing inductive proof. You write out the inductive hypothesis, then you write out, "Show P of n implies P of $n + 1$." So it's there in front of you. And then you, somehow start with the left hand side of P [of] $n + 1$ and then you make logical deductions and you use the inductive hypothesis. And you work at it, and you arrive at the conclusion of P [of] $n + 1$.

On writing of proofs in general:

P5: I gave them some very... general guidelines for proofs. [...] One mathematical claim per line. Um... and, and at the very beginning, to write every logical step from one argument to the next. So that you can go back and check if the logic is correct.

On restricting the use of certain words or phrases:

P3: For example, never use the word "any," because in mathematics it tends to mean too many things. "Any" can sometimes mean "some," it sometimes means "all".... Now the good student would know how to use the word "any" correctly. But a not very good student is likely to use it in a way that is ambiguous, so better never to use it.

P2: It's different from high school English, in which one is warned never to say "he said" and "she said," you have to use a much more precise synonym than "said." Whereas in mathematical

> English I'd like them to just use "implies," and stay away from all the synonyms for implies.

6.5. Reflections on Strategies for Structural Thinking. The professors expressed a variety of opinions about the use of guidelines and rules to support structural thinking. Professors 1 and 3, in particular, had long debated the nature of the course, with Professor 3 believing strongly in a systematic, rule-based approach and Professor 1 believing the approach should be more investigational, but feeling compelled to introduce guidelines in response to student errors. Epp (2003) discusses similar debates. Each professor, however, had seriously considered the arguments for the opposing view, and the other professors also viewed this as a complex issue. This section has a different tone as at this point I believe it is more illuminating to move from a synthesis to a contrast of opposing points of view.

Professor 3 introduced the idea of a very systematic approach, and cited one of its advantages, as follows.

> *P3:* My own idea as I said is to try to, to tell them... to try to get them to operate in a very systematic way, and systematically tell them that it's for their own good. For example, one of the purposes of having all these rules, these very precise rules, is that you know that a proof is supposed to be an argument that you give, that should convince me.

He considers it reasonable to ask for this level of rigidity, since students have not previously learned to communicate with the precision needed for mathematical proofs.

> *P3:* And I insist very much on minimizing the number of rules because I have become convinced that as an antidote to the general... amorphousness of their work, of the work of their ideas, there is a need to put a lot of rigidity. [...] Maybe it doesn't hurt to go to the other extreme for a while. Once, in one course.

He acknowledged that mathematicians do not think entirely in this way, but considers it unrealistic to hope to convey this in the time allowed.

> *P3:* Doing proofs, mathematical proofs, is not a matter of analyzing logical structures and statements, and applying the logical rules of inference. [...] That may be the frame, the, the thing that supports the proof. But the real proof and the real mathematics is not that. And that should also be conveyed in the course. But of course it's too much. That's wishful thinking.

He also acknowledged arguments in favor of a more investigation-based approach, but considers this unrealistic given students' tendency to give inadequate consideration to their claims and justifications.

> *P3:* I have doubts, because even when I put moderately challenging problems, what I always find is... that the main obstacle is not that they don't know what to do, the main obstacle is that they will do anything... and don't seem to think that it needs justification.

Professor 1, in contrast, originally began the course with the intent of structuring it around investigations of "interesting problems where the answer is not obvious." He considered a rule-based approach to be unrepresentative of the way that mathematicians think when they write proofs.

> *P1:* Although we know in principle that it can be done syntactically, we can bring it down to some axiomatic system... that, in fact, you think very semantically. I mean you know sort of... you're thinking about the objects themselves, and you are confident with your ability to do... to do legal reasoning, on this thing. [...] And then you have some pictures of it, and you are confident of your ability to translate those mental pictures into things that we all accept as real mathematics.

He argued against treating proving as a syntactic enterprise.

> *P1:* [Professor 3] wants them to understand that proof is basically a syntactic thing which we then, sort of, as we become more sophisticated we take shortcuts, but he wants them to really understand the syntax of what they're doing, and to work on writing it. My problem is that it's more of the same. In other words, it encourages the students to think of mathematics as a syntactic enterprise, rather than trying to assign meaning to what they're doing.

However, Professor 1, too, has felt compelled to issue guidelines in response to student errors (such as that quoted in the previous subsection regarding correct introduction of objects to be used in a proof, which he hopes will head off "scope" errors in which a student uses the same letter to refer to two potentially different objects in a single proof). While he hopes these rules will help students to write good mathematics, he considers their proliferation to be a problem.

> *P1:* Every time the students demonstrate to me that they, that some completely natural logical thing to me, that I couldn't imagine how anybody could fail – could make this kind of logical error. But they make it. Okay. So what that does is it causes me to institute some rule that I announce to the class. Not being able to do this, or must do this or something like this. [...] The problem is the more, the longer that I teach the course, the more of these I end up putting in.

As a caveat, it should be stressed that neither of the above professors applies their rules with absolute rigidity. Both allow for more flexible use of mathematical language, provided that a certain level of competence is demonstrated.

> *P1:* And I tell them that, as you demonstrate the maturity to me that you understand these, then we'll let you relax those rules. And since it's a small enough class, well I know your work, and I know which students have earned the right to relax the rules.

> *P3:* So I give you the rules, and that means that I am making a commitment to accept then what you do, without questioning it, as long as you obey the rules. [...] If you don't do it that way, then it might still be okay, but now you are taking a risk.

Other professors, too, expressed mixed feelings about a rule-based approach. There was a concern that proof templates can be useful but that students can come to rely excessively on them. Professor 5 commented on this.

P5: I don't think it's an open and shut case. I think that it's useful sometimes to bring in [...] a set of guidelines, or... almost a template, for writing a proof. And there is a kind of a template for proof by contradiction, and there is a template for induction, and there is a template for... proving things using this method of smallest counterexamples and the well-ordering, property. [...] I think that it's very instructive to... to lay these out and repeat them, so that the students have a starting point, and are not completely intimidated and put off and, to the point where they can't even start. But if you overdo it, you're going to end up with... sort of robots.

Similarly, Professor 2 expressed concerns about giving the impression that there is a unique right way to proceed in writing a given proof.

P2: I really got worried about the fact that I was giving the one – the impression that there was only one right way. Especially in proofs where if you follow your nose there is a, a single energy-minimizing path.

In addition to these philosophical issues, there were also concerns about the effectiveness of a teaching approach that relies heavily on writing guidelines. One problem is choosing and sticking to rules, as in this comment from Professor 2:

P2: One of the things is to show students what it means to articulate a bunch of starting facts, and a scheme for deductive reasoning, and to use those facts to investigate problems. [...] Inevitably we end up telling them they're allowed to use everything they know about some previous material. And "everything they know" is not a well-defined body of facts. So we get promptly confused as to what, what have the status of axioms.

Another problem is the difficulty of sticking to precise use of terms in one's own speech.

P2: For mathematicians, much of the time, when they say "In general you can't count on something being true," it means there exists at least one counterexample. Whereas "In general you can't count on blah blah" in the ordinary language means that, somehow more than half the time it's wrong. And... I mean I've found myself making that mistake even though I've noticed that it's a dangerous one.

Finally there were questions about the degree to which students are able to use the rules effectively in constructing proofs. Professor 2 noted that what is learned in abstract discussions of predicate calculus may not readily transfer to contexts with more mathematical "substance."

P2: What I'm learning is difficult is moving from stuff which has just the form of predicate calculus, to real live sentences in mathematics. Where the structure of the sentence is hidden,

> a little – the student gets distracted by the substance of the mathematics, from seeing the structure.

In all, it seemed that all the participants use some rules or guidelines as scaffolding to help their students to avoid logical errors and to write with the required level of precision. However, they had both philosophical and practical concerns about the efficacy of this approach.

7. Discussion

So far, I have treated the four modes of thinking separately, covering their meanings, associated teaching strategies, and professors' concerns about their teaching and learning in the classroom. In this section, I argue that the four modes may sensibly be viewed as interdependent and that successful provers switch flexibly between all four according to their changing goals within a proof attempt. I therefore suggest that in teaching proof we should balance attention to the four modes, highlighting not only how a person is thinking at a given time, but also what they hope to accomplish by thinking in that way.

7.1. Interdependence of the Four Modes. First, instantiation is arguably the basis for any semantic work. One needs to instantiate in order to have objects to think creatively about and in order to subject inferences to counterexample checks. Instantiation is not strictly necessary for structural thinking, but I would argue that it can help to clarify the structure of a statement, especially in cases involving suppressed quantifiers (Selden & Selden, 1995). For example, in a recent real analysis class, I had a student argue that the statement "Every bounded sequence contains a convergent subsequence" is false because the sequence $1, 0, 1, 0, 1, 0, 1, \ldots$ has a subsequence $1, 1, 0, 1, 1, 0, \ldots$ that does not converge. Discussing the fact that it also has a subsequence $1, 1, 1, 1, 1, \ldots$ that does converge helped to clarify the fact that the statement contains an implicit existential quantifier.

In the opposite direction, some of the skills associated with structural thinking are necessary for instantiation. One has to decide what to instantiate, and doing so involves being able to identify what objects a statement is about and what it says about those objects. Structural thinking also supports creative and critical thinking in similar ways. Further, when writing down the results of creative thinking, one needs to be able to formulate these in appropriately precise language.

Creative thinking does not so much directly support the other modes as fill gaps in what those modes can achieve. Doubtless there are cases in which a competent structural approach could generate an entire proof and critical thinking could adequately check its validity. However, as Weber (2001) argues, in many situations there are numerous correct deductions that could be made from a given premise and numerous reasonable approaches that one could try (direct proof, indirect proof, various proofs by cases, etc.), but few that would lead to a proof. In general, an individual needs some mechanism for deciding which direction to pursue. Examining instantiations can facilitate such decisions.

Critical thinking is also not strictly necessary for any given proof attempt. It might be possible to produce a proof either structurally or creatively without doing any sort of checking. However, mathematicians would surely advise against this, especially for students who are still learning a mathematical topic. Particular risks are that creative thinking can easily lead to over-generalization because instantiations often incorporate properties that do not hold for the whole set under consideration. Similarly, formal rules can easily be confused; for a true conditional statement, the contrapositive is always true but the converse need not be. Critical thinking can help to catch potential errors from such sources.

7.2. Pedagogical Implications. Based on the reasoning of the previous section, I argue that each of the four modes of thinking described in this paper is important in the construction of proofs and should be taught in introductory proof courses. However, this leaves open the question of how this should be achieved. Should we teach instantiation first, in order that students develop facility in meaningfully understanding statements? Should we teach structural thinking first, since students will clearly need to parse statements correctly in order to work sensibly in the other modes? The latter is implicit in much of the teaching that is discussed in this study, as well as in the majority of textbooks designed for such courses. However, I would suggest that in separating out any of the modes and attempting to teach it in isolation for some length of time, we are neglecting an important part of the thinking of successful provers – that is, the flexible use of all four modes in response to changing demands during a proof attempt. To illustrate the significance of this, I offer the following outline of a hypothetical proof construction for the statement that if the sum of two primes is odd, then one of the primes must be 2 (a question from the course, as mentioned by Professor 5). This is not intended to be a prescription or a description of an actual attempt. Not every step is necessary and an experienced mathematician or student might skip steps or move very quickly through some, might make similar steps in a different order, etc. However, I believe that it is a reasonable characterization of how a competent student at this level might sensibly respond.

Step 1, Structural: Identify the statement as a conditional with premise "the sum of two prime numbers is odd" and conclusion "one of the primes must be 2."

Step 2, Instantiation: Experiment with examples, confirming that adding two primes other than 2 gives an even number ($5 + 7 = 12$, $3 + 11 = 14$, etc.). Perhaps also try adding 2 to other primes, to get a feel for the fact that this gives odd numbers ($5 + 2 = 7$, $3 + 2 = 5$ etc.).

Step 3, Critical: Consider larger primes to confirm that this still holds ($29 + 19 = 58$, $53 + 17 = 70$, $53 + 2 = 55$ etc.).

Step 4, Structural: Set up the first line(s) of a proof, writing "Let x and y be prime numbers, and suppose $x + y$ is odd, that is, for some integer n, $x+y = 2n+1$." Experiment with manipulating this equation ($x = 2n+1-y$ etc.) and decide that there is no obvious way forward from here.

Step 5, Creative: Try to construct a case in which neither x nor y is 2 but their sum is odd. Notice that this is impossible because all primes other than 2 are odd, and that adding two such together will always give an even number. Notice also that 2 is even so that when added to another prime the result is odd.

Step 6, Critical: Check that there are no other primes that are even, so that no others would also allow an odd sum.

Step 7, Structural: Formulate the first part of argument in appropriate mathematical language, writing "If $x + y$ is odd, then x and y cannot both be odd." Perhaps justify this by adding "since if x and y are both odd, then $x = 2a + 1$ and $y = 2b + 1$ where a and b are integers, so $x + y = 2a + 1 + 2b + 1 = 2(a + b + 1)$, which is even."

Step 8, Structural: Formulate the second part of the argument in appropriate mathematical language, writing "So one of x and y must be even, so one of x and y must be 2, since 2 is the only even prime." Perhaps justify the last statement by adding "since if $c \neq 2$ is even then 2 divides c, so c cannot be prime."

Step 9, Critical: Check each deduction in the proof.

An initial focus on the skills needed for structural thinking does not seem to be misguided for several reasons: students need to learn the linguistic conventions of mathematics, there are many simple proofs at this level that can be tackled entirely in the structural mode, it is impossible to correctly prove any statement without understanding its structure, and teaching syntactic validation of proofs may avoid potential confusion about when checking a small number of examples is an appropriate action. However, if the classification into the four modes of thinking is taken as a reasonable basis for further discussion, then the argument above raises the question of the degree to which, in the longer term, any approach to teaching proof:

(a) is balanced with respect to the four modes and
(b) fosters the integrated and flexible use of all four.

It certainly seems reasonable to claim that collaborative classroom environments, in which students investigate, refine, and prove mathematical conjectures, address both of these issues reasonably well. Indeed, such environments may directly impact a number of the failings cited by the professors here because research indicates that in such situations students will find meaningful ways to understand statements and to check their own and others' assertions (Alibert & Thomas, 1991; Blanton & Stylianou, 2003; Rasmussen et al., 2005; Yackel et al., 2000). However, this does not mean that the potential is always maximized in such classrooms. Nor does it mean that a lecture-based presentation could not provide effective instruction in all four modes. I view this as important because my position as a mathematics educator is not that I would like to see all mathematicians drastically alter their teaching practices. Rather, I would prefer to see people think carefully in evaluating the ways in which their own preferred instructional methods can be used to maximum advantage and to consider alternatives if and when they identify something that cannot be done well within their current range of approaches. I hope that this framework provides a coherent way of making this evaluation of introductory proof courses. I also suggest that it could provide a basis for metacognitive discussions with students about what they need to do in order to successfully construct and validate mathematical proofs, and about how the tasks that they are asked to complete should contribute to this. In particular, I suggest that an emphasis on the purposes of the modes might help students to distinguish between stages of proving activity at which instantiating examples can be useful and stages at which it is insufficient.

8. Directions for Future Research

In addition to the above pedagogical questions, this study also raises questions for future research. For me the most interesting of these is the question of what might account for a perceived split in the student population, as described by Professor 4:

> *P4:* Basically the class consists of two groups. There are groups that understand it, and probably hardly need it, and then there are those who... ah... really need it, and are not learning it.

This is a controversial comment, but I consider it important to include it as several of the professors spoke of similar observations. Neither I nor they take it as a reason to "give up," and as the evidence above suggests, all those interviewed were committed to finding ways to help all of their students improve their mathematical reasoning as much as possible. However, it is a reality that students do fail transition courses despite having succeeded in previous mathematics, and even the most dedicated teacher may suffer from doubts regarding what they are actually able to teach:

> *P3:* I think there are many people in this department and in every math department who think that those things [how to write definitions correctly, that statements may be proved but noun phrases may not, etc.] are somehow so obvious that anybody who is not a complete moron and has some grasp, some ability to do mathematics should be able to see that. Therefore it is a waste of time to teach that because, the people who need to be taught are the people who are hopeless anyhow. [...] And I would have been vehemently opposed to that idea a few years ago. And now I'm less sure because you know... I'm certainly trying to teach that and I cannot... say that it works, so... maybe these people have a point!

It is clearly important that we continue to improve our understanding of factors that contribute to student difficulties with mathematical proof. My own hypothesis on completing this study was that some students who normally might fail this course could improve their performance considerably by learning to use instantiations effectively. Not doing so would be consistent with research suggesting that in earlier mathematics, students can be successful by manipulating mathematical notation without thinking about its meaning (Sfard & Linchevsky, 1994). At this level, attempting such an approach would automatically cut a student off from all three of the semantic modes of thinking described above. With this in mind, a colleague and I conducted research to investigate whether a tendency to instantiate forms a consistent distinction between the work of those who fail and those who succeed in Introduction to Mathematical Reasoning. Preliminary results from this research (Alcock & Weber, 2005; Weber, Alcock, & Radu, 2005) suggest that the situation is more complicated: there were both successful and unsuccessful students among those who tended to instantiate and those who did not. We are continuing our work on this data in order to better identify factors that contribute to successful use of both syntactic and semantic modes of thinking during proof construction.

References

Alcock, L. J., & Simpson, A. P. (2002). Definitions: Dealing with categories mathematically. *For the Learning of Mathematics*, *22*(2), 28–34.

Alcock, L. J., & Weber, K. (2005). Proof validation in real analysis: Inferring and checking warrants. *Journal of Mathematical Behavior*, *24*(2), 125–134.

Alibert, D., & Thomas, M. (1991). Research on mathematical proof. In D. O. Tall (Ed.), *Advanced mathematical thinking* (pp. 215–230). Dordrecht: Kluwer.

Blanton, M. L., & Stylianou, D. A. (2003). The nature of scaffolding in undergraduate students' transition to mathematical proof. In N. A. Pateman, B. J. Dougherty, & J. T. Zilliox (Eds.), *Proceedings of the 27th conference of the International Group for the Psychology of Mathematics Education held jointly with the 25th annual conference of PME-NA* (Vol. 2, pp. 113–120). Honolulu, HI: University of Hawai'i.

Cerulli, M., & Mariotti, M. A. (2003). Building theories: Working in a microworld and writing the mathematics notebook. In N. A. Pateman, B. J. Dougherty, & J. T. Zilliox (Eds.), *Proceedings of the 27th conference of the International Group for the Psychology of Mathematics Education held jointly with the 25th annual conference of PME-NA* (Vol. 1, pp. 181–188). Honolulu, HI: University of Hawai'i.

Chazan, D. (1993). High school geometry students' justification for their views of empirical evidence and mathematical proof. *Educational Studies in Mathematics*, *24*, 359–387.

Dahlberg, R. P., & Housman, D. L. (1997). Facilitating learning events through example generation. *Educational Studies in Mathematics*, *33*, 283–299.

Dubinsky, E., Elterman, F., & Gong, C. (1988). The student's construction of quantification. *For the Learning of Mathematics*, *8*(2), 44–51.

Dubinsky, E., & Yiparaki, O. (2000). On student understanding of AE and EA quantification. In E. Dubinsky, A. H. Schoenfeld, & J. Kaput (Eds.), *Research in collegiate mathematics education. IV* (pp. 239–289). Providence, RI: American Mathematical Society.

Durand-Guerrier, V. (2003). Which notion of implication is the right one? From logical considerations to a didactic perspective. *Educational Studies in Mathematics*, *53*, 5–34.

Edwards, B. S., & Ward, M. B. (2004). Surprises from mathematics education research: Student (mis)use of mathematical definitions. *American Mathematical Monthly*, *111*, 411–424.

Epp, S. S. (1998). A unified framework for proof and disproof. *The Mathematics Teacher*, *91*, 708–713.

Epp, S. S. (2003). The role of logic in teaching proof. *American Mathematical Monthly*, *110*, 886–899.

Gholamazad, S., Lijedahl, P., & Zazkis, R. (2003). One line proof: What can go wrong? In N. A. Pateman, B. J. Dougherty, & J. T. Zilliox (Eds.), *Proceedings of the 27th conference of the International Group for the Psychology of Mathematics Education held jointly with the 25th annual conference of PME-NA* (Vol. 2, pp. 437–444). Honolulu, HI: University of Hawai'i.

Glaser, B. (1992). *Emergence vs. forcing: Basics of grounded theory analysis.* Mill Valley, CA: Sociology Press.

Harel, G. (2002). The development of mathematical induction as a proof scheme: A model for DNR-based instruction. In S. R. Campbell & R. Zazkis (Eds.), *Learning and teaching number theory: Research in cognition and instruction* (pp. 185–212). Westport, CT: Ablex.

Harel, G., & Sowder, L. (1998). Students' proof schemes: Results from exploratory studies. In A. H. Schoenfeld, J. Kaput, & E. Dubinsky (Eds.), *Research in collegiate mathematics education. III* (pp. 234–283). Providence, RI: American Mathematical Society.

Johnson-Laird, P. N., & Hasson, U. (2003). Counterexamples in sentential reasoning. *Memory & Cognition, 31*, 1105–1113.

Leron, U. (1985). A direct approach to indirect proofs. *Educational Studies in Mathematics, 16*, 321–325.

Moore, R. C. (1994). Making the transition to formal proof. *Educational Studies in Mathematics, 27*, 249–266.

Pinto, M., & Tall, D. O. (2002). Building formal mathematics on visual imagery: A case study and a theory. *For the Learning of Mathematics, 22*(1), 2–10.

Raman, M. (2003). Key ideas: What are they and how can they help us understand how people view proof? *Educational Studies in Mathematics, 52*, 319–325.

Rasmussen, C., Zandieh, M., King, K., & Teppo, A. (2005). Advancing mathematical activity: A practice-oriented view of advanced mathematical thinking. *Mathematical Thinking and Learning, 7*(1), 51–73.

Recio, A. M., & Godino, J. D. (2001). Institutional and personal meanings of mathematical proof. *Educational Studies in Mathematics, 48*, 83–99.

Rowland, T. (2001). Generic proofs in number theory. In S. R. Campbell & R. Zazkis (Eds.), *Learning and teaching number theory: Research in cognition and instruction* (pp. 157–184). Westport, CT: Ablex.

Segal, J. (2000). Learning about mathematical proof: Conviction and validity. *Journal of Mathematical Behavior, 18*(2), 191–210.

Selden, A., & Selden, J. (1999). *The role of logic in the validation of mathematical proofs* (Tech. Rep. No. 1999-1). Cookeville, TN: Tennessee Technological University, Department of Mathematics.

Selden, A., & Selden, J. (2003). Validation of proofs considered as texts: Can undergraduates tell whether an argument proves a theorem? *Journal for Research in Mathematics Education, 34*, 4–36.

Selden, J., & Selden, A. (1995). Unpacking the logic of mathematical statements. *Educational Studies in Mathematics, 29*, 123–151.

Sfard, A., & Linchevsky, L. (1994). The gains and pitfalls of reification: The case of algebra. *Educational Studies in Mathematics, 26*, 191–228.

Simpson, A. P. (1995). Developing a proving attitude. In *Proceedings of Justifying and Proving in School Mathematics* (pp. 27–38). London: Institute of Education.

Skemp, R. R. (1976). Relational understanding and instrumental understanding. *Mathematics Teaching, 77*, 20–26.

Toulmin, S. (1969). *The uses of argument.* Cambridge: Cambridge University Press.

van Dormolen, J., & Zaslavsky, O. (2003). The many facets of definition: The case of periodicity. *Journal of Mathematical Behavior, 22*(1), 91–106.

Vinner, S. (1991). The role of definitions in teaching and learning mathematics. In D. O. Tall (Ed.), *Advanced mathematical thinking* (pp. 65–81). Dordrecht: Kluwer.

Watson, A., & Mason, J. (2002). Extending example spaces as a learning/teaching strategy in mathematics. In A. Cockburn & E. Nardi (Eds.), *Proceedings of the 26th annual conference of the International Group for the Psychology of Mathematics Education* (Vol. 4, pp. 378–385). Norwich, UK: University of East Anglia.

Weber, K. (2001). Student difficulty in constructing proofs: The need for strategic knowledge. *Educational Studies in Mathematics, 48*, 101–119.

Weber, K. (2004). Traditional instruction in advanced mathematics courses: A case study of one professor's lectures and proofs in an introductory real analysis course. *Journal of Mathematical Behavior, 23*(2), 115–133.

Weber, K. (2006). Investigating and teaching the processes used to construct proofs. In F. Hitt, G. Harel, & A. Selden (Eds.), *Research in collegiate mathematics education. VI* (pp. 197–232). Providence, RI: American Mathematical Society.

Weber, K., & Alcock, L. J. (2004). Syntactic and semantic production of proofs. *Educational Studies in Mathematics, 56*, 209–234.

Weber, K., & Alcock, L. J. (2005). Using warranted implications to understand and validate proofs. *For the Learning of Mathematics, 25*(1), 34–38.

Weber, K., Alcock, L. J., & Radu, I. (2005). Undergraduates' use of examples in a transition to proof course. In G. M. Lloyd, M. Wilson, J. L. M. Wilkins, & S. L. Behm (Eds.), *Proceedings of the 27th annual conference of the North American Chapter of the International Group for the Psychology of Mathematics Education.* Roanoke, VA: Virginia Polytechnic Institute and State University. (Published electronically at `http://www.allacademic.com/meta/p24783_index.html`)

Winicki-Landman, G., & Leikin, R. (2000a). On equivalent and non-equivalent definitions: Part 1. *For the Learning of Mathematics, 20*(1), 17-21.

Winicki-Landman, G., & Leikin, R. (2000b). On equivalent and non-equivalent definitions: Part 2. *For the Learning of Mathematics, 20*(2), 24–29.

Yackel, E., Rasmussen, C., & King, K. (2000). Social and sociomathematical norms in an advanced undergraduate mathematics course. *Journal of Mathematical Behavior, 19*(3), 275–287.

Zepp, R., Monin, J., & Lei, C. L. (1987). Common logical errors in English and Chinese. *Educational Studies in Mathematics, 18*, 1–17.

Appendix A. Sample Transcript, Conceptual Descriptions, and Memos

Throughout, words and phrases likely to become key to the analysis were highlighted in bold in order to facilitate quick re-reading of the large amount of text generated. This is reproduced below as it appeared in the researcher's notes.

A.1. Sample Transcript and Conceptual Descriptions

	Interview transcript (Professor 1, Interview 1, Lines 118-120)	*Researcher's conceptual description*
P1:	Yes so that's... I mean... the thing that I'm still wrestling with in this course, is there's a lot of stuff that I'm asking them to master really very early on. There's a whole group of skills. Propositional logic, all of these things about, the number of skills before you can even process a new mathematical definition.	**Teaching problem** that **wrestling** with is asking them to master a lot of **skills** – **propositional logic** e.g. – early on, because needed before can **process a definition**.
I:	Right.	
P1:	What does a mathematician do, you know? So one of the things, again, that's second nature to me but it's not to them, is that if I see a definition, I immediately instantiate it. You know, just try some examples of this definition, and try to fit it in. And they don't even think, you know – what happens is you know that you describe a new definition, you say, "let f be a function, let x be a real number, we say that..." and then some relationship between f and x holds if, blah blah blah. So then what they have to do, they have to realize that this definition only makes sense in the context of, I have to have a function in mind and then I can make sense out of this definition, but if I don't, then.... And what they'll do is typically if you have a sequence, you know, if I have a sequence definition to use in the rest of the problem, and they don't understand the definition, they'll just skip that sentence and go on. I will, they will come in for help on a problem, and five or ten minutes into the discussion I'll realize that, that they never bothered to process this particular definition, they have no idea what this means.	Here we are with **instantiation** as a **thing that mathematicians do**. **Structure** of **definition** appearing here again as something mathematician can just rattle off. What **instantiation** does is **make sense** of **definition** by having an **object in mind**. Typically students **don't understand the definition**, but will **skip it** and go on. They come for help and he realizes a few minutes in that they didn't **bother** to **process** the **definition**. So they have no idea what it **means**.

A.2. Examples of Researcher's Summary Memos

P1 I1 114-128

Summary memo: **teaching problem** of asking them to master a lot of **skills** needed to **process definition** and need these **early on**. **Process a definition** then seems to mean **instantiate examples**, which is a thing that **mathematicians do**. Idea that you **make sense** of a definition by having **objects in mind**, and students don't spontaneously do this. They will **skip a definition** if they don't understand it and try to go on.

P1 I1 114-128

Summary memo: sophisticated **tasks** set for **definition processing**. **Generate examples** then generate more that have **some properties but not others** (Mason ref). **Partitions** as example.

P1 I1 114-128

Summary memo: two more examples of **facility** on part of professor. Both sort of **spontaneously** generating example that **models structure** of what want to describe a **generic definition** and **generic statement that would be hard to negate**. Also **spontaneously restates** more **concisely**.

A.3. Examples of Researcher's Theoretical Memos

P1 I1 114-128

I just don't believe that you need to have these skills in prop logic etc. before you can even begin to process a definition. I believe that you could do it badly and then learn them by means of fixing what you are doing. I believe that he's more on productive lines with example generation as a means of beginning to process what is going on, rather than the meta-math. No arguments that that is useful, of course, but having it as pre-emptive makes me a little uncomfortable because where is the motivation? Unless you're doing that teaching about mathematics thing, but even then I'd say you want something to relate it to. Ah, he says that in the next line. Well, maybe you do need some parsing skills to even do that. I don't know.

P1 I1 114-128

Got a slight implicit laziness here in saying that they didn't bother to process the definition, but this sounds very much like Dahlberg & Housman in that they anticipated being able to do something with the statement without doing that. It's like operating with the statements rather than any objects. Which, of course, on some level is what we want because properties are where it's at. The problem, again, is that having an idea of the objects seems essential to the mathematician, probably for helping you with the decision of what to do. And just it being about something.

P1 I1 114-128

Ah, of course, example generation/instantiation as a way of attacking the abstraction problem head-on. You have the students make it less abstract for themselves. This is sort of obvious really. Of course we consider the task so obvious that we assume they're thinking in that way anyway.

Current address: Department of Mathematical Sciences, University of Essex, Wivenhoe Park, Colchester, Essex, CO4 3SQ, UK

E-mail address: `lalcock@essex.ac.uk`

CBMS Issues in Mathematics Education
Volume **16**, 2010

Referential and Syntactic Approaches to Proving: Case Studies from a Transition-to-Proof Course

Lara Alcock and Keith Weber

Abstract. The goal of this paper is to increase our understanding of different approaches to proving in advanced mathematics. We present two case studies from an interview-based investigation in which students were asked to complete proof-related tasks. The first student consistently took what we call a referential approach toward these tasks, examining examples of the objects to which the mathematical statements referred, and using these to guide reasoning. The second consistently took what we call a syntactic approach toward these tasks, working logically with definitions and proof structures without reference to examples. Both students made substantial progress on each of the tasks, but they exhibited different strengths and experienced different difficulties. In this paper we: demonstrate consistency in these students' approaches across a range of tasks, examine the different strengths and difficulties associated with their approaches to proving, and consider the pedagogical issues raised by these apparent student preferences for reasoning in certain ways.

1. Introduction

A central unresolved issue in mathematics education is that of how to help students develop their conceptions of proof and ability to write proofs. There has been a considerable amount of research on proof in collegiate mathematics education, much of which has focused on why students cannot construct proofs. Causes of undergraduates' difficulties include possessing different standards of justification than those held by mathematicians (Harel & Sowder, 1998; Recio & Godino, 2001), an inability to use, or a lack of appreciation for, definitions in formal mathematics (Alcock & Simpson, 2002; Moore, 1994; Vinner, 1991), difficulties with formal mathematical notation (Dubinsky, Elterman, & Gong, 1988; Dubinsky & Yiparaki, 2000; Selden & Selden, 1995), a poor understanding of important mathematical concepts (Moore, 1994), and a lack of proving heuristics and strategies (Schoenfeld, 1985; Weber, 2001). For a more complete review of this literature, see Harel and Sowder (2007), Selden and Selden (2007), and Weber (2003). This large body of work has yielded valuable insight into why students cannot construct proofs. However, the important issues of how students *can* construct proofs and what types of reasoning are desirable while students are producing proofs are still important

The authors wish to thank the students who took part in this research, along with Annie Selden and Adrian Simpson for their valuable comments on earlier drafts.

©2010 American Mathematical Society

open questions. In this paper, we shed some light on these issues by presenting case studies of productive reasoning by two individuals in a transition-to-proof course.

2. Theoretical Background

One reason that it is difficult to delineate productive proof processes is the complex nature of proving practice. From one perspective, producing a proof can be viewed as a formal exercise. Griffiths (2000) asserts that "mathematical proof is a formal and logical line of reasoning that begins with a set of axioms and moves through logical steps to a conclusion" (p. 2). It is certainly the case that there are logical rules and formal conventions that arguments must obey to be considered valid proofs by the mathematical community. Furthermore, the use of formalism, precise mathematical language, and logical inference are not only necessary in producing a correct proof; in some cases, they can also be sufficient. In other words, there are cases where one can write proofs solely by starting with definitions and assumptions and logically manipulating these assumptions until one deduces the theorem to be proven. If in doubt, consider the processes that most mathematicians would use to verify trigonometric identities or to establish that the sum of two continuous functions is continuous.[1] Thus an individual might attempt to write a proof by focusing entirely on logical syntax, without thinking about representations of mathematical concepts. In this case, we say that the individual has taken a *syntactic* approach to proving (Weber & Alcock, 2004).

From another perspective, assertions to be proven are not viewed solely as strings of symbols or well-formed formulae. Rather, these assertions can and should be understood as meaningful statements about mathematical concepts, and these concepts can be thought about both in terms of logical symbols and in terms of other intuitive representations such as diagrams and prototypical examples (for instance, see Lakoff and Núñez, 2000). From this perspective, the act of proving a statement can be understood as forming a convincing explanation for why mathematical concepts have the properties ascribed to them by that statement. The formation of this explanation deals not with logical symbols themselves, but with meaningful representations of the mathematical concepts to which the symbols relate. Obviously this reasoning needs to be formalized to meet the conventions of proof agreed upon by the mathematical community, but the formalism of proof in this case is more characteristic of the product, rather than the process, of proving. Such a viewpoint can be seen in the writings of Hanna (1991) who, while stressing the need for proofs to have a legitimate formal and logical structure, refers to this structure as more of a "hygiene factor" and claims mathematicians believe that the importance of a proof lies in the intuitive ideas used to create it. We say that approaches to proving in which individuals make use of meaningful representations of relevant mathematical concepts are *semantic* or *referential* approaches to proving (Weber & Alcock, 2004).

We believe that what we have described above is a false dichotomy. We see deep connections between the two perspectives. To borrow the language of Anna Sfard (1991), syntactic and referential approaches to proving are complementary and represent different sides of the same coin. Examining representations of mathematical objects can indicate logical assertions that are true about those objects and suggest what formal inferences should be drawn in a proof. Conversely, syntactic

[1] We are indebted to John Selden for these examples.

logical manipulations of definitions can lead to meaningful assertions that can be used to develop new representations for mathematical objects and modify existing ones (Pinto & Tall, 1999). However, while syntactic and referential approaches to proving are interrelated and complementary, Tall (1997) and others have argued that some mathematicians tend to predominantly choose one approach over the other. For instance, Tall cites Poincaré (1913), who claimed that Riemann was an intuitive thinker who brought geometric imagery to his aid while Hermite was a logical thinker who never evoked such imagery in mathematical conversation.

Exploratory studies in mathematics education have indicated that students may also prefer and consistently employ one approach to proving over the other. For instance, Pinto and Tall (2002) investigated the learning trajectories of students in a first real analysis course. They found that some students consistently used visual imagery to make sense of the formal definitions that they were given, while others learned concepts by memorizing definitions and deducing properties of the concepts from the definitions without recourse to imagery. Similarly, Alcock and Simpson (2004, 2005) conducted a qualitative study of the behavior of students in a real analysis course. They found that of the 18 students in the study, half regularly displayed multiple indicators of visual reasoning while the other half rarely used visual reasoning in their work. Neither of their studies focused specifically on the activity of proving, but both attest to students' learning and reasoning in a definition-based, proof-oriented advanced mathematical domain.

This paper provides detailed evidence of similar student preferences in the different context of a transition-to-proof course. It contributes theoretically in two ways. First, it allows us to see a contrast between syntactic and referential reasoning in an area in which there are fewer obviously "visual" ways to represent examples of the concepts involved than there are in real analysis. Second, it allows us to examine instances of students' productive reasoning in a transition course. Despite the significance of this course for mathematics students in the United States – it is typically the course in which they are first exposed to rigorous proof (Moore, 1994) – there has been relatively little research on either students' proving processes or their performance after completing this course. Most previous research has focused on skills and concepts that the students lack. Moore (1994), for example, closely followed five students as they progressed through a transition-to-proof course, finding that each of these students had serious difficulties constructing proofs throughout the course. Causes of students' difficulties included their inability to state definitions, not knowing how a proof should begin, inadequate concept images, and an inability or disinclination to generate and use examples. Selden and Selden (2003) presented eight undergraduates in such a course with arguments purported to be proofs and found that these students were unable to differentiate between valid proofs and invalid arguments. The work presented here allows us to see different processes that students use when they are not so hampered by these basic difficulties.

We present data on the reasoning of two students in a transition-to-proof course. Both students were somewhat successful in the course, but one student consistently took a syntactic approach to proving while the other used a referential approach. The goal of this presentation is to: (a) illustrate the processes used by undergraduates while taking a syntactic or referential approach to proving, (b) demonstrate that students can be at least somewhat successful in transition-to-proof courses applying either of these approaches, (c) highlight the strengths and weaknesses within each of these approaches by looking at the students' successes and difficulties, and (d) to argue that neither of these approaches should be used exclusively by students and both syntactic and referential approaches to proving are necessary for proving competence.

3. Research Context

In this exploratory study, 11 students participated in individual interviews near the end of a one-semester course entitled Introduction to Mathematical Reasoning. In the course, students practiced applying standard proof techniques and were introduced to content on logic, sets, relations, functions, and some elementary group, number, and graph theory. In the study, we observed the students as they completed a collection of proof-oriented tasks. Our overarching goals were to: (a) investigate the degree to which students at this level tended to consider examples of mathematical concepts (either using specific mathematical objects or generic ones) while working on proof-oriented tasks, (b) explore whether there was a connection between students' tendency to use examples and their success in writing proofs, (c) identify purposes for which students used their examples, and (d) explore the difficulties that some students have with using examples to construct proofs.

3.1. Interview Structure. Each participant met individually with the first author of this paper for approximately one hour. In each case, the student was asked to:

(1) Describe their experience in the transition-to-proof course, including their study habits, aspects of the course they found difficult, and their course performance.
(2) Complete two proof production tasks (the material used for specific tasks is reproduced below). They were presented with these tasks one at a time on separate sheets of paper and asked to describe what they were thinking about as they attempted to answer. They worked without assistance from the interviewer until they either completed the task to their own satisfaction or reached a point where they could no longer make progress. The interviewer then enquired about their proving processes, asking about why they had taken specific actions and, if appropriate, about why they found it difficult to proceed. These questions focused on the student's choice of actions and on their conceptions of their own difficulties rather than on conceptual understanding or logical reasoning.
(3) Talk about their answers to two questions on the mid-term examination they had taken the previous week, including any difficulties they had experienced and whether they understood why these had been graded as they had.
(4) Read a proof (see below) and explain this proof; to describe what they would do to explain it to someone who had not yet taken the transition-to-proof course, and to illustrate the proof using examples.
(5) Describe any general strategies that they used when writing and reading proofs, and then to answer more specific questions about whether they focused on examples or "rules" during such work.

The aim of having the interview structured in this way was to gather information about students' use of examples in a variety of different proof-related contexts. We included both proof production and proof reading tasks since these might reasonably be expected to elicit different behaviors. We included the examination questions in order to allow discussion of at least some mathematics completed outside the interview and without the requirement to think aloud. Again, this would help us to investigate whether students were consistent in their use (or otherwise) of examples in their reasoning. Finally, we deliberately postponed asking the participants for reflections upon their own use of examples until the end of the interview, in order to minimize the chance that they would be influenced by this line of questioning while they were addressing the mathematical tasks.

In this paper we present data from the last four sections of the interviews. Reproduced below are the two proof production tasks, one examination question, and the statement for which the participants were asked to read a proof (the proof itself appears in the appendix).

Relation task (proof-production)

> Let D be a set. Define a relation $\approx$ on functions with domain D as follows. $f \approx g$ if and only if there exists $x \in D$ such that $f(x) = g(x)$. Prove or disprove that this is an equivalence relation.

Function task (proof-production)

> Definitions:
> A function $f : \mathbf{R} \to \mathbf{R}$ is said to be **increasing** if and only if for all $x, y \in \mathbf{R}$, ($x > y$ implies $f(x) > f(y)$).
> A function $f : \mathbf{R} \to \mathbf{R}$ is said to **have a global maximum at a real number** c if and only if for all $x \in \mathbf{R}$ such that $x \neq c$, $f(x) < f(c)$.
> Suppose f is an increasing function. Prove that there is no real number c that is a global maximum for f.

Examination question

> Let S be a set with at least two elements so that there are subsets other than S itself and the empty set. Consider the power set of S, $P(S)$, that is, the set of all subsets of S, under the inclusion relation $\subseteq$.
> Determine whether or not this relation on the power set is: i) reflexive, ii) symmetric, iii) antisymmetric or iv) transitive. Justify your claims.

Proof-reading task

> Definition:
> A function $f : \mathbf{R} \to \mathbf{R}$ is said to be **increasing** if and only if for all $x, y \in \mathbf{R}$, ($x > y$ implies $f(x) > f(y)$).
> Theorem:
> Suppose f and g are increasing functions. Suppose that $f(0) = g(0) = 0$. Prove that for all $x \in \mathbf{R}$, $f(x) \cdot g(x) \geq 0$.

All of these tasks involve producing or reading proofs that involve general assertions about classes of mathematical objects with certain properties. The relation task involves a general set and (possibly) a general element within that set, though the relation itself is fully specified. The function task involves a general increasing function. The examination question involves a general set with certain simple properties, and once again a fully specified relation. The proof-reading question involves two general increasing functions. In all cases, definitions of relevant properties of these general objects are either provided or (in the case of the examination question) are expected to be familiar to the student. Hence, in all cases, an individual may reasonably approach the question either by working syntactically with the appropriate definitions or by examining particular instances of the general objects discussed.

The reader will note that we did not include any questions of the types "prove that this statement is false," "decide whether this statement is true or false" or "give a counterexample." This was a deliberate choice, made in line with our aims for the interview as stated above. Where work on all of the included tasks could reasonably begin with either a syntactic or a referential approach, we were concerned that any task that strongly suggested consideration of (counter)examples might affect the participants' inclination to talk about examples in their other reasoning or in their reflection upon this. We are interested in responses to such questions and in the issue of counterexamples and their role in mathematical learning and reasoning, so we will return to this briefly in our discussion at the end of the paper. However, we judged that it was not appropriate to include a question of this type in this particular study.

We note that it remains possible that any of our findings about example-use in interacting with proofs may be an artifact of the particular questions we asked about the particular types of mathematical object involved. This is one limitation of this study, and indeed elsewhere the first author has speculated that the degree to which an individual tends to refer to object-like visual representations may depend upon the accessibility of these within subject matter (Alcock & Simpson, 2002). However, these interview tasks involve a variety of objects. These range from real functions, for which all students had long experience of interacting with graphical representations, through inclusion relations, which were newer but may also be represented visually in a number of straightforward ways, to equivalence relations, which the students were less likely to think of as "objects" and for which they were certainly unlikely to have assimilated any substantial non-algebraic imagery. We therefore suggest that this range of tasks went some way towards allowing each participant to display a range of strategies if they were so inclined.

3.2. Analysis. The interviews were transcribed and the transcripts were analyzed by the two authors. First, each author independently read through two transcripts and identified episodes in which the student used an example. Codings were compared and any disagreements were discussed. Once it was clear that there was a high level of agreement about these codings, we then constructed a scheme for analysis in which each author would independently:

- Identify the interview context in which the example was used.
- Note whether the use was prompted by an interviewer question.
- Characterize the purpose for which the example was used.
- Describe the example.
- Note any difficulties with using the example.

Through this analysis it became clear that some students took a consistently referential approach to proving. Six students used examples in their attempts to answer multiple questions, used their examples for multiple purposes, and made reflective comments indicating that they regularly made use of examples in their proof-writing. Four students took a consistently syntactic approach to proving. These students did not refer to an example on a single task that we provided for them, and made no comments that they regularly made use of examples in their usual work for the course. There was one remaining student who was less consistent in her behavior. A more complete discussion of all the students' behavior can be found in Weber, Alcock, and Radu (2005).

3.3. Case Studies. The distinction between referential and syntactic approaches was particularly sharp in the more successful students, two of whom provide the case studies that form the remainder of this paper. These two students, Brad and Carla, attained an A- and an A respectively in the midterm examinations that they had taken the week prior to the interviews (grades are on the scale A, B, C, D, F, where D and F are considered failing grades). In their interviews, neither student was perfectly productive or efficient in their proof attempts, but both were coherent and both took actions that were mathematically sensible and likely to lead to progress. Brad, however, spontaneously referred to examples in response to all of the interview tasks, while Carla never did so.

We note that this occurred despite the fact that they had attended the same class and had been exposed to the same lectures, the same homework assignments, and the same earlier examination. We do not have information on the students' previous experiences with proof (though these are unlikely to be extensive) or mathematics in general, so we cannot comment on how these tendencies may have arisen. For instance, it is conceivable that Brad at some point had a mathematics teacher who made a point of referring to examples, that Carla had a teacher who stressed systematic writing, or that either student simply had one success early on in this particular course with their preferred strategy and so stuck to it. However, whatever the source of these tendencies, we can say that they appear to have resulted in these two students interacting very differently with the material as it was presented to them in their transition course and in the interview. We return to implications of this finding in our concluding discussion.

In addition to addressing the interview tasks competently, Brad and Carla were both articulate in reflecting upon their own strategies, and so provide good material for us to see how referential and syntactic approaches have distinct advantages and disadvantages. They also provide an opportunity for us to recognize that students with nominally similar levels of attainment might be achieving this in strikingly different ways, and that those who are successful might become even more so if we can recognize which of their skills they are using well, and which would benefit from further development. At the end of the paper we will also consider ways in which study of the good-but-imperfect performance of these students might inform our teaching of those who are doing less well in transition courses.

4. Referential Approach: Brad

Brad used examples in his reasoning consistently across all of the interview tasks, and the examples he chose were appropriate in each context. His reflective comments indicated that this referential approach was something he used regularly and often deliberately. His behavior and his reflections indicated that it was natural for him to use examples specifically to:

(1) Get an initial understanding or grasp of the concepts in a given question.
(2) Illustrate a proof to someone else.
(3) Decide on a type of proof to use.
(4) Fall back on for more ideas when he could not decide how to proceed.

Illustrations of the first and second of these are provided in Section 4.1, where we see that his referential approach seemed to serve Brad reasonably well in that it afforded him a sense of understanding of a given mathematical situation. Illustrations of the third and fourth are provided in Section 4.2, in which we examine in more detail the ways in which Brad's referential approach helped him and the ways in which it turned out to be less effective. In particular, though it allowed him to decide on a type of proof to pursue, and provided him with a way to search for more ideas, it did not afford him the ability to use this insight to write a full and correct general argument. Finally, in Section 4.3 we provide a short discussion to conclude this section, including further illustrations of Brad's descriptions of his own strategies.

4.1. Strength: Using Examples to Understand Concepts or Questions. Brad spontaneously invoked examples as a first response to all of the tasks discussed in the interview. On reading the relation task he made an initial comment that he was "trying to think what the question's asking," then announced,

> *B:* Alright, I'm just going to like write out some examples. To try and... like, set a D. And then... yes, write out a function or two. I don't know if that's going to help me.

He wrote the following on his paper:

$$D = 1, 3, 5 \qquad f(x) = x^2 \qquad g(x) = x$$

When asked later why he had done this, he said,

> *B:* Um, I guess it just... gives you something concrete [...] Because this is really general. And you can't really put your hands on this. You know I can't like, get a grasp of it.
>
> *B:* Um, I think just to get an idea of what the relation was. [...] When we're given a relation, a lot of times it's a little more like, you know "greater than," you know or "equal to." Or something like that. Or... it's easier. Like this is kind of, a little like harder to see.

Similarly, on reading the function task, Brad again commented that he was trying to understand the question, and stated:

> *B:* And I'm going to take an example to make sure I'm doing it right.

He wrote the following, along with a small sketch of a graph for $f(x) = x$, on his paper:

$$f(x) = x \qquad x = 2 \quad f(2) = 2 \qquad y = 3 \quad f(3) = 3$$

On completion of the question, he once again indicated that he was using examples to grasp the concepts:

> *B:* I didn't, I never heard of a global maximum. I don't think we learned about increasing, but I'm not sure. I don't remember learning about it. So I wanted to teach it to myself first. And, I want to teach myself by examples, you know. And I was kind of starting to understand

a bit more when I was trying to, in trying to grasp – I grasped increasing, it seemed like, okay. But then I was trying to grasp the global max.

Brad's written examination solutions and his interview comments about them also followed this pattern. In response to the examination question, Brad's first line of writing read:

$$S = \{a, b\} \qquad P(S) = \{\{a\}, \{b\}, \{a, b\}\}$$

When asked in the interview to talk through his thinking while working on this question, Brad explained:

B: ...we hadn't dealt with power sets in a while, so it kind of threw me off at first. But then I was like okay, it's not that hard. Calm down.
I: Okay.
B: So then, I set up a set S, and then I... um... I told what the power set was. So I could kind of look at... like I could see the reflex – the relations? Like, for myself.

Finally, after reading the proof-reading question, Brad was asked how he would explain the proof to someone who had not taken the course. He responded,

B: Um ...I would most likely try to relate it to something that they come and do understand. Um, maybe ah, I'd probably also like draw a graph. Like first, introduce the idea of an increasing function. And then show, you know, you know, g of... that ah, if x is greater than y, f of x must be greater than y, by the terms of the increasing function.

When asked to show the interviewer what he would do, he decided to "take the easiest f of x is equal to x," and also looked at particular numbers in order to illustrate the increasing property:

B: So we'll take a to be 1 so b must be greater than that, b to be 2. Um... and, so, since f of – oh, I'd probably show them that f of a is equal to a. And f of b is equal to b. From f of x, or whatever. [...] So then, ah... f of a, is also 1 and f of b is also equal to 1 – ah, 2, for all cases. And then show them on the graph, you know 1,1 and 2,2.

Throughout the interview it appeared that it was important for Brad in his mathematical work to feel that he could "grasp" or "see" the concepts in any given question, and that by examining examples he could achieve this to his own satisfaction.

4.2. Difficulties: Translating Understanding into Proof. In fact, Brad showed himself to be adept at choosing appropriate examples, at manipulating these as entities without losing track of their component parts and at using them to outline formal arguments. We will see this in this section. However, we will also see that although his insights were usually correct, he struggled to translate these into acceptable mathematical arguments. In this section we focus on the relation and function questions because these are the ones where Brad attempted to construct proofs of his own.

After deciding that he was satisfied with his example for the relation question, and recalling the required properties for the relation to be an equivalence relation, Brad reasoned about reflexivity and symmetry thus:

B: So... so okay if it's reflexive, then... f of x should be equal to f of x. Or there should be x in D with, so that f of x is equal to f of x. Okay. That's all I'm going to say! *Laughs.* And... that's true. Because 1 is equal to 1. Symmetric, is um... x – f implies – f is related to g implies g is related to f. So... so this is really the g, there's an x in D such that f of x is equal to g of x. g is related to x – ah, f, when there's a x in D such that g of x is related to f of x. Pause. So... implies that g of... yes. Because if... because x [is] 1, f of x is equal to g of x, then the same x in D that g of x must be equal to f of x.

We note that Brad spoke about f and g as though these stood for general functions, but occasionally referred to his example as if to confirm his thinking. We note also that throughout his work (unlike many of the weaker students in the study), Brad apparently had no problem maintaining control of the objects in the question: he correctly considered reflexivity, symmetry and transitivity as possible properties of the defined relation, without becoming confused about the different types of objects involved (an arbitrary set, elements of that set, functions on that set and a relation between these) or about the way in which the notation represents these.

While working on this and the rest of the question, Brad made some brief notes, but did not write using complete mathematical sentences. When he had finished, the interviewer asked him whether he would write anything else if he were to hand this in for homework, and what this would be. Brad elected to demonstrate for reflexivity, "since it's the easiest part," and he wrote the following:

> Reflexive is when $f \approx f$.
> $f \approx f$ assuming there exist an $x \in D$ such that $f(x) = f(x)$.
> Since $f(x) = f(x)$ the relation reflexive.

A generous interpretation might say that he excludes (sensibly) the possibility that D is empty by assuming that there is an x in D, and that for this x we will naturally have $f(x) = f(x)$. However, even if his thinking is as clear as this, the clarity is lost in his written work, in which poor use of syntax means that there is ambiguity regarding what is assumed and what is inferred. The interviewer did not challenge what Brad had written and he did not appear aware of this ambiguity.

In his work on the relation task, Brad was able to provide written answers and feel that he had answered the question. In his work on the function task, however, this was not the case, and we can see more clearly the problems encountered while trying to apply a referential strategy during proof construction. After writing down his initial example Brad continued exploring the defined concepts. He wondered aloud, "Can f of c be repeated?" and sketched a sine curve with two local maxima marked, before rejecting this idea. He then talked quietly to himself for a moment before suggesting an overall proof tactic.

B: I think we can do this by contradiction. Assume that... assume that um... if f is an increasing function then c... ah... then there is... a c? For which there is a max. And then prove that that can't happen. And then, so that'll prove it.

He began to work on this idea, but found it difficult to make progress.

B: Alright so, if there is... a global max ... (writing, mumbling) ... f of c is greater than both f of x and f of y. I'm trying to think like, you could um, I'm taking the, I'm trying to think that is c, and see if I keep repeating the process, like make um, how I could show that if you, show that there's another x, plus, like x greater than, like if x is greater than y, f of x is greater than f of y, then there's a number greater than x, that'll show that it's greater than – that number's greater than x, and then that's like, keeps going, and going and going. But I don't know how...

After this struggle, he returned to consideration of his graph of a generic increasing function:

B: I'm just trying to see it by looking at the graph. How I can relate it. Like, the two terms interrelate. Why... because I can't even see – I want to know why, there can't be one [...] like know why it can't be and then try to prove.

He appeared ready to give up shortly after this, and when the interviewer asked whether he could talk through his thinking, he said,

B: Alright. I'm thinking that in the definition of increasing, there's never going to be one number that's the greatest. There's always going to be like, a number greater than x. Because it's, because it's increasing. So there's always going to be some number greater than the last. So if x is greater than – that's what I assumed here. x is greater than y, then there's going to be some x plus 1, that is going to be greater than y plus 1, so that f of x plus 1 is going to be greater than f of y plus 1. Or something like that. Where like, it's just going to change.... So then, there can't be some number, you know that... if it's increasing there can't be some number that's greater than all of them. Or, some f of c.

In our view Brad seemed to have some reasonable idea about the fact that for any number in the domain, one can always take a greater number, whose image under the function will be greater than that of the original. However, he did not seem to have good control over the way in which the definitions, and in particular the variables x, y and c, could be used to express this argument. We think it is worth noting that Brad did not attempt to link these variables explicitly to his sketched graph by labeling it in any way.

4.3. Discussion: Brad on His Reasoning Strategies. Overall, Brad seemed able to choose appropriate examples, both specific and generic, and to work with these with some fluency. However, he did not seem to use his examples to effectively guide his manipulation of the symbolic notation at the detailed level. In fact, his own reflective comments on his general strategies for writing a proof suggest that he was not trying to do this, relying instead on his knowledge of standard types of proof to provide structure:

B: I start out by forming an example to, you know, get a strong grasp of what they're asking me. And then, ah, probably play around with like, maybe do a few examples, so I can see what it's – actually maybe how I could prove it, which method of proof I should use. And then once I find a method, proceed from there and... just remembering, just... because it seems like in all the different types of proofs we've done, there's always some kind of structure. So once I get to know what the proof is, I feel more comfortable because I guess that's the hardest part of, of proof. You know, is figuring out which kind of proof to use, because then you can structure it the way you've normally done it before.

This use of examples for specific purposes at specific junctures in his work was further confirmed by his answers to more explicit questions about whether he tended to think about examples, rules or a combination of the two when writing and reading proofs.

B: [*about writing proofs*] Probably a combination but more the rules. I probably use the examples just as a... um, more specific, so I can get like exactly what it's telling me. Or more, probably, based on the question. And then with the proof um, try and like base it on, like on the properties or...

B: [*about reading proofs*] More examples, probably. Just so that like... I always find that, whenever I, I guess like the more I write down, the more I can see... the more clear I'll be. [...] And even – even during the proof, like maybe refer back to it, I guess like the last question but... you know I could try to use – if I got stuck somewhere... in the proof, where I'm like, I don't know where to go from here, look back at – I guess I do look back at the examples I used, I guess to see where I could go from there.

At the end of this paper we will offer further discussion of the range of purposes for which students use examples in proof-related activity and the way in which they relate these to more syntactic work.

5. Syntactic Approach: Carla

In marked contrast with Brad, Carla did not spontaneously use examples in response to any of the interview tasks. Instead, she focused on the syntax and form of the given definitions, questions, and written proof. Her reflective comments indicated that this was a consistent and deliberate approach, and is demonstrated in Section 5.1. Both Carla's reasoning in the interview and her reflective comments indicated that she was often successfully able to answer questions using this syntactic strategy; further, she appeared to have fewer difficulties than Brad in constructing written arguments. However, where Brad's referential reasoning gave him security in the correctness of his own answers, Carla sometimes expressed doubts about her arguments. We illustrate this in Section 5.2, where we also note that although Carla could generate examples when asked to do so, this did not seem to occur to her as a possible approach, and she expressed doubts about the idea of using examples as a tool for mathematical reasoning. Finally, in Section 5.3 we discuss Carla's comments about her strategies in general.

5.1. Strengths: Using Syntax and Form to Construct Arguments. The first notable feature of Carla's responses to the relation and function tasks was that she began speaking and writing immediately after reading the question. For the relation task she listed the properties of an equivalence relation, and went on to draw a quick conclusion.

C: Oh... okay. It's transitive, symmetric, and reflexive. [*Writing.*] So to prove that it's transitive... um... pause... if x is in D, f of x is... equal to g of x. f of x is equal to f of x, so f is related to g. So it's reflexive... um... symmetric is... if f is related to g, then... f of x is equal to g of x, so g is related to f as well... so... symmetric. And transitive is... f is related to g, that means f of x is equal to g of x, and g is related to... a I guess... so g of x is equal to a of x. So it's transitive as well. So... yes. It's an equivalence relation.

Notice that Carla did not give due consideration to the existential quantifier in the definition of the relation (Brad made a similar error in this part of the question); the interviewer did not attempt to draw her attention to this. Like Brad, Carla had made short notes, and the interviewer asked whether she would write anything else if she were going to hand this in for homework. Carla said yes and elected to provide an answer for symmetric. She wrote the following:

Symmetric YES if $f \approx g$, then $f(x) = g(x)$
if $f(x) = g(x)$ then $g(x) = f(x)$
thus $g \approx f$, so if $f \approx g$, then $g \approx f$ thus it is symmetric.

We note that the ambiguity in quantification persisted, but that her written answer is more clearly structured than that provided by Brad.

Carla's response to the function task began in a similar way, with reading of the question followed by immediate writing.

C: So... I'm thinking the way to prove this is using contradiction. So, I would start out by assuming... there exists... a c ... for which [*writing*]... f of x is less than f of c, when x is not equal to c. Okay. [*Pause.*] So now I'm trying to use the definition of increasing function to prove that, this cannot be. Um... so there exists a real number for which f of x is less than f of c, for all x... and there's... f... is an increasing function... for... all x... [*writing*]... y in $\mathbf{R}$, x greater than y implies f of x greater than f of y. Mm... pause... I guess what I'm trying to show is if x is in reals, and they are infinite... for all x... there will be... some function f of c greater than f of x. [*Long pause.*] So... there exists... an element... in $\mathbf{R}$... [*writing*]... greater than c. Um... for x... because... f is an increasing function... f of x will be greater than f of c. Um... a contradiction... so that... there is no c for which f of c is greater than f of x... for all x.

Carla's comments when asked to reflect upon this task provide evidence that her approach was syntactic throughout. When asked what immediately made her decide to prove by contradiction, Carla answered to the effect that she had used the form of statement to decide upon an appropriate proof structure.

C: Because, in class, whenever we have some statement which says, "There is... no such number," or "There exists no such number," then we assume there is, such number. And then we go on to prove that that would cause a contradiction, thus, it doesn't exist. So it was just, something... automatically ingrained, when I see those couple of words, I think contradiction.

The interviewer then commented that Carla had appeared to be stuck for a short time, and in response Carla described having to think about the problem once she had systematically begun her answer.

> *C:* Well the first line kind of came from me, you know just recognizing those first couple of words and recognizing the format it should be starting off with. Because contradiction, so assume this, this number exists. And then I was thinking, I should take what's previously given, that it's an increasing function. So I wrote that down. And once I was done with that, systematic part, I kind of needed then to think about, how actually... actually think about the problem.

Although one might imagine that this "actually thinking about the problem" would involve some other representation, further probing revealed only that she was thinking about what a written proof would look like. Indeed, later in the interview Carla responded to direct questions by stating clearly that she was not referring to examples while working on this task.

> *I:* Did you have any sort of picture in your head for this one?
>
> *C:* No, no... not really. I mean I know what a global maximum is from calculus... I mean I've done these sort of things so many times. But I didn't imagine any, any sort of function. Something that would have a maximum. [...] Really... I guess I did it very systematically and theoretically, because I just stepped – this is the rule, and do it through.
>
> *I:* Okay. Yes. You see [...] one thing I thought you might be doing is trying to create some kind of picture or trying to imagine how that would work with a particular example. And that appears to be not what you were doing in that case...
>
> *C:* No. No. I just, I just tried to see if it was, it was connected. Just theoretically, if the words made sense. If one thing implied the other.

Carla's responses to the examination question and the proof-reading question provided evidence that she used a syntactic approach in other situations. The first lines of Carla's response to the examination question read:

S set with at least two elements
$P(S)$ under $\subseteq$ $\qquad$ x is element of $P(S)$, so $x \subseteq S$
$\qquad\qquad\qquad\qquad$ y is element of $P(S)$, so $y \subseteq S$

We note that where Brad wrote down a generic set and its power set, Carla wrote down and elaborated a list of the given conditions of the problem. When asked about this question, Carla commented "Basically the format I used was like the same one I used up here," referring to the earlier interview question. When asked to talk through her thought processes, she gave a general description of how she thinks about such questions:

> *C:* Okay. So, I just think when, when something is reflexive I think, so, there's this one element related to itself, and, does that relation hold true?
>
> *I:* Okay.
>
> *C:* I guess that was how I think about it. I just pick two elements, if it's for symmetry, or antisymmetry I pick two elements, and say if

they're related, then what does that mean, and... from then on can I conclude some next relationship between them.

For the proof-reading task, when Carla was asked how she would explain the proof to someone who had not taken the course, she did not refer to any examples and apparently did not feel the need to add much further explanation. Indeed, when specifically asked to illustrate the proof using examples, Carla expressed uncertainty about what she was being asked to do:

C: Just... what do you mean? [*Hesitation.*] What type of examples?

We explore the apparent impact of Carla's overall lack of example use in the next section.

5.2. Difficulties: Absence of a Sense of Understanding. The most striking illustration of the disadvantages of Carla's syntactic approach occurred after her initial response to the function task. Despite her competent attempt, which stood out as unusually efficient and complete compared with those of most student participants, she commented on her proof that "it seems a bit flaky." When asked why, she said,

C: I don't know, it just doesn't make sense for me. It, it feels like, I just, it's just proved systematically, without being able to imagine what's going on. So that's why it feels flaky.

As we noted before, evidence throughout the interview indicated that although she talked about "imagining what is going on," she did not at any point consider examples or generic images of functions. This is not to say that Carla was unable to generate examples, however. This is best illustrated by her eventual response to the proof-reading question. As noted at the end of the preceding section, she initially seemed thrown by the request to generate illustrative examples. However, with an opportunity to think further she went on to address the question without apparent difficulty or errors.

I: Well, whatever you think would be appropriate, really. Take your time if you want to think about it, and do write anything on the paper if you want to.

C: Okay.... I guess I would give two simple functions, f and g that were both increasing.

I: Okay.

C: And, such that... I guess... writing... f of x, equal to x... and... g of x is equal to... some other increasing function. Like... kind of like x cubed.

Like Brad, Carla stated further that she would illustrate the increasing property by using particular elements of the domain, and showed how she would do this when invited to do so by the interviewer.

C: And then... I would just give... well first I would explain why they were increasing, and just give examples of different... xs. If I were to explain it...

I: Do you want to do that for me now?

C: Oh, okay. Example... for why it's an increasing function. Just take, I guess... x_1 is equal to 1, x_2 is equal to 2. f of x would be... x and... f of x_2 would be 2 and so it's... f of x_2 is greater than f of

> x_1. Because x_2 is greater than 1, they are increasing. So I would say that first. Then I would go on to, for each of these three cases, I would give an example for, for instance, for some x, say x is negative 2... that's for the second case...

Here we observe that although Carla's language did not perfectly capture the logical relationships between what was assumed and what was demonstrated or illustrated, she did introduce appropriate examples of both functions and domain elements, and she did coordinate general statements and particular objects correctly throughout.

In fact, although the idea of using examples in reasoning seemed rather alien to Carla in relation to the tasks discussed in the interview, it turned out in the final discussion that she did refer to examples when *reading* proofs, and gave reasons similar to those given by Brad.

> *C:* I think about examples, when I read proofs. Because, it just tends to clarify it better.
>
> *I:* Okay. And, how do you decide what examples to think about, in that case?
>
> *C:* Um... I just think any random examples, or different cases if there are only a set number of cases. You know I think of one for each case. But, other than that, any example. The simplest ones, usually. *Laughs.*
>
> *I:* Right. Can you give me an illustration of the kind of thing you would look at in a certain kind of problem?
>
> *C:* Well, sometimes, like in graph theory, even when I do the proofs, I, I sometimes don't even understand what is supposed to be proven. I understand the assumptions, and then from the assumptions I have to draw an example for myself, of some simple graph that has this property. And then... have to prove to myself that that would be true, in that case. And then I go on from there, to understand that proof.

However, when asked whether she had ever tried a similar strategy when writing proofs, Carla expressed discomfort with the idea of using examples as a basis for constructing general arguments.

> *C:* Even if I have convinced myself that that proof would be true, and it would happen in certain examples, it wouldn't help me in writing out the proof itself. Because it has to hold for all graphs, and... I don't know how to explain it. I have trouble... generalizing graphs. As a whole. So those examples wouldn't work in that case.

Indeed, Carla expressed a broader discomfort with the use of examples in proving, including problems she had encountered in using counterexamples in disproof.

> *C:* I could never grasp the, just concept of giving a simple counterexample, any old thing. And those were usually the easiest problems on the exam. And I would always get zeros on them. Because I tried to disprove it in a general manner. And, I guess I'm just not, I don't trust examples, but...

It is not clear whether Carla had been over-influenced by the maxim "you can't prove by example" or whether she simply found herself unable to generate a proof based on observation of an example.

5.3. Discussion: Carla on Her Reasoning Strategies. Overall Carla took a consistently syntactic approach to all the tasks discussed in the interview, and she both confirmed and elaborated upon this in her later reflective comments. When asked about any general strategies she had for writing proofs, she said,

> *C:* Um, I just start with a claim... I usually don't have anything in my head beforehand. I start off with what I know, and then I assume, what they're talking about, that I should use, in that case. And then I just try to work off of there. And I try to imagine what my goal is, and kind of work from both sides, to the center.

When asked more specifically about the first things she would do, she stated that she "thinks of a method to use" and went on to explain how she identified an appropriate one:

> *C:* If it says, if it's something that has to be proven for all... numbers in such a set, then I use induction. And... for instance, if uniqueness is supposed to be proven, I always assume there's two different numbers that produce the same result. Or something to that extent. And use contradiction. Or, for there exists no number such that, I say yes, assume there is and then use contradiction.

This basic strategy still stood when she did not immediately know which technique to use.

> *C:* Well I would try out just different ones and see which one gets me the farthest. [...] We don't really know many methods, so it's not that difficult, to get one right.

This last comment indicates that a syntactic approach afforded Carla the ability to answer most of the questions she encountered in the transition course. As we have seen, what it did not appear to afford her was a sense of meaningful understanding of her answers, unlike that which Brad appeared to obtain by reference to examples. Interestingly, as we saw in the previous section, the use of examples to gain understanding of a situation was not alien or inaccessible to Carla. However, in her case we can see an absence of two-way links between examples and what she knew about proof structures, and a very restricted use of examples in proof-related activity.

6. Discussion

6.1. Summary: Referential and Syntactic Approaches to Proving. Compared with the majority of the interview participants, both Brad and Carla scored well in their course examinations and made substantial progress on the interview tasks. Further, at times in the interview they made similar errors relative to both conceptual correctness and the precise use of mathematical language to express logical relationships. However, we believe it is clear that they approached these proof-related tasks very differently, and that they exhibited different strengths and experienced different difficulties. Below, we review these approaches and difficulties before considering the pedagogical implications of this research.

Brad took a referential approach, referring to examples of the statements he was considering. Sometimes these referents were represented in an obviously "visual" way, as in his discussion of graphs while working on the function task and the proof-reading task. Sometimes they were represented using strings of symbols signifying mathematical objects, as in the written sets he introduced while working on the tasks involving relations. Sometimes he examined particular examples, such as the function $f(x) = x$ or the set $D = \{1, 3, 5\}$. Sometimes he examined more "generic" examples, such as the set $\{a, b\}$ and its power set. However, in all cases he treated the statements as being about mathematical objects and their properties, and he examined examples of these objects regularly in order to address the questions. This referential approach afforded Brad a strong sense of meaningful understanding – he appeared satisfied that by considering example objects he could understand the meanings of concept terms and gain insight into why a statement should be true (for more on this idea, see Alcock and Simpson (2004)).

Both the proof production tasks and Brad's own reflections suggest that the approach also afforded him a means of deciding on appropriate proof structures for given statements. What it did not seem to afford him was a consistent ability to coordinate the details of a general argument. In the interview, Brad sometimes gave reasonable example-based arguments but struggled to produce proofs in terms of the relevant formal definitions. His reflective comments suggest that a possible reason for this is that he treated understanding (via examples) and proving (via familiar proof formats) as somewhat separate parts of his activity.

Carla, on the other hand, took a syntactic approach. She did not consider the referents of the statements unless prompted to do so. Instead, she took cues from the linguistic form of the statements she was considering, using these to decide on an appropriate proof structure. She used known or given definitions to systematically write down what was to be assumed and what was to be proved, and searched for logical inferences she could make that would lead from one to the other. She explained her own reasoning in these terms, and did not appear to feel a need to add anything beyond this type of explanation when talking about a provided proof. This syntactic approach served her well in producing written answers to the interview tasks, and her reflective comments suggest that she found it generally effective in the course as a whole.

However, Carla's comments also indicated that she was sometimes able to produce such work without being able to "imagine what is going on," and that this could make her hesitant about her answers even when from an observer's point of view these were full and correct (for more on this idea, see Raman, 2003). Moreover, while Carla saw the value of looking at examples in order to understand a statement, she commented that she had struggled with tasks requiring the generation of counterexamples, and did not see how the use of examples might help her in constructing a general proof. Finally, while one may be able to use a syntactic approach to construct a fair number of proofs in the conceptually limited domains that are explored in a transition-to-proof course, we question whether one can construct a wide range of proofs in a more sophisticated, conceptually rich mathematical domain by relying exclusively on a syntactic approach to proving (see also Weber (2001) and Weber and Alcock (2004)).

6.2. Theoretical Discussion. The issue of how referential and syntactic strategies contribute to effective mathematical reasoning is both interesting and complex. Certainly a successful mathematician will be able to use both, switching from one to the other in response to the perceived demands of the situation (Alcock, 2010). For instance, a mathematician who is uncertain about the truth of a statement is likely to examine examples in a more or less systematic counterexample search, whereas one who is certain of a result may proceed directly to working syntactically with appropriate definitions (Inglis, Mejia-Ramos, & Simpson, 2007).

However, as we noted in the introduction, it has been suggested that some experienced and famous mathematicians exhibit tendencies to use one or the other strategy preferentially. Our own experience of discussing these issues in a variety of presentations with both mathematicians and mathematics educators suggests that such tendencies are widespread, and this study, along with others (Alcock & Simpson, 2004; Pinto & Tall, 1999, 2002), indicates that many students exhibit similar tendencies at the transition-to-proof level.

This leads naturally to the question of the origins of these tendencies. We are often asked whether we suppose that they are innate, or whether we think that they are learned through earlier mathematical experiences, in school or elsewhere. We acknowledge that this is a fascinating question, but we also believe it is highly intractable – it isn't feasible to trace even one individual's entire mathematical development, much less to compare many such paths through mathematical learning. We are more interested in the more practical question that may be stated as follows: Given that right now we have students in transition courses who display tendencies to favor referential or syntactic strategies in their reasoning, what should we do about it? We offer some comments on this in the final section.

6.3. Pedagogical Implications. As we have described, Brad's and Carla's respective approaches were consistent across tasks that asked for proof production, for proof explanation, and for description of reasoning undertaken in an examination context. They were also consistent with the students' reflections on the reasoning strategies they generally employed in their transition-to-proof course. If we accept that these are instances of somewhat common tendencies, then this provides a differentiated backdrop against which we can offer some comments on what it is desirable for students to learn in transition-to-proof courses, and how instruction might work toward this in the cases of more and less successful students.

First, it could be argued that Carla's syntactic approach is "better" in the important respect that it affords more success in producing written proofs and fewer errors of logical expression within these. As mathematicians, we certainly value the systematicity of such a syntactic approach and our assessment methods reflect the fact that our primary goal is that students should learn to produce correct written proofs. However, as suggested in the theoretical background, the syntactic approach corresponds to only one view of proof. The skills used in proving syntactically are only a subset of those that we might wish a student to develop. Others pertain to interpreting mathematical statements as giving meaningful information about mathematical objects and using other representations of these objects to guide proving activity. In Weber and Alcock (2004), we argue that use of such skills contributes significantly to mathematicians' ability to construct proofs, and in Alcock (2004) the first author presents interview data in which mathematicians stress the importance of these skills.

We therefore wish to recognize the value in what each student is doing, to think in terms of helping students build upon their existing strategies, and to augment these with skills they may possess but tend not to invoke without prompting. This seems eminently possible in the cases of relatively successful students like those examined here. Brad did not think only about examples, he also thought in terms of using standard proof structures. In addition, the examples he introduced were all appropriate, showing that he could correctly process the syntax of both known and given definitions. If he could be encouraged to begin his proof attempts more systematically by clearly writing down assumptions, and to spend some time linking the symbols of the definition with his examples, he might be better able to use his example-based insights to guide his proofs. Similarly, Carla did not only think about syntax, she also reported that she used examples to aid her in understanding the statements and proofs she was given in the course. In addition, when asked specifically to illustrate a given proof with examples, she was able to do so correctly and without much apparent difficulty. If she could be encouraged to make this activity a more regular part of her work, she might gain a more consistent sense of being able to "imagine what is going on," more fluency in constructing counterexamples, and perhaps more insight into how to translate from examples to properties as well as the reverse.

Of course, in a classroom setting, one is likely to have to deal simultaneously with students with different combinations of available strategies and skills. Encouragingly, we can observe a commonality in the difficulties encountered by both Brad and Carla: each seemed to have an underdeveloped notion of how to use examples and syntax together to construct a proof. Hence, we suggest that in class settings students with both syntactic and referential preferences might benefit from instruction that explicitly draws attention to links between formal statements and proofs and their referent objects and relationships, at a detailed, step-by-step level. This might help students to see more clearly how different ways of interacting with mathematical statements can be complementary and to make informed selections from a wider range of strategies than they might naturally employ.

References

Alcock, L. J. (2004). Uses of example objects in proving. In M. J. Hines & A. B. Fuglestad (Eds.), *Proceedings of the 28th annual conference of the International Group for the Psychology of Mathematics Education* (Vol. 2, pp. 17–24). Bergen, Norway: Bergen University College.

Alcock, L. J. (2010). Mathematicians' perspectives on the teaching and learning of proof. In F. Hitt, D. A. Holton, & P. Thompson (Eds.), *Research in collegiate mathematics education. VII* (pp. 63–91). Providence, RI: American Mathematical Society.

Alcock, L. J., & Simpson, A. P. (2002). Definitions: Dealing with categories mathematically. *For the Learning of Mathematics, 22*(2), 28–34.

Alcock, L. J., & Simpson, A. P. (2004). Convergence of sequences and series: Interactions between visual reasoning and the learner's beliefs about their own role. *Educational Studies in Mathematics, 57*, 1–32.

Alcock, L. J., & Simpson, A. P. (2005). Convergence of sequences and series 2: Interactions between non-visual reasoning and the learner's beliefs about their own role. *Educational Studies in Mathematics, 58*, 77–100.

Dubinsky, E., Elterman, F., & Gong, C. (1988). The student's construction of quantification. *For the Learning of Mathematics, 8*(2), 44–51.

Dubinsky, E., & Yiparaki, O. (2000). On student understanding of AE and EA quantification. In E. Dubinsky, A. H. Schoenfeld, & J. Kaput (Eds.), *Research in collegiate mathematics education. IV* (pp. 239–289). Providence,

RI: American Mathematical Society.

Griffiths, P. A. (2000). Mathematics at the turn of the millennium. *American Mathematical Monthly, 107*, 1–14.

Hanna, G. (1991). Mathematical proof. In D. O. Tall (Ed.), *Advanced mathematical thinking* (pp. 65–81). Dordrecht: Kluwer.

Harel, G., & Sowder, L. (1998). Students' proof schemes: Results from exploratory studies. In A. H. Schoenfeld, J. Kaput, & E. Dubinsky (Eds.), *Research in collegiate mathematics education. III* (pp. 234–283). Providence, RI: American Mathematical Society.

Harel, G., & Sowder, L. (2007). Toward comprehensive perspectives on the teaching and learning of proof. In F. Lester (Ed.), *Second handbook of research on mathematics teaching and learning* (pp. 805–842). Charlotte, NC: Information Age.

Inglis, M., Mejia-Ramos, J. P., & Simpson, A. (2007). Modelling mathematical argumentation: The importance of qualification. *Educational Studies in Mathematics, 66*, 3–21.

Lakoff, G., & Núñez, R. E. (2000). *Where mathematics comes from.* New York: Basic Books.

Moore, R. C. (1994). Making the transition to formal proof. *Educational Studies in Mathematics, 27*, 249–266.

Pinto, M., & Tall, D. O. (1999). Student construction of formal theories: Giving and extracting meaning. In O. Zaslavksy (Ed.), *Proceedings of the 23rd annual conference of the International Group for the Psychology of Mathematics Education* (Vol. 1, pp. 281–288). Haifa, Israel: Technion – Israel Institute of Technology.

Pinto, M., & Tall, D. O. (2002). Building formal mathematics on visual imagery: A case study and a theory. *For the Learning of Mathematics, 22*(1), 2–10.

Raman, M. (2003). Key ideas: What are they and how can they help us understand how people view proof? *Educational Studies in Mathematics, 52*, 319–325.

Recio, A. M., & Godino, J. D. (2001). Institutional and personal meanings of mathematical proof. *Educational Studies in Mathematics, 48*, 83–99.

Schoenfeld, A. H. (1985). *Mathematical problem solving.* Orlando, FL: Academic Press.

Selden, A., & Selden, J. (2003). Validation of proofs considered as texts: Can undergraduates tell whether an argument proves a theorem? *Journal for Research in Mathematics Education, 34*, 4–36.

Selden, A., & Selden, J. (2007). *Overcoming students' difficulties in learning to understand and construct proofs* (Tech. Rep. No. 2007-1). Cookeville, TN: Tennessee Technological University, Department of Mathematics.

Selden, J., & Selden, A. (1995). Unpacking the logic of mathematical statements. *Educational Studies in Mathematics, 29*, 123–151.

Sfard, A. (1991). On the dual nature of mathematical conceptions: Reflections on processes and objects as different sides of the same coin. *Educational Studies in Mathematics, 22*, 1–36.

Tall, D. O. (1997). Metaphorical objects in advanced mathematical thinking. *International Journal for Computers in Mathematics Learning, 1*, 61–65.

Vinner, S. (1991). The role of definitions in teaching and learning mathematics. In D. O. Tall (Ed.), *Advanced mathematical thinking* (pp. 65–81). Dordrecht:

Kluwer.

Weber, K. (2001). Student difficulty in constructing proofs: The need for strategic knowledge. *Educational Studies in Mathematics, 48*, 101–119.

Weber, K. (2003). Students' difficulties with proof. In A. Selden & J. Selden (Eds.), *Research sampler, 8.* Published electronically: Retrieved June 17, 2006 from `http://www.maa.org/t_and_l/sampler/rs_8.html`.

Weber, K., & Alcock, L. J. (2004). Syntactic and semantic production of proofs. *Educational Studies in Mathematics, 56*, 209–234.

Weber, K., Alcock, L. J., & Radu, I. (2005). Undergraduates' use of examples in a transition to proof course. In G. M. Lloyd, M. Wilson, J. L. M. Wilkins, & S. L. Behm (Eds.), *Proceedings of the 27th annual conference of the North American Chapter of the International Group for the Psychology of Mathematics Education.* Roanoke, VA: Virginia Polytechnic Institute and State University. Published electronically: Retrieved July 12, 2008 from `http://www.allacademic.com/meta/p24783_index.html`.

Appendix: Proof for Proof-Reading Task

Proof (by cases)

Assume f and g are increasing functions and $f(0) = g(0) = 0$.
Let $x \in \mathbf{R}$.
By the law of trichotomy, either $x > 0$, $x < 0$, or $x = 0$.

Case 1. Suppose $x > 0$.
Since $x > 0$ and f is increasing, $f(x) > f(0)$ by definition of increasing function.
Since $f(0) = 0$, $f(x) > 0$.
Since $x > 0$ and g is increasing, $g(x) > g(0)$ by definition of increasing function.
Since $g(0) = 0$, $g(x) > 0$.
Since $f(x)$ and $g(x)$ are positive, $f(x) \cdot g(x)$ is positive and hence $f(x) \cdot g(x) \geq 0$.

Case 2. Suppose $x < 0$.
Since $x < 0$ and f is increasing, $f(x) < f(0)$ by definition of increasing function.
Since $f(0) = 0$, $f(x) < 0$.
Since $x < 0$ and g is increasing, $g(x) < g(0)$ by definition of increasing function.
Since $g(0) = 0$, $g(x) < 0$.
Since $f(x)$ and $g(x)$ are negative, $f(x) \cdot g(x)$ is positive and hence $f(x) \cdot g(x) \geq 0$.

Case 3. Suppose $x = 0$.
Then $f(x) \cdot g(x) = f(0) \cdot g(0) = 0 \cdot 0 = 0$.
So $f(x) \cdot g(x) \geq 0$.
Since in all three cases, $f(x) \cdot g(x) \geq 0$, $\forall x \in \mathbf{R}$ $(f(x) \cdot g(x) \geq 0)$.

Department of Mathematical Sciences, University of Essex, Wivenhoe Park, Colchester, Essex, CO4 3SQ, UK
E-mail address: lalcock@essex.ac.uk

Graduate School of Education, Rutgers University, 10 Seminary Place, New Brunswick, NJ 08901, USA
E-mail address: khweber@rci.rutgers.edu

CBMS Issues in Mathematics Education
Volume **16**, 2010

Step by Step: Infinite Iterative Processes and Actual Infinity

Anne Brown, Michael A. McDonald, and Kirk Weller

Abstract. Students in two introduction to abstract mathematics courses were interviewed while trying to determine whether the set $\bigcup_{k=1}^{\infty} P(\{1,2,\ldots,k\})$ equals the set $P(\mathbf{N})$, where $\mathbf{N}$ denotes the set of natural numbers, and P denotes the power set operator. In their efforts to understand the infinite set represented by the union notation, the students constructed a variety of iterative processes. An APOS analysis of the data resulted in a description that identifies the role of the mechanisms of interiorization, coordination, and encapsulation in constructing infinite iterative processes and their states at infinity. This theoretical description is illustrated through a series of case studies and is compared to what is predicted by the Basic Metaphor of Infinity of Lakoff and Núñez (2000).

1. Introduction

What saves a man is to take a step. Then another step. It is always the same step, but you have to take it.

Antoine de Saint-Exupery

Aspects of mathematical infinity are ubiquitous in the collegiate mathematics curriculum. For example, students encounter references to infinity in pre-calculus and calculus courses while studying such topics as the asymptotic behavior of rational functions, infinite limits, infinite sequences and series, and improper integrals. Those who continue to upper division courses encounter Cantor's theory of infinite sets while also revisiting more formally many of the topics studied earlier.

The study reported on in this paper was motivated by a review of the existing literature on student understanding of aspects of mathematical infinity, a body of work that is concerned mainly with Cantor's theory and limits. Our study complements existing studies by focusing on aspects of infinity that fall outside the scope of previous studies, and it is the first empirical study to use APOS theory to examine student understanding of aspects of infinity.

Specifically, we provide in this report a preliminary theoretical description of learners' constructions of infinite iterative processes and their states at infinity, grounded in an analysis of interviews with college students who were working on a problem involving the infinite union of power sets. Our focus on infinite iteration

©2010 American Mathematical Society

is consistent with existing theoretical studies of the understanding of infinity. Dubinsky, Weller, McDonald, and Brown (2005a, 2005b) illustrated that, historically, infinite iteration was foundational in the development of thought about mathematical infinity, and, according to Lakoff and Núñez (2000), it forms the conceptual basis for a wide variety of topics involving infinity.

Before describing our study in more detail, we introduce some needed terminology. In this study, we use the phrase *iterate through the set of natural numbers* to refer to the situation in which one imagines starting at 1 and successively adding 1 to reach each natural number in turn. We will use the phrase *infinite iterative process* to refer to the endlessly repeated application of a transformation of mental or physical objects, involving one or more parameters that change with each repetition. For the cases discussed in this paper, one of the parameters iterates through the set of natural numbers. An iterative process may be a recursive process, but it need not be.

The overall goal of the present study was to see whether APOS theory can adequately describe the mental constructions students might make in response to mathematical situations involving actual infinity that they encounter in upper level collegiate mathematics courses. To that end, we gave students in an introduction to abstract mathematics class the following task:

$$\text{Prove or disprove: } \bigcup_{k=1}^{\infty} P(\{1,2,\ldots,k\}) = P(\mathbf{N}), \qquad (\star)$$

where $\mathbf{N}$ denotes the set of natural numbers and P denotes "the power set of," that is, the set of all subsets of the given set. One way to see that this equality cannot hold is to observe that each $P(\{1,2,\ldots,k\})$, and hence the infinite union, contains only finite subsets of $\mathbf{N}$ as elements, while $P(\mathbf{N})$ also contains infinite subsets of $\mathbf{N}$ as elements. Rather than using a formal set theory argument, however, all of the students we interviewed initially tried to make sense of the infinite union by constructing iterative processes. Consequently, we put a special focus on examining the nature of the infinite iterative processes that college students may develop and use in such problem situations. In doing so, we recognized the relevance of Dubinsky's observation:

> It is not possible to disagree with the observation that a string of symbols written on a piece of paper is static. But the nature of anything is never in the thing itself, but in the relation between it and the individual who observes it ... whatever the nature of the formal statement [is] in the absence of human thought, when an individual is using it to construct meaning then the formal expression is full of the dynamism of the several processes the individual constructs (Dubinsky, 2000, p. 235).

It is also important to note that the problem $(\star)$ has an interesting property that was a factor in our decision to use it in our study: the nth partial union equals $P(\{1,2,\ldots,n\})$ but the infinite union does not equal $P(\mathbf{N})$, so there is a "discontinuity at infinity." We hoped that by examining student reactions to this possibly counter-intuitive situation we would gain insight into how students cognitively construct actual infinity.

In the next section, we describe APOS theory and the general methods of analysis used with this research framework. Following that, we review related research literature and present our methodology. We summarize our results by presenting a preliminary theoretical description of students' constructions of infinite iterative processes and their states at infinity, and illustrate these results through five case studies from our data. We conclude with a discussion of our results and point out some future directions for research on related topics.

2. APOS Theory and Our Research Framework

Briefly, APOS theory may be summarized by the following statement:

> An individual's mathematical knowledge is her or his tendency to respond to perceived mathematical problem situations by reflecting on problems and their solutions in a social context and by constructing or reconstructing mathematical actions, processes, and objects and organizing these in schemas to use in dealing with the situations (Asiala et al., 1996, p. 7).

In more detail, according to the theory, formation of a mathematical concept begins as the individual transforms objects to obtain new objects. The individual first conceives of the transformation as an *action*. This is indicated by seeing the transformation as an explicit set of step-by-step instructions to be applied to an existing object, or objects, as a response to a perceived external cue. As the individual repeats and reflects on an action, it may be *interiorized* into a mental *process*. In contrast to an action, a process is perceived as being under the individual's control, as opposed to being a response to external cues. The individual can imagine performing the transformation without having to explicitly execute each step, and, when necessary, can mentally reverse the steps, as well as make coordinations with previously constructed processes. Full development of a process conception of a transformation includes seeing the process as *complete*, i.e., being able to imagine having performed each step. As the individual becomes aware of the process as a *totality* (i.e., sees the process as a single operation applied at a moment in time), realizes that transformations can act on that totality, and can actually construct such transformations, then we say the individual has *encapsulated* the process into a cognitive *object*. Once the process has been encapsulated, new transformations can be applied. In developing an understanding of a mathematical topic, an individual often constructs many actions, processes, and objects. When they are organized and linked into a coherent framework, one may say that the individual has constructed a *schema* for the topic.

The research methodology employed in moving from these general descriptions of the theoretical perspective to a preliminary theoretical analysis of a topic is described in more detail in the Methodology section of this paper. Although we summarize our results in a theoretical description that is presented linearly, we are not claiming that an individual's thinking necessarily proceeds in this way. Since we consider mathematical knowledge in terms of one's *tendency* to respond to a problem situation by making certain constructions, we do not assert that the actual construction of knowledge will always be amenable to the kind of precise categorization detailed in our description. Our description also does not explain what actually happens in an individual's mind, as this is probably unknowable, nor does it predict whether an individual will necessarily apply a given structure constructed in one situation to another. As such, APOS explanations tell only part of the story. However, there have been a number of empirical studies showing the viability of APOS theory in describing how students construct their conceptions of a variety of mathematical topics, including mathematical induction, functions, derivatives, limits, and group theory (see for example Baker, Cooley, & Trigueros, 2000; Breidenbach, Dubinsky, Hawks, & Nichols, 1992; Brown, DeVries, Dubinsky, & Thomas, 1997; Dubinsky, 1989). Finally, in a recent theoretical study, Dubinsky et al. (2005a, 2005b) demonstrated the potential applicability of APOS theory to a wide variety of mathematical problems involving infinity.

For a more detailed description of APOS theory, see Asiala et al. (1996) and Dubinsky and McDonald (2001). For information on the efficacy of APOS theory in the development of pedagogical strategies leading to increased student learning, see Weller et al. (2003).

3. Literature Related to This Study

As noted in the introduction, much of the existing literature on conceptions of infinity deals with students' reasoning and intuitions about Cantor's theory and limits. In a study of pre-college students, Fischbein, Tirosh, and Hess (1979) found that standard mathematics instruction did not cause significant change in the students' often contradictory intuitions of actual infinity. Tirosh (1991) subsequently developed a unit on Cantorian theory and studied its affect on students' formal and intuitive understandings of infinity. While there was little evidence that the instruction resulted in modified intuitions, it was observed that students became more wary of using intuition alone to answer questions about actual infinity and understood that using formal definitions and theorems was a more reliable approach.

Many studies have focused on the specific strategies used to compare infinite sets. The basis of comparison used in Cantor's theory, one-to-one correspondence, tends to be chosen less often by students than methods based on set inclusion or the notion of a "single infinity" (see for example Borasi, 1985; Moreno & Waldegg, 1991; Sierpinska & Viwegier, 1989). Additionally, Tirosh and Tsamir (1996) found that students' responses to comparison tasks depended on how the sets were represented. Tsamir and Tirosh (1999) proposed instructional interventions to make students aware of the contradictions caused by using more than one method of comparing sets, as did Tsamir (2001). A detailed study, using the theoretical lens of abstraction in context, of how one student dealt with these conflicts was reported in Tsamir and Dreyfus (2002), and the tendencies of prospective secondary teachers to use various strategies to compare infinite sets were studied by Tsamir (2003).

Tall (1980) argued that students' difficulties with infinity stem not from faulty intuitions but rather from an instructional focus on Cantor's theory, which is at odds with their previous school mathematics experiences that tend to emphasize notions of measurement. Tall (1992) discussed the formation of a variety of socially and individually constructed "creases in the mind," ways of viewing, understanding, and representing mathematical content, including conceptions of infinity, that could create learning obstacles for students. More recently, Tall (2001) considered two modes of reasoning about the infinite: extending experiences with the finite and using formal axiomatic methods. He argued that a variety of natural and formal infinity concepts can arise, and examined their interplay. Although not directly applicable to this report, these studies heightened our awareness of possible conflicts between intuitions of infinity and more formal mathematical notions. More relevant to the content of this study are theoretical findings offered by Dubinsky et al. (2005a, 2005b). In response to historical arguments against the existence of actual infinity, the authors explain how potential and actual infinity represent two different conceptualizations linked by the mental mechanism of encapsulation. Potential infinity, the notion of infinity presented over time, is the conception of infinity as a process. Since an infinite process has no final step, and hence, no obvious indication of completion, the ability to conceive of the completion of an infinite process is a crucial step in moving beyond a purely potential view. As one reflects on a completed infinite process, an individual may be able to conceive of the process as a totality, a single operation freed from temporal constraints. The individual can then apply the mechanism of encapsulation to mentally transform the process into a mental object, an instance of actual infinity. Encapsulation often occurs in an effort to determine the "state at infinity" of an infinite process.

A number of other researchers have raised the issue of how one conceptualizes states at infinity for infinite processes. For example, Mamona-Downs (2001) noted that many students consider the limit of a sequence $(a_n)_{n=1}^{\infty}$ to be like a last term, writing a_∞ for the limit. Lakoff and Núñez (2000), drawing on research in cognitive science and applying their method of mathematical idea analysis, hypothesized that the conceptualization of many different aspects of actual infinity is tied to a common structure called the Basic Metaphor of Infinity (BMI). They argued (p. 158) that the mechanism of conceptual metaphor enables an individual to conceptualize the "result" of an infinite process in terms of a process that does have an end. That is, they proposed that the crucial effect of the BMI is to add to the target domain, iterative processes that go on and on, the completion of the process and a final resultant state. This metaphorical final result, the state at infinity, may then be perceived as an instance of actual infinity.

Schiralli and Sinclair (2003) contended that mathematical idea analysis alone is inadequate for the task of describing the nature of mathematical concepts. In their view, individuals use a variety of cognitive approaches in connecting their internal understandings of mathematics with the more public, shared representations of mathematics, and argue that many such approaches cannot be explained through mathematical idea analysis. While agreeing that abstract concepts have a basis in sensory-motor activities, they presented several instances in which Lakoff and Núñez's "reduction of abstract concepts to more concrete ones through metaphor fails to explain fundamental processes involved in acts of abstraction" (p. 82). They noted the importance of empirical studies as a means of capturing the richness of student's individual thinking. We agree, and attempted in this empirical study to identify the roles of the mechanisms of interiorization and encapsulation in students' construction of infinite iterative processes and their states at infinity in a standard, formally presented problem situation involving actual infinity.

4. Methodology

In the first phase of the study, we conducted interviews with students enrolled in an introduction to abstract mathematics course at a regional university in the Midwestern United States. These students were sophomore or junior mathematics, computer science, secondary mathematics education, or secondary science education majors. All of the students had taken two semesters of single variable calculus, as well as at least one other course in either linear algebra, college geometry, or multivariable calculus. Finite unions, power sets, and standard proof techniques were topics in the course, but neither indexed infinite collections of sets nor infinite unions were discussed in class prior to the interviews, although their formal definitions appeared in the assigned reading in the course textbook.

The students were initially given the problem ($\star$) for homework, presented in a slightly different form:

> For each positive integer n, let $X_n = \{1, \ldots, n\}$, and let $P(X_n)$ denote the power set of X_n, that is, the set of all subsets of X_n.
>
> Prove or disprove: $\displaystyle\bigcup_{n=1}^{\infty} P(X_n) = P(\mathbf{N})$.

There were seven students in the class, and six of them attempted the homework problem. No one completely solved it. The instructor (one of the authors) asked the six students to participate in individual, audio-taped interviews. Five of the six students agreed to participate, and all five were interviewed. The students were asked to solve the problem again, with prompts given as needed by the interviewer. Typical interviewer prompts included questions intended to encourage students to resolve contradictions in their expressed reasoning, to consider the statement of definitions, and to use standard proof techniques when stymied with a particular line of reasoning. Each interview lasted less than one hour. The written work was collected, the audio tapes were transcribed verbatim, and the transcripts were checked against the original tapes to ensure accuracy.

Using the APOS research framework (see Asiala et al., 1996), all three authors individually analyzed the transcripts to identify various mathematical issues that arose in student work on the problem, shared findings, and then negotiated to consensus on the issues selected for further study. APOS theory was then used to begin to identify the mental constructions students had or had not made, relative to the most prevalent issues. In this preliminary analysis, it was agreed that iterative processes were used by all six of the students, but the nature of these processes and the role they played in students' attempts to solve the problem were not clear. For phase two of the study, we designed an expanded interview protocol that included questions that would help us distinguish difficulties with iteration in general from difficulties in forming power sets and the infinite union.

Phase two interviews took place one year after the phase one interviews. The students interviewed in this phase had just finished taking the same course from the same instructor. These students had essentially the same mathematical background as the previous interviewees. The course topics were similar, with the exception that this course included a brief introduction to the topic of cardinality of sets.

Seven of the eight students enrolled in the course agreed to participate in the interviews. These students were split into two groups of two students and one group of three students. All of the students knew each other and had often worked together in class during the semester. The instructor and one of the other authors conducted the interviews.

Small group interviews were chosen for the second phase because, in the analysis of the phase one interviews, there were several instances in which little could be learned about a student's conceptions without extensive interviewer prompting. The intent of grouping students together was to maximize articulation of student thinking while minimizing interviewer prompting. We note that, in a study using the APOS theoretical perspective to examine student conceptions of inverse functions, Vidakovic (1993) found that mental constructions resulting from group work do not differ in any significant manner from those made by students working on a topic individually.

Each interview lasted one and one half to two hours. The interviews were conducted in a distance-learning classroom and were both audio-taped and video-taped. Video images were recorded from a document camera and two wall-mounted cameras, and there were two small microphones placed near the document camera. An experienced television producer controlled the equipment from a separate room. In this way, an atmosphere close to a typical classroom setting was maintained. The fact that the interview was being recorded did not appear to distract the students in any significant way. The questions were printed on separate pieces of paper and handed to the group one at a time, in numerical order. The students discussed the problems and wrote their solutions on sheets of paper situated under the document camera. The image from the document camera was projected so that the interviewers could see what the students were writing. At times, the interviewers asked questions of one or more of the students, asked for further clarification of their ideas about the issues they were discussing, and used prompts similar to those used in the phase one interviews. The audiotapes were later transcribed verbatim. The resulting transcripts were checked against the videotapes and the written work that was collected to ensure accuracy.

The phase two interview protocol consisted of nine problems, the first eight of which explored issues related to the original problem. The ninth question was the power set question from the first interview protocol. The first four questions dealt with power sets and unions involving a finite number of sets. These were topics covered in the course, and all of the students solved the problems easily. The remaining problems, the responses to which formed the basis of our analysis, are listed below.

5. If $B_1, B_2, B_3, \ldots$ are sets, what does it mean to say that $x \in \bigcup_{k=1}^{\infty} B_k$?

6. If C and $B_1, B_2, B_3, \ldots$ are sets, what must be shown to verify that $\bigcup_{k=1}^{\infty} B_k = C$?

Comment: The purpose of Questions 5 and 6 was to emphasize the formation of a union of an arbitrary collection of sets indexed by $\mathbf{N}$, independent from the consideration of power sets, and to provide an opportunity to verify that students were aware of the formal definition of infinite union.

7. Let k be a fixed but arbitrary positive integer. Describe the set $P(\{1, 2, \ldots, k\})$, where P denotes power set. What are the elements of this set, and is the set finite or infinite?

Comment: The purpose of this question was to evoke students' conceptions of the role of the index k and to verify that the students were aware of the definition of power set apart from the context of the infinite union in Question 9.

8. Describe the set $P(\mathbf{N})$, where $\mathbf{N}$ is the set of natural numbers and P denotes power set. What are the elements of this set, and is this set finite or infinite?

Comment: The purpose of this question was to reveal the students' conceptions of the set $P(\mathbf{N})$ in a context separate from its role in Question 9.

9. Prove or disprove: $\bigcup_{k=1}^{\infty} P(\{1, 2, \ldots, k\}) = P(\mathbf{N})$, where $\mathbf{N}$ is the set of natural numbers.

Comment: There were two reasons for the change of notation in this question from what was used in the phase one interviews. In the audio-taped phase one interviews, it was sometimes difficult to distinguish (audibly) the students' references to the index n from their references to the set of natural numbers $\mathbf{N}$ (thus we changed the index to k). Also, some of the students found it awkward to use the notation X_n to represent the set $\{1, 2, \ldots, n\}$. In fact, all of the students involved in the phase one interviews unpacked the notation into this more explicit form.

The analysis of the second phase interviews proceeded similarly to the first phase: we came to consensus on the dominant mathematical issues that arose in student work, and then identified mental constructions students had or had not made related to those issues. We also re-analyzed the phase one data set. Because the expanded protocol included questions about various aspects of the main problem, we were able to further refine a description of the various infinite iterative processes students used in solving this set of problems.

5. Results

Initially, we expected that students would compare $\bigcup_{k=1}^{\infty} P(\{1, 2, \ldots, k\})$ with $P(\mathbf{N})$ on the basis of set inclusion, as emphasized in the course. As pointed out earlier, one might note that the equality does not hold since each $P(\{1, 2, \ldots, k\})$, and hence the infinite union, contains only finite sets as elements, whereas $P(\mathbf{N})$ also contains infinite sets as elements. Indeed, one might expect that if an individual knows the definition of power set and the definition of infinite union, the result should not be very difficult to see. However, despite demonstrating that they knew these definitions, none of the 12 students interviewed took a direct, formal approach to solving the problem. Instead, all of the students first emphasized process-oriented thinking about the problem, and most used formal techniques only after repeatedly being encouraged to do so by the interviewers. Thus, it is important to examine the role of the processes they constructed in their attempts to solve the problem.

Before considering the specifics, it should be remarked that all of the students' responses included the following reasoning pattern. For the infinite union, the students constructed an iterative process resulting in a sequence of partial unions whose nth term was the union of the first n power sets:

$$P(\{1\}), \quad P(\{1\}) \cup P(\{1, 2\}), \quad \ldots, \quad \bigcup_{k=1}^{n} P(\{1, 2, \ldots, k\}), \quad \ldots \quad .$$

Noting that the successive power sets are nested, that is,

$$P(\{1\}) \subset P(\{1, 2\}) \subset P(\{1, 2, 3\}) \subset \cdots,$$

the students observed that the sequence of partial unions reduces to the sequence of power sets:

$$P(\{1\}), \quad P(\{1, 2\}), \quad P(\{1, 2, 3\}), \quad \ldots, \quad P(\{1, 2, \ldots, n\}), \quad \ldots \quad .$$

The students constructed the sequence of sets as a way of understanding the infinite union, so they identified the state at infinity for this process with the infinite union. Determining the form of the state at infinity is what turned out to be most problematic. All of the students initially proposed that the state at infinity was equal to $P(\mathbf{N})$. The case studies offered below illustrate the various ways in which that conclusion was reached, and how the students reflected on and reconsidered their reasoning. Prior to presenting the case studies, we provide the theoretical description we developed during the data analysis; it introduces the constructs and terminology that we used to distinguish the approaches taken by the students.

Briefly, an infinite iterative process is one type of mental construction that falls in the category of "process" in APOS theory. We will argue that an infinite iterative process is conceptualized as a coordination of a process of iterating through **N** with a transformation that can be applied repeatedly. An infinite sequence of objects (e.g., numbers or sets) is formed, one object being adjoined to the sequence at each step. In some cases, the process consists of selecting objects in sequence from a collection of already constructed objects, while in other cases, the objects might be constructed at the same time that the iteration is being constructed.

Central to our approach is the observation that a prerequisite for constructing an infinite iterative process is the ability to construct a process of iterating completely through **N**. Certainly, modern mathematicians have this ability (i.e., we can imagine each natural number being reached in such an iteration). We can also view this process as a totality (i.e., we can think of "iteration through **N**" as a single operation). It is also commonplace in advanced mathematics to think of the complete sequence of natural numbers as being followed by an entity which we label "∞." This conception of infinity provides a method of indexing a state at infinity for other countably infinite iterative processes (see, for example, Lakoff & Núñez, 2000, p. 165).

We now provide a general description, derived from analysis of interview data, of how an individual might construct infinite iterative processes. Then, we present five case studies to illustrate the description. In showing how students used such processes and their encapsulations in thinking about the problems in our interview protocol, we hope to provide insights that may lead to future research into the conceptualizations evoked by a wider variety of problems involving actual infinity.

5.1. Understanding iteration through N. In this study, the infinite iterative processes of interest involve coordinating iteration through **N** with transformations that construct various power sets and unions. Our first step is to describe, in terms of APOS theory and based on our data analysis, what it might mean to understand iteration through **N** and its encapsulation:

Action: The individual explicitly refers to stepping through a finite segment of **N**, typically writing or speaking the values assumed in sequence. The individual may refer to repeatedly adding 1, or use other terminology that suggests passing from a natural number to its successor.

Process: The action of finite iteration through a segment of **N** is interiorized to a mental process of iterating through any finite segment of **N**; multiple instantiations of iterating through finite, but not necessarily initial, segments of **N** are coordinated to construct a complete infinite process of iterating through **N**. Being aware that each natural number has been reached, and acknowledging that only natural numbers have been reached, indicate that the individual can conceive of the process as complete.

Object: Once viewed as a totality (i.e., as a single operation that is understood to consist entirely of reaching each natural number in order), the infinite iterative process might be encapsulated as the individual attempts to apply an action of evaluation in trying to determine what is "next." The resulting object is conventionally labeled ∞. This object may be viewed as a value of an iterating variable that is beyond all natural numbers, but it must be understood that this object is not obtained in the process of iterating through **N**.

5.2. Understanding infinite iterative processes. Now we use our theoretical description of iteration through **N** as part of our description of the conceptualization of an infinite iterative process that obtains a sequence of objects, and the encapsulation of that process. In terms of APOS theory, we propose that constructing infinite iterative processes and their encapsulations might take place as follows, assuming that a transformation (or function) that accepts natural numbers as inputs is given. We also assume the description of how students might construct their understanding of function given in Breidenbach et al. (1992).

Action: Performing a small number of iterations of the transformation is considered to be an action in this situation. For example, an individual might perceive the presence of an indexing variable in the given mathematical notation as a cue to start with $k = 1$, obtain the first object using the transformation, add 1 to get $k = 2$, obtain the second object, and so forth.

Process: The action of finite iteration is interiorized to a mental process of finite iteration by coordinating a process of iterating through a finite segment of **N** with a transformation that can be applied repeatedly. Subsequently, multiple instantiations of this finite mental process (i.e., using the same transformation but stepping through different finite segments of **N**) are coordinated to construct an infinite mental process, which we will call an *infinite iterative process*. When fully constructed, this amounts to coordinating a complete process of iterating through **N** with a process conception of the transformation. The infinite iterative process obtains a (countably infinite) sequence of objects. The individual understands that an object is obtained for each natural number in order, and that objects are obtained only for natural numbers.

Object: Once it is viewed as a totality (i.e., as a single operation that associates an object to every natural number in order), the infinite iterative process might be encapsulated in response to an attempted action of evaluation. The resulting object is a state at ∞. This object is understood to be beyond the objects that correspond to the natural numbers and thus not directly produced by the process. We call this object a *transcendent object* for the process.

A comment about the subtle distinction between having completely constructed an infinite process and seeing it as a totality might be helpful to the reader. In achieving the former, the individual can imagine how to take every step and what the outcome of each step will be. In achieving the latter, the individual realizes that the process forms a single transformation that can be applied at a moment in time. For example, in considering a recursive process of forming the infinite union of an indexed collection of sets $\{B_k : k \in \mathbf{N}\}$, seeing the process as complete might entail the observation that, for each i, the step that is taken is to union $\bigcup_{k=1}^{i-1} B_k$ with B_i. On the other hand, seeing the process as a totality might entail the realization that the union operation is applied to the entire collection. According to APOS theory, seeing the process as a totality is part of the shift to object status as the process is encapsulated in response to an attempted action, such as an action of evaluation. In this example, one appropriate action of evaluation would be to determine the result when all of the elements of all of the sets are collected in a single set.

5.3. Introduction to Data Analysis. All of the data were used in developing the description above, but we present detailed analyses of just five case studies. Two are individual interviews from phase one. The other three are the group interviews from phase two, with the thinking of one student from each group highlighted. The student who was chosen from each group is the one who gave the most detail about her or his thinking processes during the interview. Taken together, these five cases address all of the major issues concerning the construction of infinite iterative processes and their encapsulations that we were able to identify as playing a role in student solutions to the main problem of the study.

In all of the interviews conducted for this study, the students constructed iterative processes in response to the formalisms of the problem situations. In addition to constructing iterative processes to deal with Question 9, the students involved in the phase two interviews constructed iterative processes for Questions 5 through 8, even though these problems were explicitly framed in terms of the formal definitions of the objects. Each case study displays the resulting interplay between informal process-oriented exploration and formal use of definitions and standard techniques. Although students' responses are not assumed to account for all possible approaches for solving these problems, we can match evidence of success with the problems with evidence of the described constructions, and show a relative absence of success in the absence of those constructions. Based on our data analysis, we conjecture that certain mental constructions help students construct states at infinity, thus facilitating their conceptualization of actual infinity.

The cases appear in this order: Tobi, Emily, Opal, Carrie, and Stan (student names are pseudonyms). Of these, Emily and Carrie were interviewed in phase one and the others in phase two. In choosing this ordering, we begin with the student (Tobi) who was most successful in solving the problem, then present two cases (Emily, then Opal) in which the students had difficulty in determining the nature of the state at infinity for the processes that they constructed, and finish with two cases (Carrie, then Stan) in which the student's view of the nature of the indexing variable appeared to cause difficulties in constructing iterative processes used in dealing with the formal aspects of the main problem.

5.4. Tobi: "This is a union of an infinite number of finite sets". Of the 12 students interviewed during the study, only two students, Tobi and her partner Steve, completely answered Question 9. Throughout their interview, Tobi led the discussion and provided most of the ideas, while Steve mainly reiterated Tobi's points or raised questions about things she said. Our analysis of their work on Question 9 will show that Tobi saw the infinite iterative process related to the infinite union of power sets as a completed totality and subsequently encapsulated it into a transcendent object. This enabled her to correctly determine its state at infinity and solve the problem using the formal definition of infinite union.

Early in their work on Question 9, Tobi and Steve both clearly exhibited an action conception as they wrote out the elements of $P(\{1\})$ and $P(\{1,2\})$. Realizing that the first is a subset of the second, they quickly interiorized their actions to construct a finite iterative process for computing and listing the partial unions of the power sets, noting that the form of the kth partial union would be $P(\{1,2,\ldots,k\})$. But, as they continued, they put $P(\{1,2,\ldots,k+1,\ldots\})$ on their list of partial unions. At that point, the interviewer asked them to look at the original statement of the problem and see whether there were any "power sets of infinite sets" among the sets to be unioned. Tobi replied:

> *Tobi*: No, because k is always an arbitrary number as you go, to infinity. But, the union of all of those go to infinity, right?
> *I*: What do you mean "the union go to infinity?"
> *Tobi*: There's always one plus whatever you go to.

Tobi pointed at $P(\{1\})$, then $P(\{1,2\})$, and noted that "those keep going till you stretch out our set, until it's the size of infinite numbers in there," suggesting that the process would directly produce a state at infinity. Continuing, she immediately discounted this idea, noting that "it doesn't ever get to infinity, 'cause it's always just one more than it was previous. But you just keep going and going and going. So this is the union of an infinite number of finite sets." Although she referred here only to the finiteness of $P(\{1,2,\ldots,k\})$, Tobi and Steve had observed earlier that such finite power sets contain only finite sets as elements.

Tobi adjusted the notation used on their list of power sets, revising the notation of the set $P(\{1,2,\ldots,k+1,\ldots\})$ to be $P(\{1,2,\ldots,p+1,\ldots,t\})$. At that point, their work read:

$$
\begin{gathered}
P(\{1\}) = \{\ \{\ \}, \{1\}\ \} \\
P(\{1,2\}) = \{\emptyset, \{1\}, \{2\}, \{1,2\}\ \} \\
\vdots \\
P(\{1,2,\ldots,p\}) \\
\vdots \\
P(\{1,2,\ldots,p+1,\ldots,t\}) \\
\vdots
\end{gathered}
$$

The interviewer asked whether the last set she wrote is finite or infinite, and Tobi replied:

> *Tobi*: They all gotta stop, somewhere. Because they're all finite, 'cause they go up to a number. ... Whatever that number is.

Tobi showed evidence of having constructed multiple instantiations of a finite process (e.g., the written work indicating a progression from $P(\{1\})$ to $P(\{1,2,\ldots,p\})$ and from $P(\{1,2,\ldots,p\})$ to $P(\{1,2,\ldots,p+1,\ldots,t\})$) and coordinated these into one infinite iterative process, related to the infinite union, that produced a nested sequence of power sets. Evidently she could imagine completing the process since she understood that an "infinite number" of finite power sets are produced during an iterative process in which the index is "always just one more than it was previous." In addition, she remarked on the properties of the power sets produced at each step: "they all gotta stop, somewhere." In her statement that the "union of an infinite number of finite sets" is formed, she alluded to seeing the process as a totality, giving a static collection of sets present all at once on which she could act by taking their infinite union.

The importance of being able to see the process as a totality and to encapsulate it can be seen in her subsequent work in which she used both her process and object conceptions in solving the problem using formal methods. For example, later in the interview, Tobi summed up what she knew, and what she thought they should do in order to answer the question:

> *Tobi*: The problem is to prove that the union of the power sets from k equals one to infinity of $P(\{1, 2, \ldots, k\})$ is equal to $P(\mathbf{N})$. And we boiled that down, realizing that each power set was contained in the next power set. ... So that you actually have a union of an infinite number of power sets, and those power sets are of finite sets. And we want to prove that is equal to $P(\mathbf{N})$.

They had already noted in their response to Question 6 that they simply needed to show set inclusion both ways to prove that the two sets were equal and they applied it to this situation. Without much difficulty, Tobi explained that any set that is an element of the infinite union would also be an element of $P(\mathbf{N})$. To consider the reverse inclusion, Tobi first identified various elements of $P(\mathbf{N})$ that she could show are also elements of $P(\{1, 2, \ldots, k\})$ for some k. Then, she focused on the set of natural numbers $\mathbf{N}$, considered as an element of $P(\mathbf{N})$, and asked "where could I find that in a finite set?" She observed that "it's not anywhere in there" (i.e., not in the infinite union) but had trouble convincing her partner Steve that this implies that the proposed equality does not hold. The interviewer prompted them to consider the relevance of their response to Question 6, when they observed that to show $C \subseteq \bigcup_{k=1}^{\infty} B_k$, one must establish that each $x \in C$ is an element of at least one B_k. Tobi commented, "I've already said I couldn't do that." Using x to signify $\mathbf{N}$ and B_k to signify $P(\{1, 2, \ldots, k\})$, she explained:

> *Tobi*: I can't show, if x is equal to the natural numbers, I can't show that it's in a particular B_k.
>
> *Steve*: Because that's a finite set.
>
> *Tobi*: Because B_k is a finite set.

What we see here is that Tobi was again able to consider the infinite union as a static object that transcended the infinite union process she described earlier. She used the formal definition of an infinite union to determine whether it contains an infinite set as an element. In terms of mental constructions, to perform this action on the union, she de-encapsulated the union back to a process and used what she knew about the sets $P(\{1, 2, \ldots, k\})$: that each one contains only finite subsets of $\mathbf{N}$ as elements. In sum, she was able to prove that infinite elements of $P(\mathbf{N})$, such as $\mathbf{N}$ itself, are not elements of the infinite union of power sets, and thus concluded that the proposed equality is false.

5.5. Emily: "When do you ever reach infinity?". In this section, we examine aspects of the phase one interview with Emily. We present her comments about the properties of the power set $P(\{1, 2, \ldots, k\})$ obtained at each step as evidence that she was able to view her infinite iterative process for the infinite union as complete. Additional excerpts will show that she was not able to consider the total collection of these sets at once, that is, she could not see the process as a totality. Our conclusion is that, as a consequence, she was unable to construct a transcendent object for her process and apply the formal mathematics needed to consider whether each element of $P(\mathbf{N})$ is an element of the infinite union. We remind the reader that, in the phase one interviews, the notation X_n was used to denote the set $\{1, 2, \ldots, n\}$.

Initially showing an action conception of iteration, Emily wrote the following in an effort to make sense of the given infinite union:

$$\begin{aligned} P(X_1) &= \{\emptyset, \{1\}\} \\ P(X_2) &= \{\emptyset, \{1\}, \{2\}, \{1, 2\}) \qquad \text{[sic]} \\ P(X_3) &= P(X_2) \cup \{\{3\}, \{1, 3\}, \{2, 3\}, \{1, 2, 3\}\} \end{aligned}$$

She observed that, for $n = 2$ and $n = 3$, $P(X_n)$ was obtained from $P(X_{n-1})$ by adjoining to $P(X_{n-1})$ all of the subsets of X_n that contained n. Reflecting on her work, she commented that she thought the same process would yield $P(\mathbf{N})$ since,

> *Emily*: ...we're starting with, here, X sub one ... and then we're just adding to it all the combinations of what we started with plus an extra number. And then we go through all the natural numbers. And so if we had started with a set of all natural numbers, then we'd have all possible combinations of all the natural numbers.

Emily commented that while she thought it was "logical," she felt "lost on how to prove it."

Note that Emily used neither the formal definition of the infinite union nor the conventional method of showing the equality of two sets here, but instead interpreted the infinite union as a recursive process. In describing how the process continued, Emily appeared to have interiorized the action of iterating through the first few natural numbers and applying the power set transformation. Her reference to moving through all the natural numbers in the same way may indicate that she had begun to coordinate multiple instantiations of the finite process to construct an infinite iterative process. In fact, the process she described would be an appropriate process to encapsulate to construct the infinite union in question, if this process were completely constructed and viewed as a totality. Her error was in the assumption that the same process would also construct $P(\mathbf{N})$, if "we had started with the set of all natural numbers," since she had not included any way to generate the infinite subsets that are elements of $P(\mathbf{N})$.

When Emily seemed unable to progress any further, the interviewer asked her to reconsider the problem more formally, using set inclusion relationships. Emily readily explained why the infinite union is contained in $P(\mathbf{N})$, noting that for any n, "the power set of X_n is included in the power set of natural numbers because the set X_n is included in the set of natural numbers." She pointed out that she used the same principle when she observed that $P(X_1)$ is contained in $P(X_2)$, that $P(X_2)$ is contained in $P(X_3)$, and that the same thing worked "all the way up to whatever, *any n*." This established that she had constructed an iteration through $\mathbf{N}$ that extended to an arbitrary natural number n, and that she had coordinated this process with the power set transformation to construct a finite process for the

nth partial union (or nth power set, since she obtained the power sets by taking a union) for an arbitrary n.

When Emily subsequently suggested that because $\bigcup_{n=1}^{\infty} X_n = \mathbf{N}$, it follows that $\bigcup_{n=1}^{\infty} P(X_n) = P(\mathbf{N})$, the interviewer questioned her reasoning and asked her to consider the proposed equality formally. Emily explained why each set T in $P(\mathbf{N})$ must be an element of the infinite union of power sets. In this argument, she used her finite iterative process for the partial unions. For example, she explained why each singleton element of $P(\mathbf{N})$ is contained in the infinite union because it "would have been picked up by [pause] as soon as you took the union of the next X_n, or that whatever n was, if, when n equals T." Though there were errors in expression here – she said "n equals T" when she meant $\{n\} = T$, and she referred to taking the union of the "next X_n" rather than $P(X_n)$ – she corrected herself a few seconds later. Emily gave a process-oriented explanation since she spoke in terms of continuing the process far enough ("as soon as you took") to include n. This suggests that she had made some progress in coordinating multiple instantiations of the finite process, since she apparently could imagine carrying the process out to any specified point. She also used past tense, suggesting an ability to think of the steps as having been completed. With this reasoning as a basis, she subsequently extended her approach to any finite set T, applying an argument based on selecting the "highest integer" in T in order to determine how far out in the sequence one needed to iterate in order to find the first power set of the form $P(\{1, 2, \ldots, k\})$ containing T.

After Emily demonstrated an understanding that all finite subsets of $\mathbf{N}$ are elements of the infinite union of power sets, the interviewer asked Emily about the set $\mathbf{N}$ itself. In attempting to determine whether it is in the infinite union, she used her process for the union of power sets to support her conclusion that, for each n, $P(X_n)$ is finite. Specifically, Emily considered the properties of the intermediate results of the process and expressed how each power set was obtained from its predecessor $P(X_n)$: one adds 1 to n, yielding $P(X_{n+1})$ as the "next set to be unioned." She understood that, at each step, one would be adjoining a finite number of elements to a set that is finite, and concluded that "every set would be a finite set even though you'd go on infinitely doing that." Taken together with her observation that the largest subset that was an element of each $P(X_n)$ was X_n itself, it appears that Emily was able to conceive of her infinite process for the infinite union of power sets as complete, since she could consider the critical properties of the set produced at each of its infinitely many steps.

Coming to understand that all of the partial unions are finite was not as easy for Emily as it seemed to be for Tobi. Earlier in the interview, she proposed that an infinite set (which she referred to as "$P(X_\infty)$") would be produced at some step of the infinite process. Near the end of the interview, she analyzed the conflict in her thinking.

Emily: And then you keep adding one and you'll still, have finite sets but eventually you have to, I just *still* want [both laugh] to include that infinity! And include, um, if you infinitely union sets, eventually you've got to union the infinite set I would think. Maybe not. When do you ever reach infinity? That's what it is.

I: *Do* you reach infinity?

Emily: No. [pause] OK, so that's why.

Emily's process conception is revealed by the way she expressed her ideas using temporal terms: "if you infinitely union sets, eventually you've got to union the infinite set I would think" and "When do you ever reach infinity?" Such language also suggests that she was not seeing the process as a totality, or the completed union as present at a single moment in time. Thus, it was impossible for her to apply an action of evaluation and encapsulate the complete infinite process into an object, although she tried. In her attempt to evaluate the infinite union, Emily seemed to try to construct a state at infinity in a way that was analogous to what she did for the finite process (i.e., she reasoned that if $P(X_k)$ is obtained at step k, then "$P(X_\infty)$" is the state at infinity.) In the end, her belief that one does not "reach infinity" in the indexing process allowed her to reject this idea, but she could go no further. That is, although she understood that the state at infinity transcends the process, she was not able to determine what that state is.

5.6. Opal: "All we need to look at is the one farthest to the right". We turn to the phase two interview with Opal, Nikki, and Ruvi. We focus on Opal, since she tended to lead the discussion and provided the most detail about her thought processes. Of particular interest here was her reasoning that a process that recursively generated the power sets $P(\{1, 2, \ldots, k\})$ would terminate with $P(\{1, 2, \ldots\})$ or $P(\mathbf{N})$. This suggests that she had difficulty seeing her process for the infinite union as complete.

In responding to Question 7, Opal described a recursive process for generating the subsets of $\{1, 2, \ldots, t\}$ from a list of the subsets of $\{1, 2, \ldots, t-1\}$, for each t from 1 to k. Her process was based on the realization that one need only adjoin t to each of the subsets of $\{1, 2, \ldots, t-1\}$, and append the resulting sets to the existing list. Leading the group discussion on Question 8, Opal referred back to that description, and noted that generating $P(\mathbf{N})$ was very much like what they had done in generating $P(\{1, 2, \ldots, k\})$, except that "it's just going to keep going, whereas before we stopped at k." She concluded that $P(\mathbf{N})$ was "going to be an infinite set." Her use of the future tense suggests that she was viewing $P(\mathbf{N})$ in terms of an ongoing process. She continued:

Opal: We should probably define a pattern. Like we did before [in Question 7]. You'd take an element, and then combine it with the previous elements. And then once you've reached all the possible combinations of the previous ones, then you move to the next natural number and take that as a set and then add it to all the previous.

As part of her explanation, she wrote the following, indicating an iteration through $\mathbf{N}$ by writing the first two sets, pausing, then the next two, pausing, and then the next four:

$$P(\mathbf{N}) = \{\emptyset, \{1\}, \{2\}, \{1,2\}, \{3\}, \{1,3\}, \{2,3\}, \{1,2,3\} \ldots\}$$

infinite set

The process that Opal described would generate only the *finite* subsets of $\mathbf{N}$, so her written statement of "infinite set" is actually incorrect. In response to an interviewer's question, Opal stated that no infinite sets would appear, "because you're always just taking the next natural number. It's going to be huge, but it's going to be finite." When asked, she confirmed that the word "it" referred to an arbitrary element being generated by her process. Her ability to describe how to obtain an arbitrary intermediate state from its predecessor indicates that she had coordinated multiple instantiations of the finite process to construct an infinite iterative process. What is unclear is whether Opal saw the process as complete, since she did not address explicitly whether she thought the process would produce all of the subsets of $\mathbf{N}$. At the end of their response to Question 8, the interviewer asked the group whether $P(\mathbf{N})$ has any infinite sets as elements. Opal agreed that it does, citing $\mathbf{N}$ itself as an example, but she did not address the conflict between that observation and her claim that the process she described generated the elements of $P(\mathbf{N})$.

Opal's view that it was possible, by iterating k, to make the transition from $P(\{1, 2, \dots, k\})$ to $P(\mathbf{N})$ appeared again as the group began to consider Question 9. Opal immediately asked: "Does $\mathbf{N} = \{1, 2, \dots, k\}$ where k goes from 1 to ∞?" She said that, if it were true, then the set indicated by the infinite union notation would simplify to $P(\mathbf{N})$. As they tried to make sense of the infinite union of power sets, Nikki and Opal compared the role of k in Questions 7 and 9. Nikki correctly pointed out that the notation in Question 9, $P(\{1, 2, \dots, k\})$, refers to a finite set, just as it did in Question 7. Opal challenged her statement, arguing that in Question 7 "it had a definite integer, it said it was an arbitrary fixed integer" whereas in Question 9, "k isn't fixed." At this point, Opal apparently did not see the role of k in Question 9 as requiring one to fix k at each positive integer, union the corresponding power set with the preceding partial union, and then increment k by 1. In particular, she did not see k, as it appeared in $P(\{1, 2, \dots, k\})$, as representing a *fixed* integer.

Opal, Nikki, and Ruvi could not agree on any specific meaning or relevance for the notion that "$\mathbf{N} = \{1, 2, \dots, k\}$ where k goes from 1 to ∞," and it appeared that they could make no further progress on the problem. At that point, one of the interviewers asked the group whether they could rewrite the notation of the infinite union in a way that might make it easier for them to work with it. Opal wrote:

$$P(\{1\}) \cup P(\{1, 2\}) \cup \dots P(\{1, 2, \dots, k\}).$$

Opal's expression was inconsistent with their work on Questions 5 and 6, where they had re-expressed the union $\bigcup_{k=1}^{\infty} B_k$ as $B_1 \cup B_2 \cup \dots$, agreed that there is one set B_k for each natural number, that there is no last set, and that the union would "keep going." On the other hand, it is notable that Opal did apply the processes correctly here: at each step, she incremented k, formed the power set, formed the union with the previous partial union, and then repeated those actions.

Recognizing the inconsistency with their previous work, one of the interviewers asked them, "Does that stop?" Opal said no, and crossed out the last power set, leaving $P(\{1\}) \cup P(\{1, 2\}) \cup \dots$, a decision whose correctness she debated as she continued talking about "the last" power set:

Opal: Well, all of these are going to be subsets of the last one. So if these are all subsets of the last one, and all we're doing is taking the union, then really all we *do* need to look at is the power set of 1, 2, to k. Or am I taking a leap there?

I2: What's "the last one"?

Opal: The power set of, um, the power set of one comma two, to infinity. [pause]
I2: That's the last power set in the union? That's what you're saying?
Opal: I think so.
I1: So you're saying there *is* a last one.
Opal: [pause] No, because ... no, there isn't a last one.

The purpose of the interviewer's comment, "So you're saying there *is* a last one," was to draw Opal's attention to an apparent contradiction in her expressed reasoning. Specifically, she was referring to "the last one," whereas a few moments earlier she had erased a final set in the expanded union notation in her written work, and, in Questions 5 and 6, they had agreed that there is no last set B_k to be unioned. To deal with the conflict, Opal continued reviewing her reasoning, pointed out that $P(\{1\}) \cup P(\{1,2\}) = P(\{1,2\})$ and generalized her conclusion, noting that "as we move, that's why I'm saying the union, all we need to look at is the one farthest to the right." A few seconds later, she rephrased her point: "So, yeah, so really all we need to look at is the power set of 1, 2, to k, where k is, k goes to infinity." That is, she claimed that one need only look at the kth partial union, observe it equals $P(\{1,2,\ldots,k\})$, and imagine k going to infinity.

Opal and her group made little progress in answering Question 9 beyond this point. Her partners were not convinced by her reasoning but they could not offer any alternatives. Some additional light was shed on Opal's reasoning much later in the interview, when the group again compared the meaning of the symbol k in Question 7 with its meaning in Question 9. Nikki stated that all of the power sets are finite, but Opal said "I don't know." She then went on to explain her difficulty in evaluating the infinite union:

Opal: Maybe my problem is still with the notation, because it still looks like, I mean, by letting it go to k, here in Problem 7, our k was a fixed integer. But *here*, I mean, it still, maybe I'm not looking at it correctly. But it seems like it's the, what you're taking the power set of is also going to infinity.

Of course, Opal was correct in saying that "what you're taking the power set of is also going to infinity," in the sense that $\{1,2,\ldots,k\}$ can be arbitrarily large. But an arbitrarily large finite set is not infinite. Moreover, she focused on a process of forming an infinite sequence of power sets of the form $P(\{1,2,\ldots,k\})$ and determined that its transcendent object is $P(\mathbf{N})$. One possible explanation for this might be that the role of the union operation was minimized in her thinking because each partial union simplified to the corresponding power set. Thus, when she encapsulated the process, her actions were in response to asking "What power set is this?" rather than in response to a question that is more consistent with the aim of forming a union, such as "What elements have been accumulated up to this point?" To summarize, although she constructed an infinite iterative process whose steps produced the same partial results as those that Tobi and Emily constructed, she appeared to have reflected on the results of the steps of the process, rather than the process itself. As a consequence, she used an inappropriate action of evaluation and obtained a transcendent object that was not useful in this particular mathematical situation.

5.7. Carrie: "It gets confusing because of the dot dot dot". We continue with an analysis of the phase one interview with Carrie, which is notable in that it shows a different way that students may have difficulty moving beyond finite iteration. As in the earlier cases, Carrie was able to construct an iterative process for generating a finite string of partial unions of indeterminate length. However, instead of attempting to construct an infinite process of iterating completely through $\mathbf{N}$, she appeared to coordinate two processes: a finite iteration up to an arbitrary n, followed by a non-iterative, continual variation of the index "to ∞." That is, Carrie's difficulties with the problem might have been caused by her failure to construct an underlying process of iteration through $\mathbf{N}$. We now look at Carrie's responses more closely, and again remind the reader that the notation X_n was used for the set $\{1, 2, \ldots, n\}$ in the phase one interviews.

Near the beginning of the interview, Carrie discussed how to form the union of the first few power sets. She described a recursive relationship between $P(X_{n-1})$ and $P(X_n)$ in essentially the same way as Emily did, although less articulately. Having done so, she immediately jumped to the conclusion that continuing this process would yield $P(\mathbf{N})$, "because that is with all the different numbers."

When asked for more details about her conclusion, Carrie reviewed her process for generating the power sets, indicating that she understood that n was repeatedly incremented and that the first few steps of her process were representative of how it continued. She concluded by saying, "And so on and so on, so you get $P(X_n)$." She confirmed, when asked, that her remarks referred to the nth partial union that resulted from taking n steps in the recursive process described above. Thus, Carrie apparently had successfully coordinated multiple instantiations of a process for the finite partial unions, up to some arbitrary point. She used this process in her formal reasoning in the problem. For example, she used it to show, in essentially the same way as Emily did, that any finite subset of $\mathbf{N}$ is an element of the infinite union.

However, Carrie struggled when she attempted to move beyond finite iteration through $\mathbf{N}$. She expressed her discomfort with this aspect over and over again in the interview. One typical comment was "And then, then it gets confusing because of the dot dot dot, I have to worry." As she continued talking, she pointed out explicitly several ways that the syntax "$n = 1$ to ∞" of the infinite union affected her thinking, including:

- the infinite union could be written as $P(X_1) \cup P(X_2) \cup \cdots \cup P(X_n)$ because the n in the last term " ... doesn't mean it really ends, it just means it's going on."
- when asked what n represented, she pointed to the symbol $\bigcup_{n=1}^{\infty}$ and then to the symbol X_n as she said "it's *infinity*, and that would imply it's infinity here." At the same time, she also dismissed the idea of a set called X_∞ as "bad news."
- she was ambivalent about stating that the set $\{1, 2, \ldots, n\}$ was finite, explaining: "Well, I, well, it *is* but it's *not*, 'cause it keeps going." When asked directly how many elements it had, she said that it had n elements but again referred to the symbol $\bigcup_{n=1}^{\infty}$ and said, "but that's ∞ [pause] so it would keep going."

Carrie's comments provide evidence that she viewed the index n and the set X_n as processes, rather than objects. This view is present in all of her comments above, but particularly in the comments in the last bulleted item, which show her struggling to apply an action of evaluation to X_n to determine its size.

It seems that she had not constructed a complete process of iteration through $\mathbf{N}$, which would involve imagining a pause at each natural number. Instead she constructed a process of passing through $\mathbf{N}$ consisting of a mixture of iteration and continual motion. She saw n as iterating up to an arbitrary natural number, followed by a continual, non-iterative, endless mental process of n becoming larger and larger, which she referred to as "n being infinity." Carrie appeared to coordinate this non-iterative process of moving n through $\mathbf{N}$ with a process of forming the set $X_n = \{1, 2, \ldots, n\}$. As a process, she saw X_n as a varying quantity that was continually increasing in size. In fact, later in the interview, she claimed that, through continuing this process, "soon a finite set of X sub n would be X to infinity" since "you could do it that many times." That is, she appeared to think that her process for generating the X_n had as an actual result, the object $\{1, 2, \ldots, n, \ldots\}$. She then applied the power set transformation to this object and concluded that her process for the infinite union resulted in $P(\mathbf{N})$.

Note that it is reasonable to view X_∞ or $\mathbf{N}$ as the *transcendent* object of a process that generates the X_n. What is problematic is to conclude that if you "do it that many times," then X_∞ is actually produced by the process. This led, in Carrie's case, to an erroneous conclusion that the infinite union is actually equal to $P(\mathbf{N})$. Cognitively, this need not be the same as assuming that the indexing variable takes on the value ∞. In fact, early in the interview Carrie rejected the idea of n taking on the value ∞ and yet later she concluded that an infinite set that she called "X infinity" was actually produced by her process.

5.8. Stan: "It's a fixed point. It's not infinite, like the natural numbers". Finally, in this analysis of Stan and Ralph's phase two interview, we examine Stan's idiosyncratic interpretation of the index k. We present evidence that, during most of his work on Question 9, he viewed the notation "$k = 1$ to ∞" on the infinite union as meaning that "k is infinite" or, more explicitly, that $k = \mathbf{N}$. That is, he viewed k as representing all of the natural numbers simultaneously, rather than representing a process of iterating through $\mathbf{N}$.

In their work on answering Questions 5 and 6, Ralph and Stan demonstrated that they understood the formal definition of the infinite union of sets. They also considered and rejected the possibility of using ∞ as an index value in forming the infinite union, finally using the notation "$B_1 \cup B_2 \cup \cdots$" to represent the infinite union. They stated that they understood this notation to indicate the union of an infinite number of sets and an endless process of unioning.

When Stan first looked at the notation in Question 9, he said "this is the union of all the power sets of this set" [pointing to $\{1, 2, \ldots, k\}$]. This suggests that he did not immediately recognize the form of the infinite union as being the same as in Question 5 (which starts with an explicit statement concerning a *collection* of sets, not a *single* set), nor Question 9 itself as a particular example of what they examined in Question 6. Next, he represented the infinite union using notation that conventionally represents a finite union:

$$P(\{1\}) \cup P(\{1,2\}) \cup P(\{1,2,3\}) \cdots \cup P(\{1,2,\ldots,k\}).$$

When one of the interviewers asked, "Does that stop, or not?" both students agreed that it did not stop, and Stan explained that the reason was " 'cause k is ∞." Ralph suggested scratching out the last term, but Stan added an ellipsis at the end of the expression instead.

Considering how to proceed, Stan said he would start with the "finite stuff" first and "we'll go to the infinity stuff later." By listing the elements of the sets, he observed that the union of the first two power sets is $P(\{1, 2\})$ and then that the union of the first three is $P(\{1, 2, 3\})$. Reflecting on these actions, he remarked:

> *Stan*: So, the union of the, all the power sets is eventually gonna be just the same as, you know the power set of this set here [as he spoke, he wrote "$P(\{1, 2, \ldots, k\})$"]. Okay. Again where k is, you know, one to infinity.

It appears that Stan had encapsulated a finite process for the union of the first k power sets, since he identified its result. The problematic part was his statement that he had found the union of "all the power sets." The interviewer asked what k represented in his notation $P(\{1, 2, \ldots, k\})$, and Stan began by saying "k is ∞, or k is –" when Ralph interrupted with "I'd just drop the k." Stan amended their written work to show that they proposed that the union is equal to $P(\{1, 2, \ldots\})$.

The interviewer then asked whether the set $P(\{1, 2, \ldots\})$ is one of the sets that was unioned. They both responded in the affirmative, reasoning that the "$k = 1$ to ∞" on the union indicated that there was an infinite number of sets unioned, and therefore the set $\{1, 2, \ldots, k\}$ could have an infinite number of elements because, according to Stan, "k is not finite, it's infinite." A close analysis of Stan's other comments about the index k in Questions 5 and 9 indicated that when he said that k was "infinite" or "one to infinity," he was expressing his view that k represented all of the natural numbers simultaneously. Other references of this sort included:

- In Question 5, Stan listed the indexed sets as "$\{B_1 \quad B_2 \quad \cdots \quad B_k\}$" where "$k = [1, \infty)$." Ralph suggested he change the equal sign to the "is an element of" sign, but Stan expressed skepticism and did not make the change. He did remark, however, that he was referring only to the whole numbers in $[1, \infty)$.
- Stan distinguished the k in Question 7 from the k in Question 9 by saying " ... right here [Question 7] says that k is fixed. That's why we thought that this set [$P(\{1, 2, \ldots, k\})$] was finite. But this [Question 9] is not implying that it's fixed, I don't think. It's implying that it's infinite."

When the interviewer showed them an alternative notation for the given infinite union, $\bigcup_{k \in N} P(\{1, 2, \ldots, k\})$, Stan remarked that this notation indicated "not that k is necessarily infinity, but that it can be any number *in* infinity." It appears that he used the word "infinity" as a synonym for $\mathbf{N}$. Seeing the alternative notation seemed to shift Stan's interpretation of the role of k. In fact, both Ralph and Stan made comments that indicated that they read this notation as saying that they could choose a k, compute the union up to that k, and compare the result to $P(\mathbf{N})$. Working with this incorrect interpretation of the notation, they argued that the equality proposed in Question 9 was false. While they were discussing it, Ralph observed that k was not fixed, but it was "fixable." That appeared to help Stan, in that he then began to speak of individual, arbitrary values of k, rather than all natural numbers at once. He finally said "... k just represents the number of elements in the set, and you can have a set with one elements, or two elements, or three or twenty, or whatever." This comment may indicate that Stan was beginning to consider an iteration through **N** as the basis for forming and evaluating the infinite union.

While there is little evidence that Stan constructed anything beyond a finite iterative process for the infinite union during the interview, he did appear to use his emerging understanding that k represented a single number at a time rather than all natural numbers simultaneously. He argued that **N** could not be an element of a set of the form $P(\{1, 2, \ldots, k\})$, reasoning that $P(\{1, 2, \ldots, k\})$ contains only finite sets as elements, because k is "the number of elements, I mean it's an element of **N**, but it's not **N**. It's a fixed point. It's not infinite, like the natural numbers."

6. Discussion and Conclusion

We have presented the results of a study of college students' thinking about a particular set theory problem ($\star$) involving actual infinity. It was found that explanations involving the mechanisms of interiorization and encapsulation could be used to account for their thinking. The mental structures constructed via these mechanisms in response to the problem are characterized as infinite iterative processes and their encapsulations into transcendent objects. Developing this empirically-based preliminary theoretical description is the first step in designing APOS theory-based instruction intended to help college students construct concepts of actual infinity that will be useful in problem situations involving infinity that are typical in undergraduate mathematics. It is important to note that our results are consistent with the theoretical study reported in Dubinsky et al. (2005a, 2005b), in which the authors found that explanations involving the same mechanisms can be used to account for the thinking about mathematical infinity of a variety of mathematicians and philosophers throughout history.

In our theoretical description, we propose that a process conception of infinite iteration develops as an individual coordinates multiple instantiations of a finite iterative process. When fully constructed, the individual is able to imagine the resulting infinite process as being complete, in the sense of being able to imagine that all steps have been carried out. Reflecting on the process, the individual may come to see it in its totality, that is, as a single operation at a moment in time. This may lead to an attempt to apply an action of evaluation to the process which, if successful, results in the encapsulation of the infinite iterative process. The encapsulation yields a transcendent object that is understood to be outside of the process, and the object is identified with the state at infinity.

In attempting to find meaning in the formal statement of the problem ($\star$), all of the students gave responses that were process-oriented, even though the problem was not stated in terms of a process. Specifically, rather than relying on the formal set-theoretic definition of the infinite union of a collection of sets, students conceived of the infinite union as a potentially infinite operation of successive unioning, in more or less detail, and saw the critical question as "What is the final result?"

Attention to the interiorization of actions that results in the individual being able to imagine running through all of the steps of a process has always been a part of APOS-based analyses. Since this study deals with infinite processes, the major issue arises from the fact that one must imagine completing all the steps, even though there is no last step. While the question of what it means to *complete* an infinite process was raised by Dubinsky (1988) in an article concerned with the teaching and learning of Cantor's proof of the uncountability of the unit interval, the present study is the first APOS-based study to examine this idea empirically.

One of the central results of this study is that those who could see the process as complete grasped the fact that sets of the form $P(\{1, 2, \ldots, k\})$ are finite, and used it to make progress on the problem ($\star$), either using formal methods (Tobi) or informal methods (Emily). Those who made limited progress in solving the problem (Opal, Carrie, and Stan) were unsuccessful in seeing the infinite processes they constructed as complete. This suggests that if an individual approaches a problem concerning actual infinity by constructing a potentially infinite iterative process, the key to successfully using it to solve the problem is to be able to see the process as a completed totality and to understand that the state at infinity is not directly produced by the process.

Choosing to interpret an infinite union of sets as a potentially infinite process of successive unioning is consistent with the mode of reasoning predicted by the Basic Metaphor of Infinity of Lakoff and Núñez (2000). Furthermore, Lakoff and Núñez propose that individuals make sense of the state at infinity of an infinite iterative process by applying a conceptual metaphor that is sourced in the domain of processes that do end. Viewing the state at infinity in this way is also consistent with prior theoretical results on limits of real sequences discussed by Mamona-Downs (2001). She contends that a central obstacle in the development of a learner's conception of limits is that the individual may associate the limit with an ultimate term of the sequence. The crucial part of her argument for our purposes is her observation that when learners consider a real infinite sequence (a_n) and perceive that the terms of the sequence have a "sensible and consistent tendency" (p. 268), they are likely to imagine that a final term a_∞ is possessed by the sequence. All of our interviewees observed how the sequence of partial unions simplified to a sequence of power sets and, on that basis, proposed that the final result of the successive unioning is $P(\mathbf{N})$. However, some of the students were able to move beyond that mode of reasoning, and some were not, so other factors must also be considered.

To further this discussion, we reconsider the case of Opal. Opal's first approach to determining the result of the infinite union process was to observe that the kth partial union of the power sets,

$$P(\{1\}) \cup P(\{1,2\}) \cup P(\{1,2,3\}) \cup \cdots \cup P(\{1,\ldots,k\}),$$

equals $P(\{1,\ldots,k\})$, so the union reduces to the set "farthest to the right." She then claimed that the same is true for the infinite union. Thinking that the state at infinity is constructed in the same way as the final result of a corresponding finite process might be an application of the Basic Metaphor of Infinity. Perceiving the consistency of the results of the finite processes leads Opal to assume the same holds for the "limit" or state at infinity, which is consistent with the results on real sequences noted by Mamona-Downs. However, that approach has a fatal flaw: the state at infinity for an infinite process need not be of the same form as what is produced after any finite number of steps of the process. The specific fallacy here is that the corresponding finite processes all have final results that are power sets, while the infinite union is not the power set of any set. Thus, while this example provides some support for Lakoff and Núñez's contention that people think about infinity by applying conceptual metaphors, it also indicates that solving problems that require determining the state at infinity for an infinite process can require more than metaphorical thinking. Consequently, this study supports the assertion of Schiralli and Sinclair (2003): by itself, Lakoff and Núñez's method of mathematical idea analysis does not always provide explanations that account for the richness of mathematical thinking.

The crucial question therefore is this: if students often conceive of situations involving actual infinity in terms of infinite processes, and there is a tendency to make fallacious conclusions of the sort we saw here, how can instruction be used to help students make constructions that are consistent with conventional mathematics? Our view is that APOS theory provides some direction. An object is obtained from a process when the individual attempts to apply an action to a process that is conceived as a totality. The key then is to determine what action should be applied so that an appropriate transcendent object results from encapsulating a completely constructed infinite iterative process. At least one possibility should be clear by examining the mathematics that the students are intended to learn. For example, in the case of the infinite union, the key idea is that the infinite union contains all of the elements of all of the sets. So, an appropriate action would be motivated by the question "What has been accumulated once all of the steps of the process are done?" rather than the question "What is the form of the final result?" The difference between these two actions, as Opal's case shows, is that the former has its roots in how the result at each finite numbered step is found (through a process of unioning), while the latter has its roots in the form of the result at each step (the object is a power set). Thus, choosing a correct action of evaluation for the infinite process depends on reflecting on the infinite iterative process and determining its nature, emphasizing *how* the objects are produced at each step, rather than focusing just on the form of the objects.

The phenomenon of students using process-oriented approaches, rather than the formal definition, to make sense of the infinite union can be viewed as an example of "reducing the level of abstraction" (see Hazzan, 1999; Hazzan & Zazkis, 2005). In terms of Hazzan's theoretical framework, students often unconsciously reduce the level of abstraction present in a problem situation by reinterpreting an unfamiliar situation in terms of one with which they are more familiar, by giving responses that are action or process-oriented, and by reducing the perceived complexity of the task by focusing on similar but less complex situations.

The students in our study constructed certain infinite iterative processes in order to understand the infinite union and $P(\mathbf{N})$, even though such processes are not those we might think of as underlying the formal definition of the objects involved. They are simply processes that allowed them to deal with the mathematical objects and the assertions in a meaningful way. We now give some examples from our data.

Carrie's approach to the problem bears a striking similarity to an informal method of evaluating a limit at infinity for a composition of continuous functions of a real variable. For example, a student might interchange the order of the limit and exponentiation processes to reason as follows:

$$\lim_{x \to \infty} e^{\frac{1}{x}} = e^{\frac{1}{\infty}} = e^0 = 1.$$

In the notation $e^{\frac{1}{\infty}}$, the symbol ∞ might be thought of as a process of increasing x; it is then coordinated with the process of taking a reciprocal, so that $\frac{1}{\infty}$ is viewed as a process that has transcendent object equal to 0. Following encapsulation, the exponential function is applied, and the limit of 1 is obtained as an output of the function. Note that even if the individual imagines the process represented by $\frac{1}{\infty}$ as actually producing 0, the calculation of the limit will still be correct.

Having recently completed calculus, it is likely that Carrie was familiar with this method; applying it in this case was one way she could "reduce abstraction" in a situation she found difficult. However, this method is not appropriate when the transformations involved are not continuous. Though

$$\bigcup_{n=1}^{j} P(X_n) = P\left(\bigcup_{n=1}^{j} X_n\right)$$

for all finite j, since each is equal to $P(X_j)$ itself, it is also the case that

$$\bigcup_{n=1}^{\infty} P(X_n) \neq P\left(\bigcup_{n=1}^{\infty} X_n\right).$$

That is, there is a "discontinuity at infinity." It is important to note that Carrie handled the coordination of the power set and union processes correctly when she considered specific finite cases, but failed to perform the coordination correctly when the situation required more than just finite iteration. This suggests that she shifted strategies, probably unconsciously.

Another example of the strategy of reducing abstraction by using the properties of functions of a real variable is Opal's error in assuming that because $\{1, 2, \ldots, k\}$ can be arbitrarily large, then it is "eventually" infinite. One possible explanation is that she is thinking of the convention that the limit of a function is said to be "infinity" if the function takes on arbitrarily large values (at a point or at infinity). She might have been evaluating the infinite union using reasoning similar to that used with real functions, as suggested by her comments that she was not sure how to interpret the set $P(\{1, 2, \ldots, k\})$ as "k goes to infinity" because "what you are taking the power set of is also going to infinity." Evidently she interpreted the "limit" of the set to be infinity or $\mathbf{N}$. While this is not a problem in itself, Opal like Carrie, also assumed that the power set operation is continuous at infinity.

The strategy of reducing abstraction might also play a role in the tendency to conflate the cardinal and ordinal aspects of the index k, thereby reducing complexity. In the main problem, k indicates both the number of sets that have been indexed up to the kth step and the size of the set $\{1, 2, \ldots, k\}$ whose power set is obtained at that step. The cardinal and ordinal aspects of the number k can be conflated without consequence in the finite case. But this does not carry over correctly to the corresponding infinite process: even when infinitely many sets are indexed, the sets are still finite. Several interviewees struggled with this distinction. Carrie suggested that a process of repeatedly increasing the size of a set by 1 would eventually produce an infinite set at one of the steps. Stan reasoned that since infinitely many sets are indicated by the infinite union notation, then k must be "infinite." Emily speculated that "if you infinitely union sets, eventually you've got to union the infinite set," although she was able to quickly dismiss this idea. While Piaget (1965) explains that one's conception of number develops through a coordination of its cardinal and ordinal aspects, the issue of how these mathematically equivalent, yet cognitively different aspects of number affect students' conceptions of mathematical infinity is a matter for future investigation.

Our results arise from an empirical study of the mathematical thinking of 12 college students trying to solve a particular set of related problems involving actual infinity. The main problem was presented analytically and the state at infinity is a countable set. This raises questions that must be addressed by further research. Are similar constructions made in efforts to understand infinite iterative processes in other contexts, such as a problem presented geometrically? For example, if students were asked to construct the Cantor set, would they have difficulty seeing the process as a completed totality and in determining the state at infinity? If students were given problems where the transcendent object is finite, such as finding the sum of a convergent infinite series, would the same type of constructions arise and would similar issues emerge? Additionally, how does an individual develop an understanding of $P(\mathbf{N})$, an uncountable set? Or how might one conceive of the definition of the Riemann integral given as a limit of the norm of partitions, an uncountably infinite collection?

More generally, Lakoff and Núñez (2000) offered a multitude of mathematical idea analyses in which infinite iteration and the Basic Metaphor of Infinity are thought to apply. Given that the results of the present study are consistent with, but go beyond what is predicted by their framework, it seems that there is available a large collection of hypotheses to test in empirical studies on the conceptualization of infinity. Empirical studies could be designed to examine the extent to which a given analysis matches with what is observed in students, and to identify what other mechanisms are used to construct the specific mental structures involved in the conceptualizations observed in students' thinking.

The theoretical results we offer are preliminary in nature, but there are some principles drawn from these results that we would emphasize in designing mathematical activities intended to foster student understanding of mathematical situations involving actual infinity. First, it is reasonable to anticipate that some students will conceptualize such situations in terms of infinite iterative processes. Assuming that occurs, there should be an explicit focus on the tension between processes continuing indefinitely, with no final step taken or object produced, and yet being thought of as complete. There should be activities that emphasize examining and identifying the relevant common characteristics of all the objects produced by the process as part of coming to see the process as a completed totality. Finally, there should be activities that emphasize the choice of an appropriate action of evaluation in order to develop the idea that the transcendent object is not produced by the process, but is instead outside of the process.

Although infinite iteration is but one aspect of infinity, it appears to be relevant to a wide range of mathematical situations from the most elementary experiences with counting to advanced topics involving indexed collections of sets, such as those we considered in this paper. Our hope is that these empirically based results will lead to additional studies on students' constructions and understandings of topics related to infinity, as well as the development of curricular materials focused on improving student understanding of all aspects of mathematical infinity.

References

Asiala, M., Brown, A., DeVries, D., Dubinsky, E., Mathews, D., & Thomas, K. (1996). A framework for research and curriculum development in undergraduate mathematics education. In J. Kaput, A. H. Schoenfeld, & E. Dubinsky

(Eds.), *Research in collegiate mathematics education. II* (pp. 1–32). Providence, RI: American Mathematical Society.

Baker, B., Cooley, L., & Trigueros, M. (2000). A calculus graphing schema. *Journal for Research in Mathematics Education, 31*(5), 557–578.

Borasi, R. (1985). Errors in the enumeration of infinite sets. *Focus on Learning Problems in Mathematics, 7*(3,4), 77–89.

Breidenbach, D., Dubinsky, E., Hawks, J., & Nichols, D. (1992). Development of the process conception of function. *Educational Studies in Mathematics, 23*, 247–285.

Brown, A., DeVries, D., Dubinsky, E., & Thomas, K. (1997). Learning binary operations, groups, and subgroups. *Journal of Mathematical Behavior, 16*(3), 187–239.

Dubinsky, E. (1988). Anatomy of a question. *Journal of Mathematical Behavior, 6*, 363–365.

Dubinsky, E. (1989). On teaching mathematical induction II. *Journal of Mathematical Behavior, 8*, 285–304.

Dubinsky, E. (2000). Meaning and formalism in mathematics. *International Journal of Computers for Mathematical Learning, 5*(3), 211–240.

Dubinsky, E., & McDonald, M. A. (2001). APOS: A constructivist theory of learning in undergraduate mathematics education research. In D. Holton (Ed.), *The teaching and learning of mathematics at university level: An ICMI study* (pp. 273–280). Dordrecht, The Netherlands: Kluwer.

Dubinsky, E., Weller, K., McDonald, M. A., & Brown, A. (2005a). Some historical issues and paradoxes regarding the concept of infinity: An APOS based analysis: Part 1. *Educational Studies in Mathematics, 58*, 335–359.

Dubinsky, E., Weller, K., McDonald, M. A., & Brown, A. (2005b). Some historical issues and paradoxes regarding the concept of infinity: An APOS based analysis: Part 2. *Educational Studies in Mathematics, 60*, 253–266.

Fischbein, E., Tirosh, D., & Hess, P. (1979). The intuition of infinity. *Educational Studies in Mathematics, 10*, 3–40.

Hazzan, O. (1999). Reducing abstraction level when studying abstract algebra concepts. *Educational Studies in Mathematics, 40*, 71–90.

Hazzan, O., & Zazkis, R. (2005). Reducing abstraction: The case of school mathematics. *Educational Studies in Mathematics, 58*, 101–119.

Lakoff, G., & Nunez, R. (2000). *Where mathematics comes from.* New York: Basic Books.

Mamona-Downs, J. (2001). Letting the intuitive bear on the formal: A didactical approach for the understanding of the limit of a sequence. *Educational Studies in Mathematics, 48*, 259–288.

Moreno, L. E., & Waldegg, G. (1991). The conceptual evolution of actual infinity. *Educational Studies in Mathematics, 22*, 211-231.

Piaget, J. (1965). *The child's conception of number.* New York: W. W. Norton.

Schiralli, M., & Sinclair, N. (2003). A constructive response to *Where Mathematics Comes From. Educational Studies in Mathematics, 52*, 79–91.

Sierpinska, A., & Viwegier, M. (1989). How and when attitudes towards mathematics and infinity become constituted into obstacles in students. In G. Vergnaud, J. Rogalski, & M. Artigue (Eds.), *Proceedings of the 13th annual meeting for the International Group for the Psychology of Mathematics Education* (Vol. 3, pp. 166–173). Paris.

Tall, D. (1980). The notion of infinite measuring number and its relevance in the intuition of infinity. *Educational Studies in Mathematics, 11*, 271–284.

Tall, D. (1992). The transition to advanced mathematical thinking: Functions, limits, infinity, and proof. In D. A. Grouws (Ed.), *Handbook of research on mathematics teaching and learning* (pp. 495–511). New York: MacMillan.

Tall, D. (2001). Natural and formal infinities. *Educational Studies in Mathematics, 48*, 129–136.

Tirosh, D. (1991). The role of students' intuitions of infinity in teaching the Cantorian theory. In D. Tall (Ed.), *Advanced mathematical thinking* (pp. 199–214). Dordrecht, The Netherlands: Kluwer.

Tirosh, D., & Tsamir, P. (1996). The role of representations in students' intuitive thinking about infinity. *International Journal of Mathematics in Science and Technology, 27*(1), 33–40.

Tsamir, P. (2001). When the same is not perceived as such: The case of infinite sets. *Educational Studies in Mathematics, 48*, 289–307.

Tsamir, P. (2003). Primary intuitions and instruction: The case of actual infinity. In A. Selden, E. Dubinsky, G. Harel, & F. Hitt (Eds.), *Research in collegiate mathematics education. V* (pp. 79–96). Providence, RI: American Mathematical Society.

Tsamir, P., & Dreyfus, T. (2002). Comparing infinite sets – a process of abstraction: The case of Ben. *Journal of Mathematical Behavior, 21*, 1–23.

Tsamir, P., & Tirosh, D. (1999). Consistency and representations: The case of actual infinity. *Journal for Research in Mathematics Education, 30*(2), 213–219.

Vidakovic, D. (1993). *Cooperative learning: Differences between group and individual processes of construction of the concept of inverse function.* Unpublished doctoral dissertation, Purdue University, West Lafayette, IN.

Weller, K., Clark, J., Dubinsky, E., Loch, S., McDonald, M., & Merkovsky, R. (2003). Student performance and attitudes in courses based on APOS Theory and the ACE Teaching Cycle. In A. Selden, E. Dubinsky, G. Harel, & F. Hitt (Eds.), *Research in collegiate mathematics education. V* (pp. 97–131). Providence, RI: American Mathematical Society.

Department of Mathematical Sciences, Indiana University South Bend, South Bend, IN 46634
E-mail: abrown@iusb.edu

Department of Mathematics, Occidental College, Los Angeles, CA 90041
E-mail: mickey@oxy.edu

Department of Mathematics, University of Michigan Flint, Flint, MI 48502
E-mail: wellerk@umflint.edu

CBMS Issues in Mathematics Education
Volume **16**, 2010

Teaching Assistants Learning How Students Think

David T. Kung

ABSTRACT. Teacher knowledge at the college level remains a largely an unexplored subject, despite the importance of such knowledge to college teaching and the preparation of future teachers at the high school and college level. In this paper, we investigate knowledge of student thinking in a group of current and former teaching assistants (TAs) who have taught in the Emerging Scholars workshop model. Using data from interviews with eight TAs at two different large public institutions, we report on TA knowledge of student solution strategies and difficulties, student coping skills, and student conceptions in the area of limits. When presented with a challenging problem, most participants exhibited extensive knowledge of student strategies and difficulties. They also demonstrated an awareness of the most common student misconceptions of limits found in the literature. Several less common misconceptions were not mentioned by most of the TAs. We also report on the ways the TAs described gaining this knowledge. They described acquiring knowledge of student thinking in various ways, including while observing students work on problems, writing problems themselves, and grading homework. Furthermore, analysis of the interviews indicated that these different activities produced different types of knowledge. For instance, observing students work problems gave participants very fine-grained details of problem solving methods but left out students' final conclusions - which TAs learned about through grading. We propose a framework for understanding TA learning about student thinking, hypothesizing a duality between the types of student-teacher interaction and the types of knowledge TAs have access to. We conclude the paper by discussing implications of this work for the professional development of teachers as well as for future research into teacher knowledge at the college level.

1. Introduction

Many professors' first teaching experiences come as graduate student Teaching Assistants (TAs). At large PhD-granting institutions TAs typically lead discussion sections, work problems in class, hold office hours, and grade homework and exams – leaving the professors to lecture to hundreds of students. As their introduction to the teaching profession, these experiences are likely influential in shaping the teaching philosophies of many future professors and also likely sites for their initial acquisition of knowledge of student thinking. Given the dearth of professional

This work was supported in part by grants from the Calculus Consortium for Higher Education and from St. Mary's College of Maryland. The author wishes to thank the reviewers and editors for their many helpful comments.

©2010 American Mathematical Society

development activities for faculty (National Science Foundation, 1992), the knowledge and attitudes developed as a TA may shape professors' teaching for their entire careers. Despite this critical role, little research has been done on any aspect of graduate student teaching experiences (Speer, Gutmann, & Murphy, 2005).

In contrast, there has been a recent explosion of research on teachers' mathematical knowledge at the K-12 level (for a review, see Ball, Lubienski, and Mewborn (2001)). Research on the knowledge needed for teaching mathematics has matured from a time when "number of math classes beyond Calculus" was taken as a proxy measure of teacher knowledge (e.g. Begle, 1979; Monk, 1994) to much more fine-grained, qualitative studies of the types of knowledge needed to teach various mathematical topics (Ball, 1990; Kennedy, 1997; Ma, 1999).

Here we extend this more recent approach to the college level – shedding light on what knowledge of student thinking is needed to teach calculus – by examining what knowledge of student thinking has been gained by teachers in a variety of settings, including the successful calculus instructional model offered by Emerging Scholars Programs (ESPs). We organize our work around two main research questions:

- What do experienced calculus teachers know about student thinking?
- How do calculus TAs gain their knowledge of student thinking in the course of their early teaching experiences?

We see this work as part of a larger program of research that seeks to answer more general questions about TA preparation. What knowledge of student thinking about calculus is needed to effectively teach the subject at the college level? Of this knowledge, what do entering graduate students already know? Through what types of activities do novice teachers gain that knowledge? How do experienced calculus teachers use their knowledge of student thinking in the course of their teaching? What experiences can we provide to novice teachers to help them develop their knowledge of student thinking more quickly? The ultimate goal of this program is to understand and improve the professional development of TAs during their graduate school experiences, which would in turn improve the teaching of mathematics at the college level.

The study described here reports on experienced calculus teachers' knowledge of student thinking and how it was developed. We explored the questions above by restricting our attention in several important ways. We focused on graduate students and their early teaching experiences which, we assume, are formative. Because of the vast number of topics covered in a modern Calculus course, we concentrated particularly on the foundational concept of limits. We also restricted our focus to a population of current and former TAs whose teaching experiences included both traditional discussion sections and at least two semesters in an Emerging Scholars Program (ESP). Described in more detail below, such programs have TAs facilitate cooperative learning and take on a Socratic role.

We focused on this set of TAs for several reasons. First, these TAs had experienced student interactions that went far beyond the typical question and answer format of a recitation section. Secondly, since all of the participants also had experiences in traditional classrooms, they were in a position to compare and contrast what they learned in a rich array of experiences. Also, numerous studies have shown the effectiveness of ESP workshops in helping students succeed in Calculus (see below for an overview of this research). Thus, the TAs had experience in a successful calculus program (it would be problematic to characterize traditional

Calculus courses as effective, with drop-out/failure rates in excess of 30%). Finally, anecdotal evidence suggests that for this group, the ESP experience played a critical role in shaping attitudes toward mathematics teaching, including understanding student thinking. The combination of these factors made TAs with ESP experience a source of rich data on knowledge of student understanding and the acquisition of such knowledge.

One way to investigate the effects of a rich teaching background on a teacher's development is through interview. In order to elicit TA participants' knowledge of student strategies, we piloted and refined a task-based instrument which asked them to predict and analyze a number of teaching situations. Following the tasks, we addressed the topic of student conceptions more directly. The interviews produced a plentiful array of information and reflections about TAs' knowledge and the ways they reported gaining it. Coding and analyzing these data for emergent themes gave us insight into the connections between TA interactions with students and their knowledge of student thinking.

We begin with a section on background research that includes a review of literature on teachers' knowledge, specifically concentrating on studies which looked at teachers' knowledge of student thinking and those we consider most relevant to our context. We then turn more directly to the post-secondary context and report what little is known about mathematics TAs. More is known, however, about student thinking in the calculus context; we review those studies next to give context to our later discussion of TAs' knowledge of student thinking. We conclude the background section with a review of ESPs and the existing literature on them, giving the reader an idea of the context in which our participants worked.

We then turn specifically to the case of former and current ESP TAs. The methods for this study are discussed, including the interview protocol and the backgrounds of the TAs chosen for this study. The results follow in three sections. The first reports on their knowledge of student thinking about limits. We compare their responses with the known literature on student conceptions and with our own experience on student solution strategies. The second section covers how the participants reported gaining their knowledge of student thinking. Here we categorize participants' responses by the activities they report being engaged in while learning about student thinking. We also give participants' comparisons between the knowledge they gained through different activities. This leads to the third section where we propose a duality framework for understanding the relationship between the sites where learning about student thinking takes place and the types of knowledge gained. We hypothesize that different types of interactions with students make different types of knowledge available to TAs. At the end of the results section, we present data to support the duality framework.

In closing, we discuss these results and their implications. Our duality framework could provide an important new perspective for professional development. It examines what TAs need to know – not from the front-end view of what skills are key in their first teaching experiences, but rather from the perspective of what experienced teachers have learned about student thinking through years of rich interactions with students. This perspective has important implications for graduate programs, for TA professional development, and in terms of directions for further research.

2. Background Research

While there is little work exploring the kinds of knowledge needed for teaching college mathematics, there are four areas of research which inform our work. First, there has been extensive research into teacher knowledge of students' mathematical thinking at the K-12 level. Such work typically views this knowledge as a subset of *pedagogical content knowledge* (PCK). The bulk of work in this area has been done at the elementary level, where evidence suggests that improving teachers' knowledge of student thinking can lead to improved student performance. Second, we move from the K-12 level to the college level by reviewing what research exists on mathematics TAs. While little of this work directly addresses TAs knowledge of student thinking, it provides context for understanding TAs more holistically. Third, our work is informed by several studies which shed light on the ways in which students struggle to understand the key calculus concept of limit. We present this literature to provide context for participants' thoughts about student misconceptions. Finally, we review research on ESPs to provide evidence that participants' experiences included settings in which calculus instruction was particularly effective.

2.1. Teachers' Knowledge of Student Thinking. Trying to bridge the gap between content and pedagogy, Shulman (1986) coined the term "Pedagogical Content Knowledge (PCK)" to mean "the particular form of content knowledge that embodies the aspects of content most germane to its teachability." For him, this included "an understanding of what makes the learning of specific topics easy or difficult: the conceptions and preconceptions that students of different ages bring with them..." (p. 9). While authors from different theoretical perspectives have conceptualized PCK in different ways, all agree with the basic premise that an understanding of student thinking is vital to the teaching of mathematics with understanding (see, for example, Ball & Bass, 2000; Borko & Putnam, 1996; Ma, 1999).

Related to the concept of PCK is a branch of research on teacher knowledge that has focused on subject matter knowledge used while teaching. Ma's (1999) work on the nature of elementary teachers' "profound understanding of fundamental mathematics" (p. 122) and Ball and Bass' study (2000) both fall into this category. Ma found that Chinese elementary teachers understand elementary mathematics in deep ways that help them overcome students' typical errors and assist in making connections between related mathematical concepts. Ball and Bass came at these questions from a different perspective, giving a detailed examination of the mathematics itself and "unpacking" the many subtleties a teacher would have to understand to teach a particular topic in a constructivist manner.

As for effects on student learning, several different programs of study have shown that, at least at the elementary level, improving teacher knowledge of student thinking can have positive impacts on student learning. Cognitively Guided Instruction (CGI) studies have used research-based findings on children's addition and subtraction strategies (Carpenter & Moser, 1984) to inform professional development activities for in-service teachers (Fennema et al., 1996; Fennema, Franke, & Carpenter, 1993). Improving a teacher's knowledge of the relationships between problem types and student solution strategies improved that teacher's ability to assess children's knowledge and adapt instruction for each child. The CGI researchers developed an assessment tool to measure teacher use of student thinking in the classroom. Level I teachers characterized student thinking only in terms of procedures students had been taught. The remaining levels describe teachers with increasing attentiveness to student thinking, culminating with Level IV teachers who let children's thinking drive instructional decisions and create opportunities to build greater knowledge of their particular students' thinking (Franke, Fennema, & Carpenter, 1997). Level IV teachers had created a feedback loop, using knowledge of student thinking to plan student activities which helped them learn more about student thinking, which further informed their instructional decisions. Using this framework, the CGI researchers were able to connect teacher understanding of student thinking with student success.

The CGI studies provide strong evidence that knowledge of children's thinking is a powerful tool that enables teachers to transform this knowledge and use it to change instruction. These findings, when viewed in conjunction with those of other studies, provide a convincing argument that one major way to improve mathematics instruction and learning is to help teachers understand the mathematical thought processes of their students. It also appears that this knowledge is not static nor completely acquired outside of classrooms (e.g., in workshops), but is dynamic and may require the context of teaching mathematics for full development (Fennema et al., 1996).

There are many differences between the early elementary school setting and the college classroom; in particular, the complexity of the material covered in a college Calculus course makes a complete cataloging of student strategies untenable. We believe, however, that the underlying philosophy still applies: teachers who understand their students' mathematical thinking will be more effective.

2.2. Mathematics Teaching Assistants. As noted above, research on mathematics TAs is sparse, especially when compared with the numerous studies of mathematics teachers at the K-12 levels. Given the important role of early experiences in solidifying beliefs and developing practices (Brown, 1985; Lacey, 1977), studying TAs might prove to be extremely important in understanding and improving mathematics instruction at the post-secondary level. Recent years have seen some progress in this direction, which we review here.

Among the many differences between K-12 teachers and graduate TAs is that most incoming TAs have little or no formal teacher preparation. Although the situation is improving, many graduate students receive only a day or two of professional development before entering the classroom and much of that professional development is not mathematics-specific (Speer et al., 2005). Many institutions are implementing lengthier, more in depth professional programs, and the mathematics community has responded with materials designed to be used with such programs (notably, the volume of case studies by Friedberg et al., 2001). Recent work by Seymour and colleagues suggests that programs designed to bring innovative teaching methods to undergraduates can serve as important professional development opportunities for graduate TAs in mathematics and the sciences (Seymour, Melton, Wiese, & Pedersen-Gallegos, 2005).

Several qualitative studies have also begun to paint a more complete picture of mathematics TAs. DeFranco and McGivney-Burelle (2001) explored TAs' beliefs about the nature of teaching and learning mathematics. Speer (2001) studied how these beliefs manifest themselves in a TA's teaching decisions in the classroom. Herzig (2002) has reported that an absence of institutional support for teaching can negatively impact graduate students' decisions to stay in graduate school. This raises the possibility that mathematics graduate education might simultaneously prepare the next generation of mathematics professors and select against those who are most interested in teaching. Finally, Belnap (2005) has noted that while TA training programs can influence TA practices and views on teaching, issues of course structure, time management, and pedagogical knowledge limit those influences. While all of this work improves our understanding of mathematics TAs, little of it addresses their knowledge of student thinking and how that impacts current (or future) teaching.

2.3. Student Thinking About Limits. The concept of limit is fundamental to every topic in calculus. However, limits have proved to be extremely difficult for students to learn, especially the modern epsilon-delta definition. Several researchers have grappled with understanding why this is the case. Tall and Vinner (1983) worked on this issue in the early 1980s and developed the term *concept image* to describe "the total cognitive structure that is associated with the concept" (p. 152). Further work indicated that students may have several misconceptions (or naive conceptions) which inhibit their full understanding of limits (Bezuidenhout, 2001; Przenioslo, 2004), that it is very difficult to teach limits in ways that avoid these misconceptions (Davis & Vinner, 1986), and that these misconceptions are intimitely connected with naive ideas about the infinite (Tall, 2001) and the metaphors used to describe limits (Lakoff & Núñez, 2000). Subsequent work showed that students' naive conceptions can be extremely resistant to change; a series of five interviews designed to confront students with examples which conflicted with their concept of limits produced little change (Williams, 1991). From this research we distill six conceptions of limits frequently held by calculus students, with a brief description of the possible student thinking behind each.

(1) Limit as function value at a point. Some students think a limit can be found simply by substituting the limiting x-value into the equation.
(2) Limit as bound. A function or sequence might be said to "never go past its limit."
(3) Limit as unreachable. Students can confuse "x never reaches a" with "$f(x)$ never reaches L."
(4) Limit found by discrete approximations. Some students attempt to find a limit by plugging in one (or a few) x-values and inferring the limit (for instance, by rounding).
(5) Limits as monotonic. Students sometime view the sequence $a_n = \frac{(-1)^n}{n}$ as two sequences rather than one
(6) Limits as dynamic. Students can view a limit as a dynamic process, with both the independent and dependent variable "moving" (as contrasted with the static, epsilon-delta definition).

In addition to these explicit conceptions, several authors have noted students' tendencies to avoid dealing with the conceptual difficulties limits present by developing "coping skills." In studying a large-scale calculus reform project, Smith and Moore (1990) noted this unfortunate phenomenon:

> Much of what our students have actually learned ... – more precisely, what they have invented for themselves – is a set of "coping skills" for getting past the next assignment, the next

> quiz, the next exam. When their coping skills fail them, they invent new ones. The new ones don't have to be consistent with the old ones; the challenge is to guess right among the available options and not to get faked out by the teacher's tricky questions. (p. 54)

This idea of students finding coping skills to help them get through has been noticed elsewhere as well. In studying student problem solving Schoenfeld (1985) noted similar behavior:

> The close examination of these students' problem-solving performance revealed that many of them had serious misunderstandings about mathematics... In some cases, students survived (often with good grades!) by implementing well-learned mechanical procedures, in domains about which they understood virtually nothing. (p. 13)

Whether or not student difficulties with limits are in some sense unavoidable, they do provoke an important question regarding teacher knowledge. Do calculus teachers know about typical student misconceptions and student use of coping strategies, and if so, how do they use that knowledge to help students come to a better understanding of limits?

2.4. Emerging Scholars Programs, ESP literature. To give a sense of the context in which our participants worked, we give a brief overview of ESP workshops and the research about them. In the early 1980s, concerned about the failure of many African American students to successfully complete the calculus sequence, Uri Treisman compared their study habits with those of Chinese American students who had been far more successful in navigating these gateway courses (Fullilove & Treisman, 1990; Treisman, 1985). He used this work to develop the Professional Development Program (later implemented at the University of Texas at Austin as the Emerging Scholars Program). The program was known as an honors program in which students, primarily underrepresented minorities, attended traditional lectures and took exams with the general population but met in special workshops for an additional four hours every week.

In these workshops, students worked in cooperative groups on worksheets of challenging calculus problems (Fullilove & Treisman, 1990). The worksheets were written by the graduate TAs, sometimes with the help of an extensive problem bank of previously-used problems. Student group work was facilitated by a TA and an undergraduate student assistant (typically a former ESP student), who circulated among the groups providing help in the form of Socratic questions, rarely giving straight, explanatory answers. The guiding philosophy behind the ESP model was to provide a supportive environment where students could engage in productive discourse about the important concepts of calculus. ESPs have now been implemented at many institutions across the country and now typically consist of four to six hours of workshops per week.

ESP programs have been extremely successful in many ways, and those successes have been documented and studied by several researchers (see Hsu, Murphy, & Treisman, 2008, for a review of this literature). In general, when ESP students are compared with traditionally-taught counterparts, they receive higher grades in Calculus (from 0.5 to 1.0 higher grade point average on a 4.0 scale), persist in mathematics-based majors longer, and are more likely to be mathematics majors. This improved performance is particularly impressive in light of the historical underperformance of the underrepresented groups targeted by ESPs. While attempts have been made to bring the successes of ESPs to a more general student-audience, results have been mixed (see, for example, Herzig & Kung, 2003).

Given the student success in these programs, one might speculate that the TAs leading the workshops were either already effective teachers or that in the workshop setting they became effective teachers. Sample selection difficulties, however, make it extremely difficult to differentiate between these two scenarios. What we can do is study those who have taught in the ESP setting and try to uncover what knowledge of student thinking they gained through such experience.

3. Case Study

Here we describe our study of former and current ESP teachers' learning about student thinking in Calculus. We first describe our research methods for this qualitative study and then examine the results. Using the data from our study, we propose a framework for understanding the types of knowledge gained by TAs in various situations. We then turn to the implications of this work for graduate programs and TA training before concluding with comments on the limitations of this work and possible directions for future research.

3.1. Methods. The interview participants were selected from among the current and former teaching assistants at two large public institutions that have well-established ESPs, one in the Midwest and one on the West Coast. Participation was restricted to those who had taught ESP for at least a year (eliminating one West Coast participant). Each interview lasted between 45 minutes and an hour and fifteen minutes, and all were conducted between June 2003 and April 2004 in a variety of locations. The interviews were audio taped and later transcribed.

The number of TAs teaching in ESPs is relatively small. To protect the identities of the participants, we avoid giving too much detail of their background, use pseudonyms when quoting them, and refer to all TAs with female pronouns. Their experiences and current job statuses are given in Table 1 (without any names, to further protect their anonymity). All participants had a variety of teaching experiences during graduate school, starting with standard discussion section assignments. At the Midwestern institution, a TA's primary job in discussion section classes was to answer homework questions at the board twice a week. At the West Coast institution, many of the classes had adopted an ESP-like format, using worksheets and having students work in groups. Also, the TAs from the Midwestern institution were, in general, further along in their careers – perhaps giving them a different perspective on their ESP experiences. The participants who had since gone on to become professors (all from the Midwest) had all taught Calculus in the intervening time as the instructor of record. Because of these differences in background, we distinguish between the two institutions in all quotes given below (MW for the Midwest, WC for the West Coast).

Following a brief biographical discussion, the interviews quickly turned to tasks designed to elicit thinking about student strategies and misconceptions of limits. The tasks were designed to probe the participants' understanding of student thinking, and the questions focused on how students would approach them (both correctly and incorrectly) and the difficulties students might have. The first task

TABLE 1. Participant Biographical Information

Status at time of interview	Semesters Teaching ESP	Years Since Teaching ESP	Graduate School Location
2nd year graduate student	2	0	West Coast
2nd year graduate student	2	0	West Coast
Dissertating graduate student	3	1	West Coast
Dissertating graduate student	2	2	Midwest
Dissertating graduate student	2	0	West Coast
Beginning Assistant Professor	2	4	Midwest
2nd year Assistant Professor	4	3	Midwest
3rd year Assistant Professor	2	4	Midwest

involved two limits of a piecewise-defined function. Though the function itself is discontinuous, both limits do exist. The second task involved the graphical representation of a function on an exam and asked the participants to draw a function to ask questions about.

Task 1: You plan to give your workshop students the following problem:

> Given the function
>
> $$f(x) = \begin{cases} x^2 + 1 & \text{if } x \leq -1 \\ x - 1 & \text{if } x > -1 \end{cases}$$
>
> find:
>
> $\lim_{x \to -1} f(x^2)$ and $\lim_{x \to -1} f^2(x)$.

What difficulties do you think your students will have with this problem?
What will the most common incorrect solutions be?
What methods and ideas might students use to get the correct answers?
What unusual ways (fruitful or not) would they have of approaching this problem?

Task 2: You are writing an exam on limits in a first semester Calculus course. You are in the middle of writing a problem which asks students to figure out the value of a particular limit given the graphical representation of a function, and are deciding on what function to draw. What possible graphs could you draw for this questions? For each graph, explain the choices you made. What specific part of the students' limit concept would you be testing with each part of the graph? [multiple blank axes were provided]

In early interviews, Task 2 did not elicit some of the misconceptions that are more common in a sequence/series setting, so a third task was added to the interview protocol after the Midwestern TA interviews were complete. That is, Task 3 was an additional interview prompt only for the four West Coast TAs.

Task 3: You are giving a quiz on sequences and want to know if your students understand what it means for a sequence to converge (to a finite limit). What sequence or sequences would you ask them about? Why?

All three tasks were designed for a generative, task-based analysis, as described by Clement (2000). All three also "suggest or entail strategic thinking of some complexity" (Goldin, 2000, p. 541), though in this case the focus moved from student to teacher thinking.

After the tasks and asking participants about student thinking about specific questions, the next part of the interview asked participants to reflect on their responses, recalling where and how they came to acquire their knowledge of student thinking. When appropriate, the participants were asked to compare the knowledge they had gained through their different experiences. The last part of the interview focused more specifically on other aspects of their ESP experiences. We hope to use these data in later work on ESP programs.

There are inherent limitations in having participants reflect on past events. There is the possibility that our participants' recollections did not give an accurate picture of their knowledge acquisition, which limits the conclusions we can draw. While we recognize this difficulty, we see these interviews as an important first step in addressing issues of TA knowledge of student thinking. The results found here can inform later studies which might follow TAs through their early teaching experiences.

The interviews were transcribed and then analyzed with qualitative analysis software using an iterative open coding method (Spradley, 1980; Strauss & Corbin, 1990). Categories of student misconceptions of limits were drawn from the literature discussed above. The remaining categories were inductively derived, at times informed by the literature. For instance, we originally developed the categories of "student reliance on algorithms" and "students avoiding deep thinking" which were later folded into a category of "coping skills" after we found references to such strategies in the literature. All of the categories on how participants reported learning about student thinking were inductively derived.

3.2. Results. During the course of the interviews, the participants expressed significant knowledge of student thinking and reflected on how they gained that knowledge. Here we detail their responses which provide evidence for three claims.

(1) Experienced Calculus teachers have significant knowledge of student thinking. Following the literature, we divide this type of knowledge into three categories: knowledge of student solution strategies and difficulties, knowledge of coping skills, and knowledge of misconceptions.
(2) College teachers learn about student thinking while engaged in a wide variety of activities, including observing students working problems, writing problems, and grading. However, the types of knowledge gained differ by the type of activity. This leads us to our third claim...
(3) There is a duality between teachers' activities and the knowledge of student thinking to which they potentially have access. For instance, grading provides an opportunity to learn about typical student errors while observing students working on problems leads to a much more fine-grained knowledge of student struggles. Teachers gain different types of knowledge of student thinking through different student-teacher interactions.

3.2.1. *Knowledge of Student Thinking.* Participants spoke at length about how students approach specific problems about limits and their ideas about limits more generally. Through the process of iterative coding, we organized these data into three categories. Here we provide excerpts from that data, giving evidence that experienced calculus teachers have significant knowledge of student solution strategies, common mistakes, frequently used scoping skills and typical misconceptions. For the first category, we compare the participants' responses with the strategies used by our own students; for the last category, we compare the responses to the misconceptions given in the literature.

Knowledge of Student Solution Strategies and Difficulties. The participants' knowledge of student solution strategies and difficulties with those strategies came out primarily in response to the first task. We have used this task with undergraduates (in a group work setting) for several years, and while we have only anecdotal evidence of their strategies, we present it here for comparison. Our Calculus students have found this task quite difficult, and have needed some assistance to complete it successfully. The most successful strategy has been thinking through left- and right-sided limits (which can amount to simply plugging numbers into the branches). For the $f^2(x)$ part, an algebraic approach of squaring each branch of the function has been reasonably successful; an algebraic approach to the $f(x^2)$ (plugging x^2 into the function) has been less so. Graphical approaches, typically graphing $f(x)$, are unsuccessful without significant assistance and the use of other strategies (e.g. using the graph to think through left- and right-sided limits. Some students have (incorrectly) used limit laws for $f^2(x)$. A successful, but rarely used strategy has been to approximate the answer using one point (or a small collection of them).

Our TA participants all noted that they felt the first task would be extremely difficult for students. As to the details of those difficulties, there was broad agreement on a few points. First, they all reported that students would be challenged by a piecewise-defined function.

> *WC4:* This function is already hard for them to see.
>
> *Interviewer:* What makes the function hard?
>
> *WC4:* The, having branches, that's something which is difficult, already.
>
> *MW3:* I know that one of the most difficult parts is that just beginning with the actual function is that this kind of split notation seems to cause infinite amounts of difficulties.... So most likely students will begin by treating this function as two functions, and that will cause problems for the actual limit part of this exercise.

Second, all participants noted that some students would fail to understand the algebraic notation used in the question. Comments like this were typical:

> *MW3:* ... I mean, the distinction between $f(x^2)$ and $f^2(x)$ I think is a fairly fine one in a student's mind...

A few noted the ambiguity in the notation, for example:

> *MW1:* Well, they might decide that $f^2(x)$ really means $f(f(x))$.

Other difficulties were cited by participants, but the challenges they reported depended on the different ways they thought students would approach the problem.

Six of the eight participants reported that students might attempt an algebraic solution and discussed the challenges that might face them. Four participants thought students might plug x^2 into the formula for $f(x)$. Such a solution is possible, but it requires one to also plug x^2 into the definition of the domain of the piecewise-defined function (rendering one branch of the function irrelevant, since $x^2 < -1$ never applies). This difficulty was noted by two participants; of the other two who mentioned substitution of x^2 into the formula for $f(x)$, one didn't think this method could work and the last never came to a conclusion about this method.

Six participants reported that students might try a graphical approach (either on their own or with a little bit of prompting). However, in this case, graphing the function $f(x)$ "doesn't really help, in a sense" (MW2). This comment might have meant that students could not read the answer directly from the graph, or that the graph would not help students at all. Some of these six participants used a graph to find a solution while others did not.

Interestingly, none of the participants reported that students might plug in a discrete set of values near -1, despite the fact that half of them later mentioned this as a typical misconception students have of limits. This difference in what knowledge of student thinking is evoked in different situations illustrates the challenges inherent in trying to understand teacher knowledge.

Knowledge of Coping Skills. The participants reported extensive knowledge of student coping skills. We use this term to mean the ways students find to complete tasks competently or correctly without fully understanding the underlying concepts. For instance, a typical algebraic coping skill would be knowing how to FOIL without understanding why. In describing how students would approach the first task, seven of the eight participants suggested that students might rely on a procedure without understanding the foundational ideas of the limit.

> *WC1:* [Task 1] is a type of problem they see in [Calculus I], and that they, you know, a lot of them will know how to do, um even without understanding really why.

The TAs suggested that students would rely on coping strategies, including to "plug in" points:

> *MW1:* Another mistake they would make regarding that, which is a subtle thing having to do with the definition, is they'll simply plug in -1 to one piece or the other and the better students will decide to plug it into both and see if they get answers that agree.

> *Interviewer:* What are the students going to see in this problem?
> *MW2:* They are going to see a problem that looks a little bit pointless and that the way to – and a problem that when they attack it, what they need to do is know what numbers to plug in where.

> *WC1:* I think a lot of students, you know can, can understand the process of what to do there, even if they don't really get why they're doing it that way. ... If the limit was x going to two for instance [instead of negative one] they might try to do the same thing, like plug it into both [branches] and see if they're the same.

Additional coping skills mentioned by participants included a "factor, cancel, plug-in" algorithm for limit problems and using L'Hospital's Rule:

MW2: They get to a point where they're like, "Oh, OK, I get these limit problems. I'm supposed to factor the top and factor the bottom and cancel and plug in."

MW4: They learn sort of these algebraic tricks for how to figure out what [limits] are. They learn like, "well, you plug this in" or "use L'Hospital's Rule." Like they learn these tricks, but I'm not sure they really ever get, a lot of them, an intuitive understanding of what it is they're doing.

Another type of student coping strategy reported was the use of a keyword strategy to decide how to approach a limit.

MW1: So you'll see students who, zero over zero automatically means L'Hospital's Rule, which is OK, but you'll also see students who, once they have L'Hospital's Rule at hand, when I say "zero over zero means more work" they hear "zero over zero means use L'Hospital's Rule." Which, again: OK, not great, because the work that's needed is different.

In discussing the third task, one participant mentioned the sequence $\left(1+\frac{1}{n}\right)^n$ and noted that students can get used to this sequence converging to e without understanding why.

WC3: And then sometimes it's nice to also throw in one that is not an indeterminate form. You know something, something that is literally like, this [writing]

Interviewer: That's two plus one over n all to the n.

WC3: Yeah, where it's, it's, you know looks like an e, it looks like something they want, but you know you'd have to look at it and realize, oh wait, this is two to the infinity which is, should be getting really large.

The concept of coping skills appears in the literature. However, the sort of detailed research that could confirm whether our participants' responses correspond to the coping skills actually used by students does not appear to exist, yet.

Knowledge of Misconceptions. Since research on student misconceptions of limits is well-developed, we used the categories from that literature to inform our coding. During the interviews, TAs made two types of comments that we took as evidence that they knew about a particular student misconception. Most compelling was when participants explicitly mentioned a misconception and described what they meant by it. Less compelling was when participants used an exam-writing task (Task 2 or 3) to show that they would test for a misconception, even if they didn't discuss the misconception explicitly. We take such data as weaker evidence that a participant knew about such a misconception. Also, several of the misconceptions could fall under the area of coping skills.

The most commonly cited misconception was the idea of limit as the value at a point. This came across, above, in descriptions of student coping skills (e.g., plugging in values), but all of the participants also discussed this idea explicitly, later in the interviews, when asked about misconceptions students have of limits.

WC2: They think limits and continuity are the same thing so it takes a while to give all kinds of examples like this to see that it's not quite.

MW3: I would see the more standard misconception being that it's just a point, that it's just a location and whether - and well, you see what the limit is by plugging in the point.

At the same time, several participants themselves used the language of plugging in values in the course of solving Task 1. Here we see one participant using such terminology in thinking through Task 1 (making an arithmetic error in the process). Clearly the participant is aware of how such shorthand might be misinterpreted:

MW4: OK, well I'm just going to try plugging in negative one here, and see what we get... and I'm going to totally use inappropriate notation. So, negative one squared is one, and one plus one is two. So it's two from that way and zero from that way, so the limit does not exist.

The fact that experienced calculus teachers see the idea of limit as the value at a point as a misconception, but use similar (or closely related) thinking and notation may indicate why this is such a popular misconception among students.

A less naive misconception held by many students is to approximate the limit by using domain values *near* the limiting value and plugging in one point, or a sequence of points, which are close (but not equal to) the limiting value. Four participants described this type of misconception, two in talking about a finite limit and two in discussing limits at infinity.

MW2: You know, and even when we plug in 2.9, 2.99, 2.999 and we're getting numbers that are getting closer to 6 we still don't really know that the answer is 6... I don't think students ever really get that.

Yet, this participant also described preferring this strategy to a strategy involving plugging in a single number, a possible source of the misconception:

MW2: One thing that students screw up is that they think they can just plug in one number that's really close to, say, three, and that will be – you know, they plug in 2.9999 and they get 6.001 – "Oh, the answer is 6."

Two other participants described approximating a limit with discrete points in the case of limits at infinity. One admitted that this was a very effective strategy:

MW4: Occasionally you get students that say, "Well, if it's the limit as x approaches infinity, then I can just plug in like 1000 and kind of round to the nearest whole number and then that's my answer. You do get a couple of these, and the problem is that most of the problems in the books, like, that method works.

With this misconception, we again get a picture of both knowledge of students' problematic thinking and possible reasons for that problematic thinking.

Three of the misconceptions from the literature are closely related and were frequently only implicit in the participants' discussions of student misconceptions. These are limit as monotonic, limit as bound, and limit as unreachable. These are more frequently evoked in the context of sequences. Since Task 3 was introduced only for the West Coast participants, this affected the frequency with which they were mentioned. These misconceptions were implicitly noted when participants discussed an oscillating function (e.g. $x\sin(1/x)$ as x goes to 0 or $a_n = (-1)^n/n$), which serve as a counterexamples to several misconceptions simultaneously. Five of the participants suggested such examples (either graphically in Task 2 or algebraically in Task 3), including three of the four West Coast participants.

As for explicit discussions of these misconceptions, only a few were made. One TA gave an example of a convergent, non-monotone sequence as part of Task 3. When asked what students would think about this, she responded:

> *WC2:* "They would say - a lot of them - the sequence is decreasing, which is false. And you know, for us it's a big false. But for them it's why, why - because it really looks like it's decreasing.

It remains unclear if this participant sees this as a general misconception of limits or as a misinterpretation of this particular sequence.

Just one participant explicitly discussed the misconception of limit as unreachable, and she did so in the context of horizontal asymptotes. In the section of the interview explicitly on misconceptions, she said,

> *MW1:* The idea that a function cannot cross a horizontal asymptote and still have it be an asymptote is a common misconception.

Finally, the misconception of limit as dynamic was not explicitly discussed by any of the participants. We speculate on two possible causes for this. First, the participants, themselves, might see limits as dynamic (there was some evidence of this) and thus not think of it as a misconception at all. Alternatively, it's possible that these participants would have seen this idea as a misconception if the static epsilon-delta definition had been discussed explicitly. Thus this misconception might be part of their knowledge, but it was simply not evoked in the course of the interview. In general, there is evidence that most of the participants knew of the most common misconceptions of limits, but the less common ones were frequently only implicitly reported.

3.2.2. *How TAs Gained Knowledge of Student Thinking.* In the iterative process of analyzing the data, consistent patterns emerged in terms of the ways in which the participants learned about student thinking. Here we detail their responses, organizing them by the activities participants were engaged in during their reported acquisition of that knowledge. Those categories were, in decreasing order of the frequency of reporting: observing students work, writing problems, grading, self-reflection, and miscellaneous activities. Within each category, the type of knowledge gained is detailed using quotes from the interviews to illustrate the types of comments made. Also included are quotes that indicate participants' descriptions of the limitations of the knowledge gained through that activity or indicating how the knowledge gained in that way was useful in later teaching experiences.

Observing Students Work Problems. All participants indicated that watching students work and listening in on their conversations was instrumental in their learning about student thinking in calculus. This is unsurprising given the participants' experiences in ESP workshops. However, those participants who had gone on to do other teaching reported continuing to learn about student thinking in later classes through watching and listening to students work. Also, several of those participants from the West Coast university reported these types of knowledge gains in both their ESP and standard sections (which used ESP-type methods). Most of the quotes here were in response to a general question about how they learned about student solution strategies and misconceptions.

The very nature of the ESP classroom dictates that TAs spend a fair amount of time listening and watching students work. This act of observing while students worked was instructive for all of the participants:

> *MW4:* [In ESP] I'm sort of constantly lurking and they're supposed to just ignore me even though I'm just looking at their paper. ... It's– you're just a fly on the wall and you can just watch, and that's fantastic and you learn a lot from that.
>
> *MW2:* I think that in the [ESP] section the neat thing that happens is that– or the neat thing that I really try to do in the [ESP] section is that I really do a lot of listening. I let the students talk.
>
> *Interviewer:* How did you learn about [student misconceptions]?
> *WC1:* Watching them do problems like on the board, helping them with problems.

While participants watched and listened to students in both ESP and non-ESP classes, the generous time in the workshop allowed TAs to see student thinking in more depth:

> *WC3:* With [ESP] you have a lot of time, and you have that one-on-one interaction, and yet at the same time, you're no less capable of teaching algebra than you were in the other section, you just are with the people more to see the, to see the difficulties even more pronounced...

TAs also reported being able listen to students at length without intervening:

> *MW2:* Sometimes they're off base in an interesting way and you realize, "Oh, they're, you know, a good student. They're going to be able to steer themselves back. I want to see how this plays out." And that can really give some insight into how they're thinking things.

We can only speculate if this TA would have taken the time to "see how this plays out" if it had been a more time-pressured situation – and about how much she would have learned about student thinking.

The amount of time and the attention paid to each student also improved TAs' knowledge of students thinking. Several participants noted that they learned the most from watching the struggling students.

> *MW2:* In the [ESP] class you have to pay attention to everybody. ... the C students, you would just sit there and I would learn so much from them, because they would say, "OK, I'm doing this, but I don't know what to do next." And it would just be like– I would be surprised. I would say, "Wait, how could you not know what to do next? OK, what are you thinking? Where are you at?"

All of those who had gone on to teach in other settings mentioned that, in one form or another, they had implemented group work in later classes as a way to understand student thinking. One TA later decided to give weekly group quizzes in a non-ESP Calculus class. When asked why, she gave the following explanation:

> *MW3:* Every week it gave me a time where I could see where the students were at and where their misconceptions were coming from, what they were even if I didn't necessarily know how to solve those things.

The experience of observing students work problems gave one participant knowledge about problem difficulty that she continued to use several years later.

> *MW1:* If I'm writing an exam now, I have a much better idea of how long that exam is going to take than I would have had before teaching [ESP], because I actually saw students working on problems and I saw how long it would take them to get into a problem, and how long before they would have an insight that would actually help them solve the problem.

This same participant noted that the process of watching and listening to students work had given her knowledge about calculus that she lacked about other subjects:

> *MW1:* [Because of ESP] I understand what students can do wrong and will do wrong in Calculus, but if I'm teaching Abstract Algebra, for instance I taught it for the first time last spring, the errors that I would see were completely new to me. ...I had to pick up this knowledge about how to– or what my students will and will not pick up quickly in Abstract Algebra – and in Calculus I feel like I have a better grip on that.

Writing Problems. Writing challenging problems for worksheets represents the bulk of class preparation for ESP TAs. This is in stark contrast with a standard discussion section TA appointment which involves little or no problem writing. Six of the eight participants mentioned this activity, along with reflecting on what students did with the problems, as important in learning about student thinking.

While all the participants had access to banks of previous ESP-type problems, all reported writing their own problems at least some of the time. When they talked about problem writing and what they learned from that process, they largely focused on understanding how difficult students found different problems. One participant discussed finding a balance between problems that were too hard and too easy, indicating that she was learning about how difficult students found various problems:

> *MW1:* For me, the biggest thing was writing the worksheets, making sure that the worksheets were at once challenging but not impossible so that the students could make progress during the two hours but not zip through it and be bored after half an hour.

In talking about worksheet writing, one participant described the differences between what she intended students to get out of a problem and what they actually did get out of it:

> *MW4:* So, there was some of that, some of, you know, some problems that maybe I thought were really interesting and the students managed to do the problems, but they never really quite got the interesting thing I was trying to get at. You know, that maybe needs to be rewritten or have some extra parts or something, or maybe I need, you know, a different approach to try to get them to see what's going on.

They also noted that the process involved a fair amount of trial and error. In this sense, the activity of writing problems was intimately linked to the activity of observing students working problems, creating a feedback loop.

> *Interviewer:* You said earlier that ESP taught you a lot about what a good assignment is, could you explain more about that?
>
> *MW4:* Writing the worksheets and then seeing what happened with them. It's like instant feedback. You know, three times a week you write problems and then you give them to the students and you find out if they worked or not. And they all work to some extent and I think it's sort of good to take risks and try new things and see, but some of them, you know, maybe I wouldn't try again and some of them I would, like they're surprises.
>
> *MW4:* I think that thing where you're constantly writing problems and then giving it to them over and over again is ...instantly seeing how the problems worked, is really good.

At times this feedback was even directed at a single student who struggled with some topic:

> *MW3:* They would say, "Ok, I'm doing this, but I don't know what to do next." And it would just be like– I would be surprised. I would say, "Wait. How could you not know what to do next? Ok, what are you thinking? Where are you at?" And I would really sort of turn those things over in my head and sometimes I would have to wait until the next time and say, "Ok. Now I think I have a way to try to get you to understand that." Or, "I put a problem on this week's worksheet just for you." I wouldn't say that to them, but I'd be thinking to myself, "This problem is on this worksheet just for you. This is your own personal worksheet problem to try to address that misconception."

This attention to individual thinking mirrors the feedback loop of the Level IV instructors described in the CGI literature (Fennema et al., 1996).

Grading. Grading exams and/or homework was mentioned by five of the eight participants as a way of learning about student thinking. Most TAs spend a significant amount of time grading, and the process of carefully examining student work appears to be important in developing knowledge of student thinking.

> *WC1:* I learned a lot from grading midterms, I mean that's 'cause, I need, I'm like, even students who I'd watched in class, like I'll grade their quizzes or midterms and I'll see them doing, you know something completely different.

For most, a desire to learn about students' thought processes colored the way they approached grading. Grading was clearly not just for assessment; rather it involved struggling to understand what students had been thinking.

> *MW1:* [Grading is] not simply assigning point values and pressing on, but sort of looking at problems that are not worked out correctly and saying, "What in the world was this student thinking here?" ...if you see a student go completely off in another direction, usually that's not random. If you look closely and think about

it you can figure out what it is that might have caused them to go in that direction.

Grading homework can also be vital in assessing student progress during the course of the semester. The importance of grading as such a tool became clear to one participant after a later teaching experience:

MW4: Last semester I also had a really bad semester last semester teaching, in my opinion, and part of that was because I was teaching for a class that had no homework and no quizzes and no feedback of any kind.

For another participant, the desire to understand student thinking informed her assessment strategies:

MW2: I learn a lot when I grade things. I learn a lot when I grade exams. I try to write exams in a way where I'm going to try to learn something about what they know.

Here we see the participant make an implicit distinction between a procedural skill (i.e., a coping strategy which might have been described as "what they can do") and the deeper, more profound understanding (i.e., "what they know"). Clearly for many of the participants, grading played an important role in building knowledge of student thinking.

Self-Reflection. As one might expect, a certain amount of knowledge of student thinking can be gained by thinking about one's own thought processes. While only three of the eight participants discussed self-reflection explicitly as a way of understanding student thinking (one in a negative sense – that one does not learn much through self-reflection), several of the others implicitly did so in talking about how students would approach the first task. One participant was initially confused by the Task 1, tried a method that failed, and then noted that students might be similarly confused.

MW2: Does this actually work? Now I'm confused. Yeah, 'cause this doesn't actually work, does it? ... (laughs) ... So, they might try this and then fail.

Those participants who mentioned self-reflection generally noted the insufficiency of this method for understanding student thinking. For these participants, the thought that self-reflection produced only limited knowledge might have been prompted by the contrasts with the rich knowledge they were gaining in other ways. For some, this insufficiency was implicit in their discussion of the other ways of gaining knowledge, but others made it explicit.

Interviewer: How do you know where they might go wrong?
MW4: Yeah, I mean, some of it is just like imagination like, "If I were a student, where could I possibly go wrong on this?" I mean, you don't know. Right? That's the thing. Like I don't claim to know exactly where everybody is going to go wrong on any particular problem, but part of it is just through imagining, part of it's through experience.

MW1: Um, you know, it's different from actual math. With actual math if you think hard enough about the answer you'll come to the answer and you'll come to the same answer that everyone who

> thinks hard about the problem will come to. It's convergent in that sort of sense, but learning what the pitfalls are, you can't do that in isolation.

One of the participants with the least total teaching experience noted the limited nature of self-reflection in teaching epsilons and deltas. In fact, she noted that her particularly strong knowledge of the subject may actually inhibit understanding student difficulties. She discussed how she learned about the difficulties students have with the rigorous definition of limits:

> *WC3:* So it's hard for me particularly, as an analyst, [I] bat around epsilons and deltas and limits and approximate identities and approximate solutions and just, at this point, without blinking, you know I mean it's just [snap, snap]... So in many ways you have to retrain yourself to believe that it is not second nature.

Her description echoes Ball and Bass's (2000) idea that "unpacking" complicated math concepts is a vital part of the teaching process.

Finally, one participant discussed self-reflection as insufficient to understand student thinking in a later class:

> *MW1:* I taught [Abstract Algebra] for the first time last spring, the errors that I would see were completely new to me because I'd never taught the class before and I'd never made those errors myself before. I'd made other errors and managed to find my own and so I had to deal with each one.

Miscellaneous. Several activities that led to knowledge of student thinking were mentioned only by a single participant. Whether or not other participants also benefited from these activities would be purely speculative. However, it seems reasonable to assume that the effect of these activities was, at least in retrospect, minor compared to the above activities for the other participants.

Two participants each contributed two unique activities to this list. One participant mentioned office hours as a way to learn about student thinking but did not speak at any length about what types of knowledge she gained in that setting. It is possible that the structure of the ESP workshop, with longer classes and closer TA-student interactions, reduced the importance of office hours as a means for communication (and thus learning about student thinking.) The same participant mentioned reading materials on student thinking, presumably as part of the professional development program for new TAs. What she said about this experience provides potential insight into the nature of TA thinking, for those who run professional development programs:

> *WC4:* I think there are some materials which come and tell you what students think about certain problems so I read some of those, but no, I don't remember them because you know, when you read, you don't ... one thing is just to experience it, obviously you remember better than just read[ing] "students [have] these types of problems."

Another participant also mentioned two ways of learning about student thinking which were unique. She alluded to learning about student thinking by paying close attention to their actions during traditional discussion sections,

> *WC2:* But, even when I am lecturing you know, I also see that, because I encourage them to ask questions, and you see it in their questions. ... and I also know exactly which student[s] follow me and what, who didn't follow me, so I'm also - very good at this."

For this TA, this sometimes led to another unique way of learning about student thinking: looking at their lecture notes.

> *WC2:* So, then when we discuss problems, I will ask them for their notes, you know, ok, what did you take notes on, from me, from the professor, let's see your notes...

While some of the other participants may have learned about student thinking through these activities, none reported in during the course of the interviews.

We should note that in addition to the specific activities mentioned above, there were many others discussed in the course of the interviews. However, they were not mentioned in the context of providing insight into student thinking. For instance, in discussing general questions about teaching, several participants noted that they learned a significant amount from talking with their peers. In every case, however, they couched this knowledge in general ways (e.g., from *MW1:* "Being able to work with other people who are [excellent teachers] is very beneficial because you get to learn all kinds of wonderful things about how to do a better job teaching.") It is possible that this learning included knowledge of student thinking, but from the interviews alone we were unable to determine if that was their intent.

4. Duality Framework

Participants reported learning about student thinking in myriad ways. In fact, if one examines all the ways in which TAs typically interact with students (in a lecture setting, during group work, in office hours, as well as through homework, quizzes and exams, and in social settings), at least one participant reported learning something about student thinking of calculus in each setting (save the last). The fact that the participants learned something in each setting where calculus was discussed is not surprising.

Also unsurprising is the finding that observing students working on problems was most prominent among these participants' ways of learning about student thinking. This is, after all, the primary activity in an ESP workshop, and having engaged in such close contact with a relatively small number of students for such long periods of time, these participants were bound to have learned a significant amount about student thinking in that setting.

However, a closer look at the data reveals interesting patterns in the types of knowledge TAs reported gaining through the various activities. In conducting the first set of interviews, it became clear that participants were distinguishing between the types of knowledge they gained in differing settings: different activities led to different types of knowledge. Because of this, later interviews included questions that asked participants to compare directly the knowledge they had gained in a setting that they had mentioned. For instance, while participants reported learning something through both grading and watching/listening to students work problems, they described the content of that knowledge in very different terms. The former led to knowledge of what students thought the answer should be (i.e., the end result, or what the grader wanted to see), and one could sometimes "decipher" the intent of the student. The latter gave more detailed knowledge of problem difficulty, where students struggled, and the many missteps they would make.

Somewhat surprisingly, the differences were not simply a matter of grain size, with some activities giving a more detailed picture than others. Rather, there were certain qualities of the knowledge gained in one setting that were not duplicated by that gained in another – and vice versa. In the language of mathematics, there appears to be a certain duality between the ways in which the TAs interacted with the students and the types of knowledge about students' calculus thinking that they gained. This leads us to propose this duality as a framework for understanding such TA knowledge. Graphically, we represent it as follows:

Types of Interactions Between TAs and Students	$\Longleftrightarrow$	Types of Teacher Knowledge About Student Thinking

We start by giving some evidence for this framework from our interview data. When participants were asked to compare what they learned in two different settings, they invariably drew rather contrasting pictures. The most common comparison was made between grading and watching/listening to students work problems. In this case, observing students working problems gave a more detailed, nuanced view of what students were thinking – a view not accessible through grading.

MW4: Well, wandering around watching them is just going to give you more, because you can see what they're writing down and you can see sort of what order they're writing it in. You can see when they're erasing something. You know, there's just more to see, and you can maybe hear them talking to their neighbors. You can hear the weird things they're saying. Like they may be writing something correct, but then saying something a little weird. ...You can also just sort of see like how long it takes them or where they pause – you know, where do they get stuck and think for a while before moving to the next step and where they are just like writing. You can't always see that from the homework.

WC1: Well, so grading their tests you would, I mean you know it's sort of, sort of like very superficial, you can't really see anything about their thought process. I mean you can try to piece it together... You can speculate about what was going through their mind, but, I mean, all you see is what they thought the answer should be.

However, the process of grading gave some participants important information that they didn't have access to in the watching/listening phase.

MW3: And actually [not grading much work] is something I missed with the [ESP] program is that – because you never really look at what they've written down as their answers ...

This participant concluded that neither activity gave a complete picture of student thinking; that each gave her knowledge of student thinking that the other could not:

MW3: So in some sense I think it's important to see both sides, because on the one hand you see what kinds of thoughts are going through

TABLE 2. Duality Framework

Types of Interactions Between TAs and Students	$\Longleftrightarrow$	Types of Knowledge of Student Thinking Gained
Watching/Listening to students	$\longleftrightarrow$	Knowledge of solution strategies and mistakes. Fine grained, but missing students' conclusions.
Writing problems	$\longleftrightarrow$	Knowledge of problem difficulty, how question wording affects student response.
Grading homework & exams	$\longleftrightarrow$	Knowledge of students' "final answers." Only indirect knowledge of their thinking.
Student questions during traditional sections	$\longleftrightarrow$	Knowledge of where students have trouble, generally without details of the nature of that trouble.
No interaction (self-reflection)	$\longleftrightarrow$	Limited, speculative knowledge.

> the student's mind and hopefully you can help reinforce the right ones and on the other hand you can see what is being reinforced and what thoughts are actually, you know, being translated into "Yes, this is what I know."

She went on to tell an anecdote about using group quizzes in a later, non-ESP setting, illustrating why observations and grading yield different types of knowledge.

> *MW3:* And there's one group which, they just did fabulous work. ... They had fabulous discussions, got into large arguments about, you know, which way they should approach this question, often came up with the right solution, but when I'd actually look at what they'd written down on their quizzes, because I had them write up a– each write up their own quiz ... it was completely not at all what had gone into the actual discussion, so somehow I think one of the greatest difficulties that students have is figuring out which of their thoughts that they had are the right ones to put down...

Though other comparisons were not made directly by participants, their responses included contrasts. For instance, during the process of writing problems, one TA reported learning "how much adding one more layer of complexity adds to the time that it takes to do the problem ... but you don't really believe it until you see it in action, until you see a problem that you've actually written ..." The act of writing the problem and then getting feedback about it gave her qualitatively different information than getting feedback about someone else's problem. We encapsulate the dual nature of the TA actions and the types of knowledge available to the TAs in the course of that action in Table 2.

5. Implications

This work has a number of important implications, both in terms of professional development and future research. First, it provides a window into teacher knowledge of student thinking at the college level which might inform professional developers. Instead of examining from the front end what skills TAs will need for their early teaching experiences and building materials around those skills, this work looks from the opposite end of the time scale at the types of knowledge gained through years of various teaching experiences and how that knowledge might be gained earlier. If, as the K-12 literature indicates, improving knowledge of student thinking can be transformative for teachers and lead to improved student understanding, this new perspective has the potential to greatly improve the professional development opportunities we provide to graduate students.

Our duality framework could serve as a guide for early professional development activities. Typically, initial activities are focused on pedagogical knowledge such as writing on the board, speaking loudly and clearly, dealing with difficult students. If opportunities to build knowledge of student thinking are to be included in such programs, professional developers need to know what types of activities to include. The duality framework can provide guidance by indicating the types of knowledge that might be gained through engaging in (or possibly simulating) various activities.

The duality framework also provides an important lens through which to structure future research. For instance, do simulated grading experiences (such as those suggested in the case studies by Friedberg et al., 2001) lead to the same knowledge of student thinking gained by TAs in their grading experiences? Could video case studies, combined with reflection and discussion, provide the rich opportunities to observe students working problems described by former ESP TAs? Would novice TAs' knowledge of student thinking benefit from these types of simulated experiences? Can the benefits of writing problems be gained without observing students working on them, but perhaps by having other TAs comment on them? These are all important questions that are more clearly articulated through the use of the duality framework.

Beyond the implications for structured professional development, this study has implications in terms of the ways in which graduate programs make teaching assignments. First, while all TAs typically get plenty of practice grading homework and exams, only a select few have the rich experiences of observing students working problems or of writing and implementing new problems. This work suggests that without those types of experiences, TAs might be left without important knowledge of student thinking. Making sure that all graduate students have some experience observing students working problems might enhance graduate students' understanding of student thinking, and, if the lessons learned at the K-12 level apply to the collegiate level, improve the learning experiences of their calculus students. And if early teaching experiences are formative, as other research indicates, this might have positive effects on college teaching for future generations as well.

Finally, we see this research as part of a larger program to understand and find ways to build TA knowledge of student thinking – and make sure that knowledge is usable. In the setting of this larger program, the research reported here raises several important questions that could be answered by other studies:

- Can the self-reported findings here be confirmed by more detailed studies that follow graduate students through their early teaching experiences?

- What aspects of the knowledge reported here are known to TAs who did not have similar experiences (e.g., observing students working problems)?
- Exactly how do calculus teachers use their knowledge of student thinking in their teaching?
- Is it possible to simulate the activities of watching and listening to student work and writing problems in a manner appropriate for professional development? Do participants still gain knowledge about student thinking through these simulated activities?
- What knowledge of student thinking do veteran calculus teachers have? How do they use that knowledge in their own teaching?

We look forward to answering these questions and invite other researchers to help address them.

References

Ball, D. L. (1990). Prospective elementary and secondary teachers' understandings of division. *Journal for Research in Mathematics Education, 21*(2), 132–144.

Ball, D. L., & Bass, H. (2000). Interweaving content and pedagogy in teaching and learning to teach: Knowing and using mathematics. In J. Boaler (Ed.), *Multiple perspectives on the teaching and learning of mathematics* (pp. 83–104). Westport, CT: Ablex.

Ball, D. L., Lubienski, S. T., & Mewborn, D. S. (2001). Research on teaching mathematics: The unsolved problem of teachers' mathematical knowledge. In V. Richardson (Ed.), *Handbook of research on teaching* (4th ed., pp. 433–456). New York: Macmillan.

Begle, E. G. (1979). *Critical variables in mathematics education: Findings from a survey of the empirical literature.* Washington, DC: Mathematical Association of America and National Council of Teachers of Mathematics.

Belnap, J. K. (2005). *Putting TAs into context: Understanding the graduate mathematics teaching assistant.* Doctoral dissertation, University of Arizona, Dissertation Abstracts International, A 66/06, p. 2142.

Bezuidenhout, J. (2001). Limits and continuity: Some conceptions of first-year students. *International Journal of Mathematics Education in Science and Technology, 32*(4), 487-500.

Borko, H., & Putnam, R. (1996). Learning to teach. In D. Berliner & R. Calfee (Eds.), *Handbook of educational psychology* (pp. 673–708). New York: Macmillan.

Brown, C. A. (1985). *A study of the socialization to teaching of a beginning secondary mathematics teacher.* Doctoral dissertation, University of Georgia, Dissertation Abstracts International, A 46/09, p. 2605.

Carpenter, T. P., & Moser, J. (1984). The acquisition of addition and subtraction concepts in grades one through three. *Journal for Research in Mathematics Education, 15*, 179-202.

Clement, J. (2000). Analysis of clinical interviews: Foundations and model viability. In A. E. Kelly & R. A. Lesh (Eds.), *Handbook of research design in mathematics and science education* (pp. 547–589). Mahwah, NJ: Erlbaum.

Davis, R., & Vinner, S. (1986). The notion of limit: Some seemingly unavoidable misconception stages. *Journal of Mathematical Behavior, 5*, 281-303.

DeFranco, T. C., & McGivney-Burelle, J. (2001). The beliefs and instructional practices of mathematics teaching assistants participating in a mathematics pedagogy course. In R. Speiser, C. A. Maher, & C. N. Walter (Eds.), *Proceedings of the 23rd annual conference of the International Group for the Psychology of Mathematics Education – North America,* (pp. 681–690). Snowbird, Utah.

Fennema, E., Carpenter, T., Franke, M., Levi, L., Jacobs, V., & Empson, S. (1996). A longitudinal study of learning to use children's thinking in mathematics instruction. *Journal for Research in Mathematics Education, 27*, 403-434.

Fennema, E., Franke, M. L., & Carpenter, T. P. (1993). Using children's mathematical knowledge in instruction. *American Educational Research Journal, 30*(3), 555-583.

Franke, M. L., Fennema, E., & Carpenter, T. P. (1997). Teachers creating change: Examining evolving beliefs and classroom practice. In E. Fennema & B. S. Nelson (Eds.), *Mathematics teachers in transition* (pp. 255–282). Mahwah, NJ: Erlbaum.

Friedberg, S., Ash, A., Brown, E., Hallett, D. H., Kasman, R., Kenney, M., Mantini, L. A., McCallum, W., Teitelbaum, J., & Zia, L. (2001). *Teaching mathematics in colleges and universities: Case studies for today's classroom.* Providence, RI: American Mathematical Society. (Issues in Mathematics Education series, Vol. 10)

Fullilove, R., & Treisman, P. (1990). Mathematics achievement among African-American undergraduates at the University of California, Berkeley: An evaluation of the mathematics workshop program. *Journal of Negro Education, 59*, 463–478.

Goldin, G. (2000). A scientific perspective on structured, task-based interviews in mathematics education research. In A. E. Kelly & R. A. Lesh (Eds.), *Handbook of research design in mathematics and science education* (pp. 517–545). Mahway, NJ: Erlbaum.

Herzig, A. H. (2002). *Sowing seeds or pulling weeds?: Doctoral students entering and leaving mathematics.* Doctoral dissertation, University of Wisconsin – Madison, Dissertation Abstracts International, A 63/04, p. 1282.

Herzig, A. H., & Kung, D. T. (2003). A study of cooperative learning in calculus reform. In A. Selden, E. Dubinsky, G. Harel, & F. Hitt (Eds.), *Research in collegiate mathematics education. V* (pp. 30–55). Providence, RI: American Mathematical Society.

Hsu, E., Murphy, T. J., & Treisman, U. (2008). Supporting high achievement in introductory mathematics courses: What we have learned from 30 years of the Emerging Scholars Program. In M. P. Carlson & C. Rasmussen (Eds.), *Making the connection: Research and teaching in undergraduate mathematics education* (pp. 205–220). Washington, DC: Mathematical Association of America.

Kennedy, M. (1997). *Defining optimal knowledge for teaching science and mathematics.* Madison, WI: National Institute for Science Education.

Lacey, C. (1977). *The socialization of teachers.* London: Methuen.

Lakoff, G., & Núñez, R. E. (2000). *Where mathematics comes from: How the embodied mind brings mathematics into being.* New York: Basic Books.

Ma, L. (1999). *Knowing and teaching elementary mathematics: Teachers' understanding of fundamental mathematics in China and the United States.* Mahwah, NJ: Erlbaum.

Monk, D. (1994). Subject area preparation of secondary mathematics and science teachers and student achievement. *Economics of Education Review, 13*(2), 125–145.

National Science Foundation. (1992). *America's academic future: A report of the Presidential Young Investigator Colloquium on U.S. Engineering, Mathematics, and Science Education for the Year 2010 and Beyond* (Tech. Rep.). Arlington, VA: Directorate for Education and Human Resources, National Science Foundation.

Przenioslo, M. (2004). Images of the limit of function formed in the course of mathematical studies at the university. *Educational Studies in Mathematics, 55*(1), 103–132.

Schoenfeld, A. H. (1985). *Mathematical problem solving.* Berkeley, CA: Academic Press.

Seymour, E., Melton, G., Wiese, D. J., & Pedersen-Gallegos, L. (2005). *Partners in innovation: Teaching assistants in college science courses.* Boulder, CO: Rowman & Littlefield.

Shulman, L. S. (1986). Those who understand: Knowledge growth in teaching. *Educational Researcher, 15*(2), 4-14.

Smith, A., & Moore, L. (1990). Duke University: Project CALC. In T. W. Tucker (Ed.), *Priming the calculus pump: Innovations and resources* (pp. 51–74). Washington, DC: Mathematical Association of America.

Speer, N. M. (2001). *Connecting beliefs and teaching practices: A study of teaching assistants in collegiate reform calculus courses.* Doctoral dissertation, University of California, Berkeley, Dissertation Abstracts International, A 63/02, p. 533.

Speer, N. M., Gutmann, T., & Murphy, T. (2005). Mathematics teaching assistant preparation and development. *College Teaching, 53*(2), 75–80.

Spradley, J. P. (1980). *Participant observation.* New York: Rinehart and Winston.

Strauss, A. L., & Corbin, J. (1990). *Basics of qualitative research: Grounded theory procedures and techniques.* Newbury Park, CA: Sage.

Tall, D. O. (2001). Natural and formal infinities. *Educational Studies in Mathematics, 48*(2), 199–238.

Tall, D. O., & Vinner, S. (1983). Concept image and concept definition in mathematics, with special reference to limits and continuity. *Educational Studies in Mathematics, 12*, 151–169.

Treisman, P. (1985). *A study of the mathematics performance of Black students at the University of California, Berkeley.* Doctoral dissertation, University of California, Berkeley, Dissertation Abstracts International, A 47/05, p. 1641.

Williams, S. R. (1991). Models of limit held by college calculus students. *Journal for Research in Mathematics Education, 22*(3), 219–236.

Department of Mathematics and Computer Science, St. Mary's College of Maryland, St. Mary's City, Maryland 20686-3001

E-mail address: dtkung@smcm.edu

CBMS Issues in Mathematics Education
Volume **16**, 2010

An Examination of the Knowledge Base for Teaching Among Mathematics Faculty Teaching Calculus in Higher Education

Kimberly S. Sofronas and Thomas C. DeFranco

ABSTRACT. In 2001, the International Commission of Mathematics Instruction issued a report outlining the need for studies in mathematics education that examine the pedagogical knowledge of college and university mathematics faculty. This paper responds to that call for action and reports on a qualitative study designed to explore the knowledge base for teaching among 7 college and university mathematics faculty teaching calculus at 4-year institutions in the Northeastern United States. Data were collected through 3 extensive semi-structured individual interviews, observation field notes, and video-recorded segments of instruction. Data were analyzed through methods of categorical content analysis. A model of the knowledge base for teaching among the participants in the study was developed and included 5 main categories of knowledge: context-specific knowledge, knowledge of the nature of disciplinary mathematics, personal practical knowledge, general pedagogical knowledge, and knowledge of research. Within each category, themes across cases are reported. The knowledge base for teaching framework offered here provides a ground-level tool for conducting more in-depth research on this topic while the themes and patterns that emerged highlight the character of each category of knowledge important in interpreting instructional practices.

1. Introduction

The twenty-year national movement within the mathematics community to reform the teaching and learning of calculus has emphasized the applications of calculus and the need for students to gain a conceptual understanding of the subject. Initially, research conducted on calculus reform programs focused on the relationships between various instructional approaches – using technology, working cooperatively, writing – and student learning or attitudes toward mathematics (Darken, Wynegar, & Kuhn, 2000; Ganter & Jiroutek, 2000; Herzig & Kung, 2003; Robert & Speer, 2001). Since then, it has become clear that "the impact of the reform is perhaps not so dependent upon what is implemented, but rather the educational environment that is created in which to implement it" (Ganter, 1999, Implications of Findings, para. 1). Educational environment is a function of several factors, including the pedagogical knowledge of mathematics faculty. In shifting the focus to the learning environment in the calculus classroom, studies examining the knowledge base for teaching (KBT) among mathematics faculty in higher education

©2010 American Mathematical Society

become an essential complement to existing research on reformed and traditional programs.

Research on the KBT conducted at the K-12 level provides a solid foundation for examining related issues in higher education. Among the first to argue for the existence of a KBT, Shulman (1987) claimed teaching was more than "personal style, artful communication, [and] knowing some subject matter" (p. 5-6). His framework of the KBT outlined the specific content and character of the types of knowledge required for teaching: content knowledge, general pedagogical knowledge, curriculum knowledge, pedagogical content knowledge, knowledge of learners, knowledge of educational contexts, and knowledge of educational ends, purposes, and values. Many researchers have built upon Shulman's research by adding to or refining his framework of the KBT or by offering alternative frameworks (Abd-El-Khalick & BouJaoude, 1997; Ball, 1988; Borko & Putnam, 1996; Carter & Doyle, 1987; Clandinin, 1985; Cochran-Smith & Lytle, 1999; Connelly & Clandinin, 1985; Elbaz, 1983; Fennema & Franke, 1992; Grossman, 1990; LaBerge, Zollman, & Sons, 1997; Leinhardt & Smith, 1985; Rahilly & Saroyan, 1997; Sherin, Sherin, & Madanes, 2000).

Mathematics faculty in higher education vary considerably in their knowledge of pedagogy. It is well-documented that most have little or no formal training in pedagogy, though teaching frequently is a significant component of their professional responsibilities (Rahilly & Saroyan, 1997). Nevertheless, college and university mathematics faculty do develop a knowledge base upon which they draw in teaching graduate and undergraduate courses. Little is known about the content and character of that knowledge base. Selden and Selden (2001) have noted more needs to be learned about the knowledge for teaching held by mathematics faculty in higher education.

This paper summarizes the results of a study that examined what college and university mathematics faculty know about teaching. It offers an overview of the components of knowledge identified as critical to participants' KBT and highlights significant themes that emerged within each category of knowledge.

2. Research Methodology

This qualitative study was designed to examine the KBT among college and university mathematics faculty teaching calculus. Early in the study, pilot work was conducted to refine the procedures for collecting and analyzing the data. A review of relevant research literature allowed for the development of a preliminary KBT framework. The categories of that framework were then used to design interview questions which were field-tested through the pilot work. The preliminary KBT model was refined as new categories and subcategories of knowledge emerged from pilot data that could not be coded by preexisting categories. Interview questions (see Appendix A) were revised to reflect changes made to the preliminary KBT model.

Participants in the larger study were PhD mathematicians, five male and two female, teaching Calculus I, II, or III at four-year institutions in the Northeastern region of the United States. Participants' experience teaching college-level mathematics ranged from 15 to 41 years. Six participants had tenure. None had participated in any formal training in pedagogy. English was the native language for all.

Participants Jennifer, Ben, and Kevin (pseudonyms) were from Baccalaureate Liberal Arts Colleges emphasizing research. Participants Jeff, Ron, Austin, and Carrie (pseudonyms) were from Doctoral Granting Research I institutions. Participants were selected from a list of 30 PhD mathematicians from four-year colleges and universities in the Northeast generated from the American Mathematical Society's (AMS) and the Mathematical Association of America's (MAA) membership directories and from professional recommendations. On-line course schedule information made it possible to identify those teaching Calculus I, II, or III. Mathematicians were contacted until seven agreed to participate in the study.

Data were collected through three extensive, semi-structured, individual interviews, observation field notes, and video-recorded segments of instruction. The piloted preliminary KBT model and field-tested interview questions were used to gather information related to participants' knowledge and beliefs about the teaching and learning of mathematics within the higher education setting. All participants were asked the same questions. However, the course of the interviews was partly determined by participants' responses, with additional questions asked as needed to follow-up on or clarify comments made by interviewees. To collect data on classroom environment and teaching practices in their respective calculus courses, participants were each observed teaching three lessons, one of which was video-recorded.

Audio-recorded interviews were transcribed and analyzed through categorical-content analysis (Leiblich, 1998). According to Leiblich, categories represent the various themes or perspectives – in this case, types of knowledge – that provide a means for classifying the words, sentences or groups of sentences in a transcript. Categories can be predefined or emergent. In this study, a priori categories and subcategories from the preliminary KBT model were used to code the data. Each sentence or complete thought from the transcribed interview data was coded according to the category or categories of knowledge it represented and placed on an index card (see Appendix B). Data that could not be coded by the preexisting categories were reexamined to refine, add, or remove framework components. The extensiveness of the categories and subcategories, according to Leiblich, "...retain[s] the richness and variation of the text but require[s] meticulous sorting of the material" (p. 113). Coded data were sorted, read, and reread to identify the major recurrent ideas within each category and subcategory of knowledge for each participant. Emergent themes were entered into a text matrix (knowledge base second-level subcategory x participant) for the purpose of within and cross-case analyses of the data. The process yielded rich knowledge base profiles for each participant and also allowed for themes to emerge across cases. Cross-case themes were defined as those ideas expressed by three or more participants of the study. Two doctoral students in mathematics education were trained in the methods for data analysis to independently code and sort sections of the data to establish trustworthiness.

3. Results and Discussion

Many frameworks of teacher knowledge base exist at the K-12 level; however, frameworks still need to be developed to examine the KBT among mathematics faculty in higher education (LaBerge et al., 1997; Rahilly & Saroyan, 1997; Selden & Selden, 2001). Through the analysis of the data collected in this study and a review of findings from relevant research literature, a model of the KBT among college and university mathematics faculty teaching calculus was developed. The five primary categories of knowledge and the first- and second-level subcategories identified in the model are depicted in Table 1. A discussion of the themes and patterns that emerged within each category of knowledge follows. As with any model of teacher knowledge, "...all knowledge is highly interrelated, [and since] the categories of teacher knowledge within a particular system are not discrete entities, [the] boundaries between them are necessarily blurred" (Borko & Putnam, 1996, p. 675). While discussion is organized around the primary and first-level subcategories, some indirect discussion of second-level subcategories is apparent.

3.1. Context Specific Knowledge. Mathematics faculty at the college and university level function within a cultural context comprised of ideologies, values, and norms that guide their professional behavior (Beyer, 1996). New mathematicians tend to adopt the values and norms of the particular mathematics community to which they belong (Goroff, 1999; Schoenfeld, 1989). Views of teaching and learning are shaped by culture. The context in which learning takes place is directly linked to most knowledge development, which underscores the importance of establishing a sense of community to build trust and cooperation in undergraduate mathematics classrooms (Millet, 2001; Schoenfeld, 1989; Seeger, Voigt, & Waschescio, 1998; Smith, 2001).

Examining what college and university mathematics faculty know about classroom culture and their undergraduate students is a first step in identifying the characteristics of mathematics classroom environments "...in which the 'right' kinds of interactions take place, so that students develop the right sense of what mathematics is all about, as well as mastering the formal mathematics they need to know" (Schoenfeld, 1989, p. 83). Changes in student demographics have resulted in increasingly diverse student characteristics and learning needs. The influx of client students – students majoring in other fields, such as psychology, required to complete courses in mathematics – has brought attention to a mathematically under-prepared undergraduate population whose instructional needs cannot go unattended (Hillel, 2001; Saunders & Bauer, 1998). In part, context specific knowledge shapes the pedagogical practices of faculty.

As outlined in Table 1, the following first-level subcategories are subsumed under context specific knowledge: knowledge of the institutional cultures and norms and the mathematics department setting, knowledge of the classroom culture, and knowledge of undergraduate students. Cross-case analyses of the data revealed themes within each of these first-level subcategories that are italicized and followed by supporting data. Quotations from participants were purposefully selected to offer evidence for the themes, defined as ideas expressed by at least three participants, which emerged across cases.

Institutional Cultures and Norms and Mathematics Department Setting

3.1.1. *In both liberal arts and doctoral-granting institutions, participants were generally satisfied with the academic freedom they were afforded and the support they received to conduct their research. Satisfaction translated into positive work environments at both institutional and departmental levels. Implications for pedagogy included reasonable teaching loads and significant faculty control in the classroom.* Academic freedom has been defined as the rights of faculty to free inquiry, unconstrained communication in the classroom, decisions on the hiring and promotion of colleagues, and collective self-governance (Slaughter, 1996). At the micro level,

TABLE 1. Model of the KBT among College and University Mathematicians Faculty Teaching Calculus

Categories	First-Level Subcategories	*Second-Level Subcategories*
I. Context specific knowledge		
	A. Institutional cultures and norms and mathematics department setting	
		1. *Institutional mission*
		2. *Expectations of faculty*
		3. *Opportunities for faculty*
		4. *Systems of reward and recognition*
		5. *Campus environment*
		6. *Composition of mathematics department faculty*
		7. *Faculty interactions, collaboration, and behaviors*
		8. *Department guidelines*
		9. *Curriculum development and reevaluation*
		10. *Roles and responsibilities*
		11. *Evaluation of faculty*
	B. Classroom culture	1. *Universal meanings*
		2. *Mathematics learning environment*
	C. Undergraduate students	1. *Students' backgrounds, interests, strengths, weaknesses, beliefs and work ethic*
		2. *Strategies for getting to know students*
II. Nature of disciplinary mathematical knowledge		
	A. Subject matter knowledge	
		1. *Content*
		2. *Structures*
		3. *History of mathematics*
		4. *Aesthetics of mathematics*
	B. Pedagogical content knowledge	
		1. *Purpose of the discipline*
		2. *Learners' cognitions*
		3. *Calculus curriculum*
		4. *Instructional strategies*
		5. *Representation*
		6. *Mathematical discourse*
III. Personal practical knowledge		
	A. Images of teaching	1. *Dimensions of teaching images*
		2. *Practical principles*
		3. *Rules of practice*
IV. General pedagogical knowledge		
	A. Learners and learning	1. *Theories of learning*
		2. *Meaningful and rote learning*
		3. *Conceptions of learning*
	B. Principles of instruction	1. *Planning*
		2. *Classroom routines*
		3. *Assessment and reflection*
V. Knowledge of research		
	A. Relationships between research and teaching	
	B. Knowledge of research in mathematics education	

participants of this study varyingly defined academic freedom to mean selecting their own calculus textbook, designing their own syllabus, choosing a particular instructional technique or, as in Ben's case, being "trusted" in the classroom by his department. In Ben's department, senior faculty did not observe junior faculty and, according to Ben, held the belief that "snapshots" often did not accurately portray the instructional practices of faculty.

While participants agreed that imposing heavy guidelines in higher education contexts where faculty genuinely care about teaching takes away freedom and prevents innovation, more curricular and pedagogical guidelines were imposed at larger institutions to allow for continuity across multiple course sections. Ron related, "[We follow] a very specific syllabus. This is a long list of topics out of a prescribed textbook. We all agree that we will use the same textbook and we will try to cover a body of material that's listed in all those sections. We will assign the very problems that are set forth on that syllabus. So, that's more than guidelines, that's almost marching orders, right?" (Q6, I-1).

3.1.2. *All participants reported institutional or departmental expectations of faculty that included research, teaching, and service. All believed accomplishments in research and teaching were recognized by their institutions, though three participants believed research was promoted more heavily than teaching and service work received little, if any, recognition.* Jeff noted disproportionate emphasis on and rewarding of accomplishments in research. "Research universities reward research the most. Do I think that's reasonable? Yes. Do I think they should reward good teaching? Yes. Do I think they should reward it at the same level? No. Honestly,...I think it's easier to be a reasonably good teacher than I think it is to do reasonably good research in most fields" (Jeff, Q3, I-1). Jeff's view echoes the position of many mathematicians which, according to (Bass, 1997), "...implicitly demeans the importance and substance of pedagogy" (p. 20).

Participants also distinguished between official and unofficial administrative positions on criteria for reward and promotion at their respective institutions. Kevin stated, "Officially, [research and teaching] count the same. In my take, it is clearly teaching that is first among the two. That's not where I place it though. I'm more interested in research, actually. I don't think of myself as a teacher. I'm a mathematician" (Q3, I-1).

Austin agreed official and unofficial positions exist but believed teaching was less weighted at his institution. He expressed, "...the university will not recognize this... they pay a lot of attention to [teaching] when they hire you, but once you are hired I don't think...that [teaching] makes a real big difference in how much money one makes and things like that" (Q3, I-1). Austin's views align with research literature and policy reports showing rewards for promotion and tenure are often more strongly tied to accomplishments in research than accomplishments in teaching or service work (Antonio, Astin, & Cress, 2000; Frost & Teodorescu, 2001; National Research Council, 2003; National Science Foundation, 1996). Pressure to demonstrate sufficient research productivity has led many faculty in higher education to view research and teaching as conflicting activities, often to the neglect of teaching and service commitments (Barnett, 1996; Colbeck, 1998; Fairweather, 1996; Frost & Teodorescu, 2001; Johnston, 1997; Kline, 1977; Knapper, 1997; Michalak & Friedrich, 1996; Weimer, 1997).

3.1.3. *Participants did not believe they had ever succumbed to the "either-research-or-teaching" polarity that exists among some faculty in higher education. Moreover, six participants who had been granted tenure reported their professional activities had changed little as a result.* Participants' curriculum vitæ revealed they were generally well-published. Observations of their teaching suggested all were deeply invested in the teaching and learning of their students. This finding is inconsistent with the research of Fairweather (1996), who identified the "exchange relationship," a negative correlation between time faculty spend on teaching and on research activities: the more time spent on one activity, the less on the other.

Ron did not believe tenure affected his commitment to research and teaching, but he acknowledged the kind of mathematical problems he examined in his research changed significantly following his tenure appointment: "If you were subject to the same kind of review process that a person has when they come up for tenure, you would probably be very concerned not to tackle really hard problems where you were not assured of publishable results within, say, six months to a year...[Tenure] frees you to really focus your attention on something...that won't maybe yield for some period of time, but this is how big progress is made" (Q4, I-1).

Classroom Culture

3.1.4. *Participants claimed to value the ideas, thought processes, and learning of their students.* Jennifer was clear about her goal to give students a voice in the classroom, "The more actively [students] can be engaged in [mathematics], the more learning goes on. Even when I am lecturing, [students] are interrupting me and asking me questions and I see that as a very positive thing. That is the strategy – it is to get them to say as much as they possibly can" (Q23, I-2). Classroom culture, or the universal meanings that teachers and pupils bring to the learning environment, governs the interactions between and among students and teachers. In most settings, learning is a communal activity, a sharing of culture. In the mathematics classroom, the environment nurtured by the teacher constitutes the context in which the processes of sense-making occur among students (Nickson, 1992; Seeger et al., 1998).

3.1.5. *Observations of participants' classrooms revealed largely teacher-directed learning environments. Although participants were receptive to questions from students, communication in their classrooms was predominantly one-way, from professor to student.* Jeff, for example, described the communication in his own calculus classroom as uni-directional. Kevin viewed student-to-student communication as something that occurred outside of the calculus classroom. This is somewhat inconsistent with participants' professed goal to encourage active participation among students during instruction.

According to Zevenbergen (2001), the formal and abstract nature of mathematics in higher education has implications for patterns of interaction in the undergraduate mathematics classroom:

> In formal teaching contexts, including the formal lecture situation as well as the less formal tutorial situation, there are particular unspoken rules about who speaks to whom, when and how. Where the lecturer stands in front of a large group of students and works through mathematical problems or examples, there tends to be very little interaction (p. 20).

Literature and policy documents at the K-12 level have suggested students do not intuitively know how to talk about mathematics and tend to remain quiet in the mathematics classroom without encouragement from the teacher. Consequently, teachers need to help children learn to communicate their ideas about concrete and theoretical aspects of mathematics (Cobb, Wood, & Yackel, 1993; National Council of Teachers of Mathematics, 2000). It is possible that undergraduate mathematics students require similar kinds of support to engage in mathematical discourse.

The efficacy of mathematical discourse as a pedagogical tool for students to share ideas, identify misconceptions, and clarify understanding is well documented (Brendefur & Frykholm, 2000; Cazden, 1988; Pirie & Schwarzenberger, 1988; Wells, 1994; Wertsch & Toma, 1995). In mathematical discourse, students have something to say beyond the ideas and answers already known by the teacher (Moll, 1990). Voicing their thinking in class offers students opportunities to analyze and critique their mathematical work, learn effective communication, and bring their own background, personality, and beliefs into the construction of mathematical knowledge (Hiebert & Wearne, 1993; Leonard, 1999; National Council of Teachers of Mathematics, 2000).

> Teachers need to listen as much as they need to speak. They need to resist the temptation to control classroom ideas so that students can gain a sense of ownership over what they are learning. Doing this requires genuine give-and-take in the mathematics classroom, both among students and between students and teachers. The best way to develop effective logical thinking is to encourage open discussion and honest criticism of ideas... Honest questions by teachers are rare in mathematics classrooms. Most teachers ask rhetorical questions because they are not so much interested in what students really think as in whether they know the right answer... Classroom activities must encourage students to express their approaches, both orally and in writing (National Research Council, 1989, pp.59-61).

Ron imagined a mathematical learning environment for his calculus classes that mirrored the experience of the research mathematician, but had not found effective ways for making his vision a reality:

> One of my ideas that has never really quite made it is to try and turn the classroom experience into something more like the research experience... The normal situation [for the research mathematician] is that you work on a problem and you will get stuck at some point... usually [you] go to a colleague... and bounce some ideas around with that person... I've picked my sabbatical sites with the idea of working with people who are in the same subfield [as me] so that we could really get into a conversation that is ongoing... In the classroom, by contrast, ... there is a reporter in the front of the room reporting [on] the final polished outcome of [some]thing from three centuries ago... Most instruction aims to package the idea as attractively as possible..." (Q25, I-2).

Ron's sentiments echo comments by Selden, Selden, Hauk, and Mason (2000) in their study on non-routine problem-solving, which called for changes in the classroom culture of undergraduate calculus courses: "Students often expect to be told precisely how to work problems. Thus, a change in the prevalent classroom culture that prefers tedium over struggle and reflection might be required" (p. 150).

Kazemi (1998) made the distinction between social norms and sociomathematical norms that exist within the culture of a mathematics classroom. Social norms transcend subject matter and include practices such as encouraging students to explain their thinking, share strategies, or collaborate with peers. For example, Jennifer encouraged her students to make mathematical conjectures in class. While examining parametric equations of the form $x = \pm 2\cos t$ and $y = \pm 2\sin t$, a student in Jennifer's calculus class formulated a conjecture about the direction of the path of the parameterized curve: "If one of them is negative and the other positive, will it automatically go clockwise; and if both of them are negative or both positive, will it automatically go counterclockwise?" Kazemi (1998), however, maintained social norms may not be sufficient to advance students' conceptual thinking in mathematics. Sociomathematical norms "identify what kind of talk is valued in the classroom, what counts as a mathematical explanation, and what counts as a mathematically different strategy" (Kazemi, 1998, p. 411). In essence, sociomathematical norms guide the quality of the mathematical discourse in the classroom. Yackel, Rasmussen, and King (2000) studied social norms and sociomathematical norms at the college level and found students in an introductory differential equations course furthered their mathematical development through classroom discussions that emphasized making sense of the reasoning of their peers.

3.1.6. *Three participants used humor in their teaching to create a light-hearted, comfortable mathematics learning environment in their classrooms.* Carrie emphasized, "It helps to share stories, anecdotes, and jokes...[students] like that. They want to be entertained in some way when they are learning. I think that relaxes them and opens them more to the learning itself" (Q29, I-2). This finding supports the growing body of literature related to the positive effects of using humor as a pedagogical tool in the college and university classroom (Bartlett, 2003; Garner, 2006; Torok, McMorris, & Lin, 2004). "Humor can help an individual engage in the learning process by creating a positive emotional and social environment in which defenses are lowered and students are better able to focus and attend to the information being presented" (Glenn, as cited in Garner, 2006, p. 177). Results of a study conducted by Schacht and Stewart (1990) supported the use of humor as a powerful tool in reducing mathematical anxiety among students enrolled in an introductory undergraduate statistics course. The appropriate use of humor has also been shown to increase student interest in the subject matter, understanding of concepts, retention of content, attention in class, and perceptions of instructor credibility (Garner, 2006; Schacht & Stewart, 1990; Torok et al., 2004).

Kevin's engaging teaching delivery included running up and down the aisles of the lecture hall, jumping up on desks, and varying the intonations in his voice in ways students found entertaining. He reflected that his class "... doesn't have the appearance of having a tight structure... there are lots of jokes, not that I plan those, but it's just kind of like you talk and people fall asleep, so you've got to wake them up" (Q38, I-2). The use of humor in the teaching of introductory mathematics courses is especially encouraged, given students' inclination to perceive these courses as dry or difficult (Garner, 2006; Torok et al., 2004). Thus, college and university mathematics faculty may benefit from expanding their theoretical and practical knowledge base related to the use of instructionally-appropriate humor in the teaching and learning of undergraduate students.

Undergraduate Students

3.1.7. *Most participants expressed a desire to know more about their students but acknowledged knowing little about them. Learning students' names, beginning courses with verbal introductions, and collecting written biographies or personal data sheets were some of the strategies faculty used to learn about their students. Participants were most familiar with students who attended their office hours.* Making informed pedagogical decisions requires faculty in undergraduate mathematics classrooms to be aware of the needs of their students in order to adapt their instruction to meet those needs. A one-size-fits-all approach to college-level mathematics instruction is unlikely to achieve the best result, particularly in light of changing undergraduate demographics. Tucker (1999) articulated a need to bridge the gap between the knowledge and experiences of college students and mathematics faculty. Many students who experienced differentiated instruction at the K-12 level encounter what Kirst (2004) has termed the "high school-college disconnect," to which he attributes the significant percentage of college freshman who fail to complete a bachelor's degree. Further, large class sizes and course pacing, both difficult for many first-year college students, contribute to the disconnect between high school and college contexts for learning mathematics (Chronicle, 2003; Hillel, 2001; Saunders & Bauer, 1998; Wood, 2001; Zevenbergen, 2001). Austin found his large class size of approximately 130 students made it difficult for him to get to know them in any meaningful way. Other participants' class sizes were smaller: Kevin ~60 students; Carrie ~45; Jeff and Ben ~30; Ron ~22; Jennifer ~15.

3.2. Nature of Disciplinary Mathematical Knowledge. Knowledge of the nature of disciplinary mathematics includes both subject matter knowledge (SMK) and pedagogical content knowledge (PCK). SMK is composed of knowledge of the content, structures, and nature of mathematics: the facts, concepts, and procedures within mathematics, the organization of the discipline, the questions that guide further inquiry, the ways in which truth is determined in mathematics, and the certainty and source of mathematical knowledge (Ball, 1991; Gfeller, 1999; Shulman, 1986; Wilson, Shulman, & Richert, 1987). Shulman (1986) defined PCK as "the most useful forms of representation of [mathematical] ideas, the most powerful analogies, illustrations, examples, explanations, and demonstrations - in a word, the ways of representing and formulating [mathematics] that make it comprehensible to others" (p. 9). PCK includes an understanding of the relevance of mathematics to everyday life, the value and societal impact of mathematics, and the ways mathematical understanding develops in students (Abd-El-Khalick & BouJaoude, 1997). The following were emergent themes within the two subcategories of the nature of disciplinary mathematical knowledge.

Subject Matter Knowledge

3.2.1. *Participants' view of the nature of mathematics was predominantly one of Formalism.* The origins of mathematical ideas have intrigued scholars for centuries (Farrell & Farmer, 1988; Gfeller, 1999; Kulikowich & DeFranco, 2003). A number of viewpoints have attempted to define the nature of mathematics, ranging from a static discipline developed abstractly through axiomatic structures to a dynamic and constantly changing field leading to new discoveries through experimentation and a heuristical approach to solving problems. Within that continuum, the viewpoint of Formalism holds that mathematical knowledge consists of immutable truths "waiting out there" to be discovered.

There was significant agreement among participants that mathematical theorems are discovered, as opposed to created or invented. Ron stated,

> It takes an incredible amount of mental work to understand and discover the patterns that are there, but those patterns are so pervasive in so many different aspects of life that you have to say there is some objective reality there - it is not just us creating it! (Q13, I-1)

However, participants also recognized that the mathematical notation and terminology used to express ideas are human creations.

Jennifer professed to believe mathematics is primarily discovered, but not as strongly as other participants. She remarked,

> I am not a Platonist, but my thesis advisor used to say, "You sort of act like you are a Platonist, even if you are not. You act as if [mathematics] is there to be discovered, even though some of it is invention." There is a famous quote that God created the integers, and all the rest is the invention of man. I think the connections, the pattern recognition, the seeing how something works goes beyond just discovery in a laboratory sense (Q13, I-1).

In speaking to her class about calculus, Jennifer used the word "invented" (see Appendix C, lines 1, 10, and 21).

3.2.2. *Data collected from interviews with participants supported inclusion of knowledge of the history of mathematics as a subcategory of subject matter knowledge for the KBT model. Though not mathematics historians, most participants held knowledge of the timeline of events and discoveries in mathematics and the contributions of various mathematicians throughout history, which they drew upon in their teaching.* Austin, who characterized himself as more of a humanities type than a science type, used the history of mathematics to frame his entire calculus course. He stated,

> I also enjoy talking about history and philosophy... The history of calculus and the philosophies that accompanied the development of calculus ... these are the things that I find very interesting. I guess my view on mathematics, in general, [is that]... most students are more interested in the human side of the subject, which comes from history and philosophy" (Q9, I-1).

Support has increased for the integration of mathematical history into the teaching and learning of mathematics at all levels (Furinghetti, 2000). Michel-Pajus (2000) observed, "the cultural content of mathematics should not be simply sacrificed to its technical aspects" (p. 17). Siu (2000) identified four levels of the use of the history of mathematics in the classroom, which included anecdotes, broad outlines, content, and the development of mathematical ideas. An excerpt of a dialogue from Jennifer's Calculus I course (see Appendix C) demonstrates how she drew upon her knowledge of the history of mathematics in discussing the origins of calculus and derivative notations.

3.2.3. *Data collected from interviews with participants supported inclusion of knowledge of the aesthetics of mathematics as a subcategory of subject matter knowledge for the KBT model. Participants were optimistic that their love of and excitement about mathematics would result in an improved student opinion of the discipline. They perceived a connection between their mathematical research and their own appreciation for the beauty of mathematics.* Schoenfeld (1989) characterized the mathematician's aesthetic as "... a predilection to analyze and understand, to perceive structure and structural relationships, to see how things fit together" (p. 87). His view is clearly reflected in Jeff's comment:

> People always ask me what I like about doing math and what is the point of doing pure math research and I always say it's art for art's sake... You do it because it is a beautiful game you get to play. You get to invent stuff, you get to discover stuff. You get to prove things that nobody else on the planet knows... It is possible that what I do will be useful at some point, but I don't care. I do it as art for art's sake and that is really how I view it. (Q12, I-1)

The development of the mathematician's aesthetic is part of the enculturation process for apprentice mathematicians (Schoenfeld, 1989).

Pedagogical Content Knowledge

3.2.4. *Six participants relied heavily on lecture to deliver content.* The viewpoint of Formalism has implications for classroom norms and instructional practices, including an overemphasis on mathematical procedures, a polished and linear presentation of the discipline delivered primarily through lecture, and a de-emphasis on mathematics as a way of interpreting experience or as a human activity (Nickson, 1992). Observations of participants' teaching substantiated a heavy reliance on lecture, but not necessarily a de-emphasis on mathematics as a way of interpreting experience or as a human activity.

An analysis of video transcripts of a typical segment of Jeff's teaching revealed his students spoke in only 8 of 143 lines. Jeff characterized his lectures as monologues rather than dialogues and was explicit about his practice to not ask students any questions in class. Likewise, Ron lectured in Calculus II to move through the material at the pace demanded by the syllabus. He valued active student learning enough to allot one of four weekly classes to group problem solving. Carving out time for group work required him to cover larger amounts of material in his lectures. His resulting instructional approach was a ratio of one day of dynamic, interactive problem solving to three days of traditional lecture, with little or no opportunity for student participation or questions. By contrast, Jennifer did not employ traditional teaching approaches in her Calculus I course. She regularly engaged students in collaborative problem solving activities and interactive classroom discussions, knowing that many would have difficulty sitting through long lectures. Alsina (2001) has recommended greater implementation of interactive methods of teaching in undergraduate mathematics education despite the fact the discipline "is still dominated by the 'chalk-and-talk' paradigm" (Hillel, 2001, p. 64).

3.2.5. *Four participants shared the goal of getting at the root of students' confusion in mathematics to facilitate their learning, but believed students often have difficulty articulating their thinking and do not think deeply enough about the mathematical ideas during class lectures.* Studies conducted at the K-12 level support the notion that teachers' knowledge of learners' cognitions can influence classroom learning (Fennema & Franke, 1992). Experienced teachers know what their students know and are capable of learning, which topics they are likely to find difficult, and the function of mathematics in the everyday lives of students (Abd-El-Khalick & BouJaoude, 1997; Borko & Putnam, 1996; Shulman, 1986). Jeff wanted his students to actively think in class. In developing his lectures, he drew upon his knowledge of learners' cognitions to address some of the questions he anticipated from students. In preparing a lecture on intervals of convergence of power series, Jeff was aware that students often have difficulty with problems in which the radius of convergence is finite and not zero and therefore made a point to highlight this type of problem in the lesson.

Questioning the extent to which learning occurs in a lecture-based environment, Jennifer cultivated a dynamic classroom learning environment in which students had opportunities to solve problems, articulate their ideas verbally and in writing, and interact with their peers. Jennifer relied on student participation in classroom discourse to identify their mathematical conceptions and to make instructional adjustments that targeted students' conceptual and procedural difficulties. An excerpt (see Appendix D) from a problem-based lesson on related rates exemplifies how Jennifer listened to students' mathematical conversations to understand their thought processes and provided scaffolding at pivotal points in their thinking (see Appendix D, lines 39 - 44, line 46).

How other participants were able to uncover students' thought processes in class remains unclear, given the limited student participation observed during lectures. According to Wagner, Speer, and Rossa (2006), who conducted a case study on the role of teacher knowledge in the implementation of a reform-oriented differential equations course, the acquisition of PCK among university mathematicians is largely a function of the kinds of instructional experiences they foster. "What we learn about students through instruction is shaped by the nature of that instruction, so the pedagogical content knowledge supporting traditional instructional methods may not be the same as that which best supports reform-minded instructional practices" (Sherin, as cited in Wagner et al., 2006, p. 7). Non-traditional instructional methods such as collaborative activities, student-student discussions, and student-teacher discussions allow mathematics faculty to gain access to student thinking that is qualitatively different from student thinking generated in traditional lecture environments. Faculty expectations and questioning can constrain the kinds of knowledge students reveal in traditional settings (Wagner et al., 2006).

3.2.6. *Four participants expressed beliefs that students learn calculus by doing their homework, solving problems, and thinking deeply about mathematics. Three participants believed students often mistakenly think they can learn to solve calculus problems by watching the instructor.* "Mathematics isn't that superficial," Ron emphasized as he discussed students' conceptions about learning calculus,

> It's not like history, where, if you remember what year the civil war started, you are going to get that question right on an exam.

> You can't really memorize... You only get [it] by struggling with lots and lots of problems. (Q20, I-1)

According to Ferrini-Mundy and Graham (1991), many calculus textbooks provide homework sets that are nearly identical to the examples worked out in the text and thereby reinforce students' conceptions about mimicking as a means for learning calculus. With respect to the findings of this study, the teacher-directed learning environments may have contributed to or strengthened students' conceptions about learning mathematics. Moreover, it is unclear how participants were counteracting these conceptions among students.

3.2.7. *Despite participants' beliefs in the applicability of mathematics, only four reported even alluding to applications of calculus in their teaching. Participants found most applications to be too artificial, too messy, or in need of too much background knowledge.* Commenting on the contrived nature of application problems in calculus textbooks, Ron recalled the example of a problem in which its author stated a colony of bacteria had one hundred cells. He observed,

> My wife, who is a biochemist, would really like to know what kind of breakthrough discovery they have made in Canada, where they can count the number of cells in a colony of bacteria, because if we could get our hands on that, we could probably cure cancer. (Q27, I-2)

Similarly, Jeff responded,

> Do I really know what applications they'll use those for in their later engineering courses and do we address those? No. First of all, I don't know just because I don't know. Even if I did know, we wouldn't do them to any serious extent [because] the problems that those address are typically so complicated that we don't want to address them in a calculus class. (Q27, I-2)

Calculus reform initiatives have supported the inclusion of more applications in the teaching and learning of calculus. However, some disagree with the position that mathematicians must motivate and justify everything they teach, "Applications involve other disciplines. Don't let them browbeat us into teaching every subject but our own! We have calculus for engineers, calculus for biological sciences, calculus for business – why can't we have calculus for mathematicians?" (Kleinfeld, 1996, p. 230). A de-emphasis on the applications of mathematics might explain Hillel's (2001) finding that students are increasingly not convinced of the relevance of pure mathematics to potential careers. Many students have turned to fields like computer science, engineering, and finance where job opportunities are more plentiful and the relationships between those fields and academic studies are more explicit.

3.2.8. *Four participants believed that teaching methodologies continue to change, but the calculus curriculum should not. A significant goal was to communicate the "spirit" or big ideas of the calculus to students.* Curricular knowledge is central to the development of solid PCK and includes knowledge of the scope and sequence of the subject along with the various materials available for teaching the content. Faculty must be aware of topics typically presented in the calculus curriculum, how those topics might be organized for presentation, and the significance of those topics to the mathematical concepts students have studied in the past and are likely to study in the future (Grossman, 1990).

Mathematicians tend to have strong opinions about the reform or redevelopment of the calculus curriculum. Some have called for a leaner, more modern calculus curriculum, while others continue to favor the conventional calculus developed in the 17th and 18th centuries (Ferrini-Mundy & Graham, 1991; Kaput, 1997; Knisley, 1997). Participants in this study generally identified with the latter position. Ron, on the contrary, argued for mathematicians to agree to let go of certain topics in the calculus curriculum:

> I think mathematicians are going to come to understand that we are not just serving out an almost limitless supply of little factoids, but rather we are trying to promote some overall mastery of calculus as a tool... to bring to bear on real life problems. I think we are going to get more depth and less breadth, but right now we are not there. (Q23, I-2)

Questions of "what mathematics content is central and what is redundant, as well as, how present-day learning, teaching and assessment practice, can be and ought to be changed" (Hillel, 2001, p. 63) have arisen, particularly in light of the growing diversity among students in undergraduate mathematics classrooms. According to Hillel, new discoveries in mathematics and the development of new technologies should result in changes to courses that bring high visibility to some topics of study, while marginalizing others. As a result, "strong [mathematics] departments find that they replace or change significantly half of their courses approximately once a decade" (p. 61).

3.3. Personal Practical Knowledge. Personal practical knowledge is the experiential knowledge of teachers. It forms the basis for identifying teachers' images of teaching, which can help to explain the ways in which teachers draw upon past experiences in their instructional practice (Clandinin, 1986). Noting personal practical knowledge has the most significant bearing on problems central to education, Cochran-Smith and Lytle (1999) discussed the notion of "knowledge-in-practice," a view of learning in which teacher knowledge is embedded in the artistry of their practice.

> Often we cannot say what we know. When we try to describe it, we find ourselves at a loss, or we produce descriptions that are obviously inappropriate. Our knowledge is ordinarily tacit, implicit in our patterns of action and in our feel for the stuff with which we are dealing. It seems right to say that our knowledge is in our action. (Schon, as cited in Cochran-Smith & Lytle, 1999, p. 263)

Kung, Speer, and Gucler (2006) conducted a study that examined the processes by which graduate mathematics teaching assistants and novice PhD mathematicians acquire knowledge about students' mathematical thinking. Findings supported the hypothesis that "learning while teaching" represents a main method for gaining knowledge about student thinking in mathematics, especially in the absence of relevant department-supported professional development. However, Alsina (2001) cautioned too much emphasis has been placed on the experiential knowledge of undergraduate mathematics faculty. A common conception within college and university departments of mathematics is that excellence in undergraduate teaching is "just a matter of accumulated experience, clear presentation skills and a sound knowledge of the subject" (Alsina, 2001, p. 4). However, in the absence of formal training in pedagogy, many faculty develop collections of tips, tricks, and gimmicks not grounded in theories of learning to motivate students and facilitate success (Weimer, 1997).

Images of Teaching

3.3.1. *Images of teaching were unique to each participant.* Images of teaching play a central role in the thought processes that teachers have as they interact with their environment. Grounded in personal experiences, images of teaching are particular to each individual and, according to Clandinin (1986, 1986b), are shaped by a number of dimensions including, but not limited to, the moral dimension, the emotional dimension, and the private and professional dimension. "I see [teaching] as coaching more than anything because that way [the students and I] are on the same side" (Q30, I-2) says Carrie, whose image of teaching stemmed from a personal struggle with mathematics in graduate school. She recalled,

> I was out of school for two years before I went to graduate school and I had been a double major [as an undergraduate], so I had the minimum number of math classes. When I got to graduate school, it was almost impossible. It was very hard for me to get back into mathematics. I think I empathize with students who are just having a really hard time grasping the concept... I understand that because I have been in a place where I was just grinding my gears for a while. (Q29, I-2)

Embedded in Carrie's teaching-as-coaching image is a strong emotional dimension. The emotional experiences that contributed to her teaching image are evident in her characterization of graduate school and also translate into her feelings of empathy and understanding toward struggling students in her calculus course.

3.3.2. *Several participants reported negative experiences studying mathematics in their K-12 schooling.* Ron recalled,

> In high school, the most frustrating of all for me was mathematics ... a chemistry teacher filling in teaching trigonometry told me, "You have no mathematical ability whatsoever! You had better stay out of any math courses in college or you'll flunk out!" (Q31, I-2)

The idea that PhD mathematicians would have had negative experiences with mathematics at the K-12 level, while surprising on the surface, provides a possible rationale for negative attitudes toward mathematics education.

Kevin acknowledged that he did not keep abreast of current initiatives in mathematics education. He stated,

> I probably have strong biases against [mathematics education]... it comes out of my own educational experience. I didn't like school. I hated school until college, where no one who was ever

> taught how to teach taught me. In grade school, I was taught a lot by people who were taught to teach. (Q43, I-2)

3.3.3. *Those same participants reported a watershed experience that profoundly influenced their pursuit of mathematics as a field of study.* Austin said he had never been able to learn from anyone in a classroom setting. An average student, he consistently failed to complete assignments, though he was recognized as talented in mathematics. "I couldn't do what I was told to do," said Austin,

> I was very independent... I went to a program at [—] State University for eight weeks after my sophomore year [in high school] ... There was a one-hour lecture in the morning and then we spent all of our time working on the problems independently ... I have never learned so much so fast in my life as I did in that one eight-week experience where, basically, I was given interesting things to do with help to do it and I did. (Q31, I-2)

This experience impacted his philosophy of teaching and his ideas about student learning.

3.4. General Pedagogical Knowledge. General pedagogical knowledge has been defined as that which "encompasses a teacher's knowledge and beliefs about teaching, learning, and learners that transcend particular subject matter domains" (Borko & Putnam, 1996, p. 675). It includes a general body of knowledge and beliefs related to learners and learning as well as general principles of instruction (Grossman, 1990).

Learners and Learning

3.4.1. *Participants' knowledge of general pedagogy was grounded in their experiences and their own styles of learning. In the absence of any formal knowledge of learning theory, participants developed implicit "self-created" theories of student learning which influenced their teaching practices.* Austin, for example, acknowledged his ideas about student learning reflect his own ways of learning mathematics: "What influenced me is how I learned. It wasn't from teachers, it was from people who prepared good experiences for me - that [has held true], by the way, throughout graduate school and, for that matter, for the rest of my life" (Austin, Q31, I-2). Austin believed students need to grapple with ideas to develop their own understanding of them – a view aligned with constructivist theories of learning. "I won't put ideas into the students' heads. That is not my job. My job is to create experiences for them so that they will come to learn these things" (Austin, Q25, I-2). Rejecting transmission or absorption theories of learning, constructivism holds that most human knowledge is actively created and that students' interpretations of ideas are shaped by their experiences and social interactions (Clements & Battista, 1990; Resnick & Ford, 1981).

Jeff believed students learned mathematics, in part, by listening to ideas and thinking deeply about them. Not a proponent of note-taking, he observed, "The best students in my classes typically don't take any notes at all. I never took any notes... You know, you try and get it as it is being said and that's certainly the easiest way to learn it" (Q34, I-2). He planned well-outlined lessons that highlighted the overall structure of the ideas to emphasize the "big picture." However, according to Jeff, his seamless one-way flow of information was so smooth that students often believed they understood the ideas even when they did not. Jeff also saw no benefit to cooperative learning, though he did not discourage it:

> I think group work is bad. I think it is time consuming. Typically in a group, the smart person ... dominates ... and explains everything to them ... in a worse way than we would explain it to the other students. The other students aren't benefiting from the group work. It's not as though they are really sharing ideas. It's not as though ... you could take a thousand people off the street, or a thousand idiots, and combine them to get an Einstein. It is not as though groups of people who do not understand series are going to typically ... reinforce each other and understand series. If that were true, all these people who work together at home on their homework, who aren't very good in your class, would get a lot better. I don't believe it works. (Jeff, Q23, I-2)

Jeff is not alone in his opposition to cooperative learning. Wu (1997) questioned the extent to which substantive mathematics can be learned when cooperative teaching methods are employed. Further, Jeff's belief that the professor's "perfect explanation" will enlighten the student is consistent with traditional assumptions about learning that claim students are able to move through the mathematical content faster when knowledge is passed on to them in the most simple and elegant forms (Ferrini-Mundy & Graham, 1991; Kleinfeld, 1996).

The tendency for subject-matter experts to draw upon their own "implicit theories" of student learning in their instructional practice can be problematic. Those implicit theories are not always aligned with the ways students actually come to understand the major ideas within the discipline – a phenomenon Nathan and Petrosino (2003) refer to as the "expert blind spot," noting that

> [experts] tend to use the powerful organizing principles, formalisms, and methods of analysis that serve as the foundation of [the] discipline as guiding principles for their students' conceptual development and instruction, rather than being guided by knowledge of the learning needs and developmental profiles of novices. (Nathan & Petrosino, 2003, p. 906)

Nathan and Petrosino argue that experts possess "the curse of knowledge," a familiarity with mathematical concepts that leads experts to underestimate the difficulty others might have in understanding them. Processes for solving problems are often so automatic that cognitive activity becomes inaccessible to experts. While the "expert blind spot theory" raises important questions, findings of this study indicate that gaps in participants' knowledge of general pedagogy, not expertise in subject matter knowledge, explain their tendency to adopt instructional methods grounded in their own learning styles.

Principles of Instruction

3.4.2. *Participants engaged most heavily in psychological planning for instruction.* Participants' most common reasons for planning were to fulfill course objectives, to provide direction and course pacing, and to increase confidence – all aligned with reasons cited by Clark and Yinger (1987) for why teachers engage in instructional planning. Participants' *psychological planning* involved thinking about the purposes of the subject matter, new directions within the field, the important topics to cover, lecture organization, and the direction of instruction in subsequent lectures. Their *actions for planning* were limited to preparing lecture notes, reviewing the text, and outlining the course. While participants claimed to have a comfort level with the material sufficient to "teach on the fly" if necessary, a primary reason cited for planning was to avoid embarrassing or uncomfortable situations.

3.4.3. *Some participants engaged in reflective thinking about their own pedagogical practices and instructional goals.* Assessment of the teaching and learning of mathematics often comes from end-of-the-course student evaluations and, in some instances, observation by a senior faculty member. Reflection is another process for evaluating student learning and instructional practices, "[it] is what a teacher does as he or she looks back at the teaching and learning that has occurred and reconstructs the events, the emotions, and the accomplishments" (Wilson et al., 1987, p. 120). Ben recorded his observations of the difficulties students had with various mathematical concepts and the strategies he used to address their problems. "I make notes to myself at the end of the course – I want to catch those things a little earlier on when I repeat the course" (Q37, I-2). Ben's actions are an important first step in improving environments for student learning. However, this kind of feedback comes too late to make adjustments while courses are still in progress (Knoerr, McDonald, & McCormick, 1999). Ideally, instructional decision-making should be informed by *regular* assessments of students' mathematical understanding (National Council of Teachers of Mathematics, 1995). The National Research Council (2003) supports the use of formative assessment among faculty in higher education to obtain information about student learning for the purpose of promptly and accurately adjusting teaching practices to promote effective learning.

3.5. Knowledge of Research. *Knowledge of research* first emerged as important to the KBT model as a result of findings from pilot work conducted earlier in the study. Mathematicians tend to develop areas of specialization that become increasingly more focused over the duration of their careers (King, 1992). The participants in this study were devoted to research to produce new ideas in mathematics that are significant, interesting, validated by the mathematical community, and publishable in a refereed journal. Findings of this study supported the inclusion of Knowledge of Research as a category in the model of the KBT for college or university mathematics faculty. There are two subcategories of knowledge of research: knowledge of the relationships between research and teaching, and knowledge of research in mathematics education.

Relationships between Mathematics Research and Teaching

3.5.1. *Four participants believed their mathematical research played a role in their teaching practice, though the nature of that role varied from indirect to direct.* As a number theorist, Austin believed,

> History plays a big role in the way that I teach. I think that is because history plays such a big role in the theory of numbers. It certainly influences my way of thinking about things. Number theory is a very empirical branch because the objects that we study are numbers, the natural numbers [and] natural means coming from nature. Numbers are things people experience. We

> all know how to count. Our first experience in mathematics is counting. (Q41, I-2)

Kevin characterized his mathematical research as having a more direct connection to his teaching practice,

> It is a heavy role. There is not a clean distinction between [research and teaching], though they are different. I'm thinking about math all of the time. When I'm teaching, my teaching is informing my research a bit and my research is what I think is important. By doing research, I'm always on the alert of not just what is out there, but what is important, what is serious. I am trying to develop an aesthetic sense of what the nature of mathematics is and that's going to inform how I teach, what I teach, what I think is important, and how I explain things. And it does change over time. Certainly, the way I think about math now is different than [it was] ten years ago. I am hoping in ten years I will know more than I do now and not just knowledge, but more of a feel for how it all fits in. (Q41, I-2)

Studies have shown research positively impacts teaching by increasing the knowledge base of the instructor, stimulating the instructor's interest in and enthusiasm for the subject matter, encouraging good intellectual habits, allowing faculty to model thinking and critical reasoning for students, and enabling faculty to offer students challenging experiences (Hilton, 1978, 1986; Johnston, 1997; Weimer, 1997). Conversely, teaching may have a positive impact on research. For example, Kevin published a paper on a mathematical problem researched as a result of a question from an undergraduate student in one of his classes.

Other researchers have found no relationship between research and teaching (Feldman, 1987; Michalak & Friedrich, 1996). Some who challenge the positive correlation argue the self-discipline and isolation often associated with research activity detracts from the development of the kinds of interpersonal skills necessary for quality teaching. Moreover, they claim the level of the instructor's research is nearly always too advanced to bring into the undergraduate classroom in any meaningful way (Alsina, 2001; Barnett, 1996; Michalak & Friedrich, 1996).

3.5.2. *Four participants attempted to heighten students' awareness of on-going faculty research in mathematics in order to depict the field as alive and fruitful.* Kevin stated,

> ... most of us talk regularly about research in class to let them know it's not a dead subject... most people believe math is important, but it's over. That's totally false! We are now in the greatest age of mathematics ever! (Q16, I-1)

The knowledge acquired by college and university mathematics faculty through discipline-specific scholarly activities needs to be acknowledged and incorporated into initiatives directed at making positive change in the teaching and learning of undergraduate mathematics.

Knowledge of Research in Mathematics Education

3.5.3. *Five participants reported reviewing no mathematics education research literature. Some did not know where to obtain it, while others expressed concerns about the validity of its findings.* Austin said he occasionally skimmed articles in the *Journal for Research in Mathematics Education.* He commented,

> I am a great skeptic about math education research. I am open-minded in the sense that I would love to see math education research produce things that I really believe in, but I see so many conflicting things that... I accept what I already believe and reject what I don't. I am not sure that the research changes me. (Q43, I-2)

Disinclination of mathematicians to review relevant research literature in mathematics education is a finding supported by earlier research (Hillel, 2001; Stevens, 1991; Weimer, 1997). One explanation for this might be that "the nature of evidence and argument in mathematics education is quite unlike the nature of evidence and argument in mathematics" (Schoenfeld, 2001, p. 221). Theories in mathematics are explicitly stated and results are obtained through proof. Absolute proofs do not exist in mathematics education research. Methods in mathematics education research are usually only suggestive of results and rest on the cumulative evidence of multiple studies to substantiate findings (Schoenfeld, 2001).

3.5.4. *Participants held differing beliefs about the effectiveness of the calculus reform movement, with some supporting it and others strongly opposed.* Ben, Jennifer, and Ron attributed significant positive changes in their methods of teaching calculus to the initiatives of the reform movement, launched by the 1986 Conference at Tulane University (Douglas, 1986). Unsatisfied with student performance in calculus, an antiquated curriculum, traditional methods of instruction, and limited use of technology, many mathematics departments began to rethink their calculus programs (Knisley, 1997). In an effort to eradicate the old calculus curriculum and practices that purportedly failed to encourage students to think, the Calculus Consortium based at Harvard University framed a revitalized curriculum that presented the graphical, numerical, and analytical aspects of calculus (i.e., the "rule of three") to help students build meaning and grasp concepts. As reformers shifted their attention toward issues of pedagogy, cooperative learning assumed a significant role in many calculus reform initiatives. "By early 1995 over two-thirds of the mathematics departments surveyed indicated that some reform efforts were underway at their institutions, most of which included some component of group work" (Herzig & Kung, 2003, p. 32).

Carrie, Kevin, and Jeff worked in mathematics departments that fundamentally disagreed with calculus reform. Jeff's opposition to collaborative learning is consistent with his more broadly-based denunciation of the calculus reform movement:

> We tried using the Harvard text [i.e., Hughes-Hallet et al. (1998)] and tried the calculus reform in a serious way for a while... Most of the people in the department seemed to think that it wasn't working to different degrees... There are a number of people in the department who never liked the Harvard text [and] I was the one who stopped us from using it. (Q7, I-1)

Opponents of the calculus reform movement have claimed the new calculus courses lack rigor, omit important ideas from the curriculum, promote an unhealthy reliance on computers and calculator technology, and over-emphasize applications (Cipra, 1996; Kleinfeld, 1996; Tucker, 1999; Wu, 1997).

4. Summary and Implications for Future Work

This study developed a KBT framework among mathematics faculty in higher education teaching calculus at the undergraduate level. At present, little is known about the pedagogical knowledge of the research mathematician. Existing KBT models are primarily directed at K-12 educators. However, research indicates that KBT models among faculty in higher education are structurally and substantively different:

> Professors draw their notions of teaching from their own experiences 'on the job' – these experiences result in a different kind of knowledge base, structured differently for teaching in higher education, than that which is found or required in other educational milieus. (Rahilly & Saroyan, 1997, p. 12)

Although a few studies have examined some aspects of teacher knowledge base (e.g., PCK only) among mathematics faculty in higher education (Wagner et al., 2006), no study has comprehensively examined the KBT among mathematics faculty in higher education. Both the KBT model developed in this study and the themes identified within each knowledge category offer a foundation upon which others may build.

The KBT framework that emerged from this study differs in number of ways from existing KBT frameworks for K-12, including, but not limited to, the following:

(1) Study data supported inclusion of *Knowledge of the Aesthetics of Mathematics* which does not appear in K-12 models.
(2) A significant component of general pedagogical knowledge among K-12 educators is knowledge of classroom management. Participants in this study equated classroom management with behavior management and did not view it as an issue in undergraduate teaching. Participants' definitions of classroom management may have been too narrow since they did not incorporate issues of instructional time, teaching routines, classroom discourse, or other variables. This finding may warrant further study.
(3) Most participants identified more strongly with the role of research mathematician, though all valued and cared about their teaching responsibilities. The research activities of PhD mathematicians contribute to a knowledge base that differs substantially from that of K-12 educators. Professional development should explore pedagogically powerful ways that mathematics faculty in higher education can draw upon their knowledge of mathematical research in undergraduate mathematics instruction.

Collaborative efforts to examine and improve the teaching and learning of mathematics at the undergraduate level must begin with an awareness of the unique character of KBT models among faculty in higher education.

On the other hand, this study identifies some common ground between existing K-12 KBT frameworks and emergent KBT frameworks in higher education. For example, a finding of this study showed participants claimed to value students' mathematical thinking and sought to uncover the sources of students' confusion in mathematics to facilitate their learning. However, they believed students have difficulty articulating their thought processes and are not always cognitively engaged with the mathematical ideas during class lectures. Knowledge of students' mathematical backgrounds, conceptions, and processes for understanding new content –

all important components of PCK – enable teachers to prepare appropriate explanations and representations of mathematical ideas. Pedagogical content knowledge has been widely studied at the K-12 level. Researchers can learn from and build upon the findings of these earlier studies as they examine PCK among college and university mathematics faculty.

Though results of this study cannot be generalized, several findings suggest the need for future study including, but not limited to, the following:

(1) Analysis of the data supported inclusion of *Knowledge of the History of Mathematics* as a subcategory of *Subject Matter Knowledge* in the KBT model. Participants presented the rich history of mathematics as auxiliary to, rather than separate from, the teaching and learning of mathematics, allowing students to examine the historical conditions under which mathematical concepts have evolved (Grugnetti, 2000). The potential for this type of knowledge to positively impact the teaching and learning of mathematics is significant. Future study should examine in greater depth the role knowledge of history among mathematics faculty in higher education plays in students' understanding of mathematical concepts.
(2) Examination of participants' *Personal Practical Knowledge* showed several participants reported negative experiences studying mathematics in their K-12 schooling. For some, those experiences appeared to have an impact on their attitudes and beliefs about pedagogical initiatives in the field of mathematics education. Calls for a greater interchange between university mathematicians and mathematics educators must take the prior experiences of the former into consideration.
(3) While most participants expressed wanting to know more about their students' backgrounds and experiences, they acknowledged knowing little about them. Some made an effort to acquire background information about their students, but were generally not familiar with students unless they attended office hours. This finding suggests the need to explore ways for mathematics faculty to learn about and connect with students, even in the large class setting.
(4) A noteworthy finding related to participants' *Knowledge of General Pedagogy* was their tendency to construct their own theories of student learning. This study found participants' knowledge of general pedagogy was grounded in their experiences and their own styles of learning. In the absence of any formal knowledge of learning theory, participants developed implicit "self-created" theories of student learning which influenced their teaching practices. Professional development opportunities should be directed at building the general pedagogical knowledge of mathematics faculty in higher education particularly as it relates to the learning needs and developmental profiles of today's undergraduate calculus students.

Though not a focus of this paper, findings of this study indicated participants' knowledge and beliefs about the teaching and learning of mathematics influenced their approaches to mathematics instruction. This underscores the need for those invested in advancing the teaching and learning of undergraduate mathematics to be well-informed about the knowledge base for teaching among mathematics faculty in higher education.

References

Abd-El-Khalick, F., & BouJaoude, S. (1997). An exploratory study of the knowledge base for science teaching. *Journal of Research in Science Teaching, 34*(7), 673-699.

Alsina, C. (2001). Why the professor must be a stimulating teacher. In D. Holton, M. Artigue, U. Kirchgraeber, J. Hillel, M. Niss, & A. H. Schoenfeld (Eds.), *The teaching and learning of mathematics at university level: An ICMI study* (pp. 3–12). Boston: Kluwer.

Antonio, A. L., Astin, H., & Cress, C. M. (2000). Community service in higher education: A look at the nations faculty. *The Review of Higher Education, 23*(4), 373–398.

Ball, D. L. (1988). Research on teacher learning: Studying how teachers' knowledge changes. *Action in Teacher Education, 10*(2), 17–24.

Ball, D. L. (1991). Research on teaching mathematics: Making subject matter knowledge part of the equation. In J. Brophy (Ed.), *Advances in research on teaching: Teachers' subject matter knowledge and classroom instruction* (Vol. 2, pp. 1–48). Greenwich, CT: JAI.

Barnett, R. (1996). Linking teaching and research. In D. E. Finnegan, D. Webster, & Z. F. Gamson (Eds.), *Faculty and faculty issues in colleges and universities: Association for the Study of Higher Education reader series* (2nd ed., pp. 397–409). Needham Heights, MA: Simon and Schuster.

Bartlett, T. (2003). Did you hear the one about the professor? *Chronicle of Higher Education, 49*(46), 8–10.

Bass, H. (1997). Mathematicians as educators. *Notices of the American Mathematical Society, 44*(1), 18–21.

Beyer, J. M. (1996). Organizational cultures and faculty motivation. In J. L. Bess (Ed.), *Teaching well and liking it: Motivating faculty to teach effectively* (pp. 145–172). Baltimore: Johns Hopkins University.

Borko, H., & Putnam, R. T. (1996). Learning to teach. In R. Calfee & D. Berliner (Eds.), *Handbook of educational psychology* (pp. 673–708). New York: MacMillan.

Brendefur, J., & Frykholm, J. (2000). Promoting mathematical communication in the classroom: Two preservice teachers' conceptions and practices. *Journal of Mathematics Teacher Education, 3*(2), 125–153.

Carter, K., & Doyle, W. (1987). Teachers' knowledge structures and comprehension processes. In J. Calderhead (Ed.), *Exploring teachers' thinking* (pp. 147–160). London: Cassell Educational.

Cazden, C. B. (1988). *Classroom discourse: The language of teaching and learning.* Portsmouth, NH: Heinemann.

The Chronicle of Higher Education Almanac. (2003, August 29). *The Chronicle of Higher Education, L(1)*, pp. 15–29.

Cipra, B. (1996). Calculus reform sparks a backlash. *Science, 27*(1), 901–902.

Clandinin, D. J. (1985). Personal practical knowledge: A study of teachers' classroom images. *Curriculum Inquiry, 15*(4), 361–385.

Clandinin, D. J. (1986). *Classroom practice: Teacher images in action.* Philadelphia: Falmer.

Clandinin, D. J. (1986b). Personal practical knowledge: A study of teachers' classroom images. *Curriculum Inquiry, 15*(4), 361–385.

Clark, C. M., & Yinger, R. J. (1987). Teacher planning. In J. Calderhead (Ed.), *Exploring teachers' thinking* (pp. 84–103). London: Cassell Educational.

Clements, D. H., & Battista, M. T. (1990). Constructivist learning and teaching. *Arithmetic Teacher, 38*(1), 34–35.

Cobb, P., Wood, T., & Yackel, E. (1993). Discourse, mathematical thinking, and classroom practice. In E. Forman & A. Stone (Eds.), *Contexts for learning sociocultural dynamics in children's development* (pp. 91–119). Oxford, UK: Oxford University.

Cochran-Smith, M., & Lytle, S. L. (1999). Relationships of knowledge and practice: Teacher learning in communities. *Review of Research in Education, 24*, 249–305.

Colbeck, C. L. (1998). Merging in a seamless blend: How faculty integrate teaching and research. *Journal of Higher Education, 69*(6), 647–671.

Connelly, F. M., & Clandinin, D. J. (1985). Personal practical knowledge and the modes of knowing: Relevance for teaching and learning. In E. Eisner (Ed.), *Learning and teaching the ways of knowing: Eighty-fourth yearbook of the National Society for the Study of Edcuation* (pp. 174–198). Chicago: University of Chicago.

Darken, B., Wynegar, R., & Kuhn, S. (2000). Evaluating calculus reform: A review and a longitudinal study. In E. Dubinsky, A. H. Schoenfeld, & J. Kaput (Eds.), *Research in collegiate mathematics education. IV* (pp. 16–41). Providence, RI: American Mathematical Society.

Douglas, R. G. (1986). Proposal to hold a conference / workshop to develop alternative curricula and teaching methods for calculus at the college level. In R. Douglas (Ed.), *Toward a lean and lively calculus* (pp. 6–15). Washington, DC: Mathematical Association of America. (MAA Notes #24)

Elbaz, F. L. (1983). *Teacher thinking: A study of practical knowledge.* London: Croom Helm.

Fairweather, J. S. (1996). Academic values and faculty rewards. In D. E. Finnegan, D. Webster, & Z. F. Gamson (Eds.), *Faculty and faculty issues in colleges and universities: Association for the Study of Higher Education reader series* (2nd ed., pp. 361–376). Needham Heights, MA: Simon and Schuster Custom Publishing.

Farrell, M. A., & Farmer, W. A. (1988). *Secondary mathematics instruction: An integrated approach.* Dedham, MA: Janson.

Feldman, K. A. (1987). Research productivity and scholarly accomplishment of college teachers as related to their instructional effectiveness: A review and exploration. *Research in Higher Education, 26*(3), 227–298.

Fennema, E., & Franke, M. L. (1992). Teachers' knowledge and its impact. In D. A. Grouws (Ed.), *Handbook of research on mathematics teaching and learning* (pp. 147–164). New York: Macmillan.

Ferrini-Mundy, J., & Graham, K. G. (1991). An overview of the calculus curriculum reform effort: Issues for learning, teaching, and curriculum development. *The American Mathematical Monthly, 98*(7), 627–635.

Frost, S. H., & Teodorescu, D. (2001). Teaching excellence: How faculty guided change at a research university. *The Review of Higher Education, 24*(4), 397–415.

Furinghetti, F. (2000). The long tradition of history in mathematics teaching: An old Italian case. In V. J. Katz (Ed.), *Using history to teach mathematics* (pp. 49–58). Washington, DC: Mathematical Association of America.

Ganter, S. L. (1999). An evaluation of calculus reform: A preliminary report of a national study. In B. Gold, S. Z. Keith, & W. A. Marion (Eds.), *Assessment practices in undergraduate mathematics* (pp. 233–236). Washington, DC: Mathematical Association of America. (MAA Notes #49. Retrieved January 9, 2008, from http://www.maa.org/saum/maanotes49/233.html)

Ganter, S. L., & Jiroutek, M. R. (2000). The need for evaluation in the calculus reform movement: A comparison of two calculus teaching methods. In E. Dubinsky, A. H. Schoenfeld, & J. Kaput (Eds.), *Research in collegiate mathematics education. IV* (pp. 42–62). Providence, RI: American Mathematical Society.

Garner, R. L. (2006). Humor in pedagogy: How ha-ha can lead to aha! *College Teaching, 54*(1), 177–180.

Gfeller, M. K. (1999). Mathematical MIAs. *School Science and Mathematics, 99*(2), 57–59.

Goroff, D. L. (1999). *The enculturation of mathematicians in graduate school.* ERIC Document Reproduction Service No. ED436403.

Grossman, P. L. (1990). *The making of a teacher: Teacher knowledge and teacher education.* New York: Teachers College.

Grugnetti, L. (2000). The history of mathematics and its influence on pedagogical problems. In V. J. Katz (Ed.), *Using history to teach mathematics* (pp. 29–35). Washington, DC: Mathematical Association of America.

Herzig, A., & Kung, D. T. (2003). Cooperative learning in calculus reform: What have we learned? In A. Selden, E. Dubinsky, & G. Harel (Eds.), *Research in collegiate mathematics education. V* (pp. 30–55). Providence, RI: American Mathematical Society.

Hiebert, J., & Wearne, D. (1993). Instructional tasks, classroom discourse, and students' learning in second-grade arithmetic. *American Educational Research Journal, 30*(2), 393–425.

Hillel, J. (2001). Trends in curriculum: A working group report. In D. Holton, M. Artigue, U. Kirchgraeber, J. Hillel, M. Niss, & A. H. Schoenfeld (Eds.), *The teaching and learning of mathematics at university level: An ICMI study* (pp. 59–69). Boston: Kluwer.

Hilton, P. (1978). Teaching and research: A false dichotomy. *Mathematical Intelligencer, 1*, 76–80.

Hilton, P. (1986). Teaching and research: The history of a pseudoconflict. In R. E. Ewing, K. Gross, & C. F. Martin (Eds.), *The merging of disciplines: New directions in pure, applied, and computational mathematics* (pp. 89–99). New York: Springer-Verlag.

Hughes-Hallet, D., Gleason, A. M., Flath, D. E., Lock, P. F., Gordon, S. P., & Lomen, D. O. (1998). *Calculus: Single variable.* New York: Wiley.

Johnston, S. (1997). Preparation for the role of teacher as part of induction into faculty life and work. In P. Cranton (Ed.), *Universal challenges in faculty work: Fresh perspectives from around the world: New directions for teaching and learning* (pp. 31–39). San Francisco: Jossey-Bass.

Kaput, J. J. (1997). Rethinking calculus: Learning and thinking. *The American Mathematical Monthly, 104*(8), 731–737.

Kazemi, E. (1998). Discourse that promotes conceptual understanding. *Teaching Children Mathematics, 4*(7), 410–414.

King, J. P. (1992). *The art of mathematics.* New York: Ballantine.

Kirst, M. W. (2004). The high school/college disconnect. *Educational Leadership, 62*(3), 51–55.

Kleinfeld, M. (1996). Calculus: Reformed or deformed? *The American Mathematical Monthly, 103*(3), 230–232.

Kline, M. (1977). *Why the professor can't teach: Mathematics and the dilemma of university education.* NY: Saint Martin's.

Knapper, C. (1997). Rewards for teaching. In P. Cranton (Ed.), *Universal challenges in faculty work: Fresh perspectives from around the world: New directions for teaching and learning* (pp. 41–52). San Francisco: Jossey-Bass.

Knisley, J. (1997). Calculus: A modern perspective. *The American Mathematical Monthly, 104*(8), 724–727.

Knoerr, A. P., McDonald, M. A., & McCormick, R. (1999). Departmental assistance in formative assessment of teaching. In B. Gold, S. Z. Keith, & W. A. Marion (Eds.), *Assessment practices in undergraduate mathematics* (pp. 261–264). Washington, DC: Mathematical Association of America. (MAA Notes #49)

Kulikowich, J., & DeFranco, T. (2003). Philosophy's role in characterizing the nature of educational psychology and mathematics. *Educational Psychologist, 38*(3), 147–156.

Kung, D., Speer, N., & Gucler, B. (2006). Teaching as learning: Mathematics graduate students' development of knowledge of student thinking about limits. In S. Alatorre, J. L. Cortina, M. Sáiz, & A. Méndez (Eds.), *Proceedings of the 28th annual meeting of the North American Chapter of the International Group for the Psychology of Mathematics Education* (Vol. 2, pp. 835–836). Mexico: Universidad Pedagógica Nacional.

LaBerge, V. B., Zollman, A., & Sons, L. R. (1997, March). *Awareness, beliefs, and classroom practices of mathematics faculty at the collegiate level.* Paper presented at the annual meeting of the American Educational Research Association. Chicago, IL. (ERIC Document Reproduction Service No. ED409758)

Leiblich, A. (1998). *Narrative research: Reading, analysis, and interpretation.* Thousand Oaks, CA: Sage.

Leinhardt, G., & Smith, D. A. (1985). Expertise in mathematics instruction: Subject matter knowledge. *Journal of Educational Psychology, 77*(3), 247–271.

Leonard, J. (1999, September-October). *From monologue to dialogue: Facilitating classroom debate in mathematics methods courses.* Paper presented at the joint annual meeting of the School Science and Mathematics Association and the North Carolina Council of Teachers of Mathematics. Greensboro, NC. (ERIC Document Reproduction Service No. ED441696)

Michalak, S. T., & Friedrich, R. J. (1996). Research productivity and teaching effectiveness at a small liberal arts college. In D. E. Finnegan, D. Webster, & Z. F. Gamson (Eds.), *Faculty and faculty issues in colleges and universities: Association for the Study of Higher Education reader series* (2nd ed., pp. 429–441). Needham Heights, MA: Simon and Schuster Custom Publishing.

Michel-Pajus, A. (2000). On the benefits of introducing undergraduates to the history of mathematics: A french perspective. In V. J. Katz (Ed.), *Using history to teach mathematics* (pp. 17–25). Washington, DC: Mathematical Association of America.

Millet, K. C. (2001). Making large lectures effective: An effort to increase student success. In D. Holton, M. Artigue, U. Kirchgraeber, J. Hillel, M. Niss, & A. H. Schoenfeld (Eds.), *The teaching and learning of mathematics at university level: An ICMI study* (p. 137-152). Boston: Kluwer.

Moll, L. C. (Ed.). (1990). *Vygotsky and education: Instructional implications and applications of sociohistorical psychology.* New York: Cambridge University.

Nathan, M. J., & Petrosino, A. (2003). Expert blind spot among preservice teachers. *American Educational Research Journal, 40*(4), 905–928.

National Council of Teachers of Mathematics. (1995). *Assessment standards for school mathematics.* Reston, VA: Author.

National Council of Teachers of Mathematics. (2000). *Principles and standards for school mathematics.* Reston, VA: Author.

National Research Council. (1989). *Everybody counts: A report to the nation on the future of mathematics education.* Washington, DC: National Academy.

National Research Council. (2003). *Evaluating and improving undergraduate teaching in science, technology, engineering, and mathematics.* Washington, DC: National Academy.

National Science Foundation. (1996). *Shaping the future: New expectations for undergraduate education in science, mathematics, engineering, and technology.* Arlington, VA: Author.

Nickson, M. (1992). The culture of the mathematics classroom: An unknown quantity? In D. A. Grouws (Ed.), *Handbook of research on mathematics teaching and learning* (pp. 101–114). New York: Macmillan.

Pirie, S. E. B., & Schwarzenberger, R. L. E. (1988). Mathematical discussion and mathematical understanding. *Educational Studies in Mathematics, 19,* 459–470.

Rahilly, T. J., & Saroyan, A. (1997, March). *Memorable events in the classroom: Types of knowledge influencing professors' classroom teaching.* Paper presented at the annual meeting of the American Educational Research Association. Chicago, IL. (ERIC Document Reproduction Service No. ED411719)

Resnick, L. B., & Ford, W. W. (1981). *The psychology of mathematics for instruction.* New Jersey: Erlbaum.

Robert, A., & Speer, N. (2001). Research on the teaching and learning of calculus/elementary analysis. In D. Holton, M. Artigue, U. Kirchgraeber, J. Hillel, M. Niss, & A. H. Schoenfeld (Eds.), *The teaching and learning of mathematics at university level: An ICMI study* (pp. 283–299). Boston: Kluwer.

Saunders, L. E., & Bauer, K. W. (1998). Undergraduate students today: Who are they? In K. W. Bauer (Ed.), *Campus climate: Understanding the critical components of today's colleges and universities* (p. 7-16). San Francisco: Jossey-Bass.

Schacht, S., & Stewart, B. J. (1990). What's funny about statistics? A technique for reducing student anxiety. *Teaching Sociology, 18,* 52–56.

Schoenfeld, A. H. (1989). Problem solving in context(s). In R. I. Charles & E. A. Silver (Eds.), *The teaching and assessing of mathematical problem solving* (pp. 82–92). Reston, VA: National Council of Teachers of Mathematics.

Schoenfeld, A. H. (2001). Purposes and methods of research in mathematics education. In D. Holton, M. Artigue, U. Kirchgraeber, J. Hillel, M. Niss, & A. H. Schoenfeld (Eds.), *The teaching and learning of mathematics at university level: An ICMI study* (pp. 221–236). Boston: Kluwer.

Seeger, F., Voigt, J., & Waschescio, U. (1998). Introduction. In F. Seeger, J. Voigt, & U. Waschescio (Eds.), *The culture of the mathematics classroom* (pp. 1–9). Cambridge, UK: Cambridge University.

Selden, A., & Selden, J. (2001). Tertiary mathematics education research and its future. In D. Holton, M. Artigue, U. Kirchgraeber, J. Hillel, M. Niss, & A. H. Schoenfeld (Eds.), *The teaching and learning of mathematics at university level: An ICMI study* (pp. 237–254). Boston: Kluwer.

Sherin, M. G., Sherin, B. L., & Madanes, R. (2000). Exploring diverse accounts of teacher knowledge. *Journal of Mathematical Behavior, 18*(3), 357–375.

Shulman, L. S. (1986). Those who understand: Knowledge growth in teaching. *Educational Researcher, 15*(2), 4–14.

Shulman, L. S. (1987). Knowledge and teaching: Foundations of the new reform. *Harvard Educational Review, 57*(1), 1–22.

Siu, M. K. (2000). The ABCD of using history of mathematics in the (undergraduate) classroom. In V. J. Katz (Ed.), *Using history to teach mathematics* (pp. 3–9). Washington, DC: Mathematical Association of America.

Slaughter, S. (1996). Academic freedom at the end of the century: Professional labor, gender, and professionalization. In D. E. Finnegan, D. Webster, & Z. F. Gamson (Eds.), *Faculty and faculty issues in colleges and universities: Association for the Study of Higher Education reader series* (2nd ed., pp. 457–475). Needham Heights, MA: Simon and Schuster Custom Publishing.

Smith, D. A. (2001). The active/interactive classroom. In D. Holton, M. Artigue, U. Kirchgraeber, J. Hillel, M. Niss, & A. H. Schoenfeld (Eds.), *The teaching and learning of mathematics at university level: An ICMI study* (pp. 137–152). Boston: Kluwer.

Stevens, T. C. (1991). Obstacles to change: The implications of the National Council of Teachers of Mathematics [NCTM] standards for undergraduate mathematics. In N. D. Fisher, H. B. Keyes, & P. D. Wagreich (Eds.), *Mathematicians and education reform 1989 - 1990* (pp. 119–126). Providence, RI: American Mathematical Society.

Torok, S. E., McMorris, R. F., & Lin, W. (2004). Is humor an appreciated teaching tool? Perceptions of professors' teaching styles and use of humor. *College Teaching, 52*(1), 14–20.

Tucker, T. W. (1999). Reform, tradition, and synthesis. *The American Mathematical Monthly, 106*(10), 910–914.

Wagner, J., Speer, N., & Rossa, B. (2006, April). *"How much insight is enough?" What studying mathematicians can reveal about knowledge needed to teach for understanding.* Paper presented at the annual meeting of the American Educational Research Association. San Francisco, CA.

Weimer, M. (1997). Integration of teaching and research: Myth, reality, and possibility. In P. Cranton (Ed.), *Universal challenges in faculty work: Fresh perspectives from around the world: New directions for teaching and learning* (pp. 53–62). San Francisco: Jossey-Bass.

Wells, G. (1994). *Discourse as tool in the activity of learning and teaching.* Paper presented at the annual meeting of the American Educational Research Association. (ERIC Document Reproduction Service No. ED371619)

Wertsch, J. V., & Toma, C. (1995). Discourse and learning in the classroom: A sociocultural approach. In L. P. Steffe & J. Gale (Eds.), *Constructivism in education* (pp. 159–174). Hillsdale, NJ: Erlbaum.

Wilson, S. M., Shulman, L. S., & Richert, A. E. (1987). '150 different ways' of knowing: Representations of knowledge in teaching. In J. Calderhead (Ed.), *Exploring teachers' thinking* (pp. 104–124). London: Cassell Educational.

Wood, L. (2001). The secondary-tertiary interface. In D. Holton, M. Artigue, U. Kirchgraeber, J. Hillel, M. Niss, & A. H. Schoenfeld (Eds.), *The teaching and learning of mathematics at university level: An ICMI study* (pp. 87–98). Boston: Kluwer.

Wu, H. (1997). The mathematics education reform: Why you should be concerned and what you can do. *The American Mathematical Monthly, 104*(10), 946–954.

Yackel, E., Rasmussen, C., & King, K. (2000). Social and sociomathematical norms in an advanced undergraduate mathematics course. *Journal of Mathematical Behavior, 19*(3), 275–287.

Zevenbergen, R. (2001). Changing contexts in tertiary mathematics: Implications for diversity and equity. In D. Holton, M. Artigue, U. Kirchgraeber, J. Hillel, M. Niss, & A. H. Schoenfeld (Eds.), *The teaching and learning of mathematics at university level: An ICMI study* (pp. 13–26). Boston: Kluwer.

Appendix A. Interview Prompts

Interview I

(1) What kind of opportunities do you feel that the university has to offer you?
(2) What are the expectations, if any, that the university has of you as a faculty member?
(3) What types of faculty activities receive recognition from the university and-or the mathematics department?
(4) What role does tenure play in your ability to accomplish your own professional goals?
(5) Can you describe the culture of the mathematics department? (e.g., make-up of faculty; collegiality; research and teaching activities; interactions with other departments; evaluation procedures).
(6) Describe the guidelines, if any, imposed by your mathematics department that relate to teaching Calculus.
(7) How is the curriculum for this course developed? How often is it reevaluated?
(8) Do you ever observe your colleagues teaching? If so, for what purpose? Is it welcomed?
(9) If I walked into your classroom on a typical day, can you describe what I would see?
(10) What personal information do you know about the student population in your Calculus class (i.e., life background; strengths, weaknesses and interests and so forth)?

(11) In what ways does the university impact the local community?
(12) Complete the sentence, "Mathematics is ..."
(13) Do you believe that mathematics is created or discovered?
(14) How do you judge the validity of new ideas in mathematics?
(15) Is it important for people to study mathematics? Why or why not?
(16) How do you believe a positive attitude toward mathematics can be developed in students?
(17) What are the mathematics backgrounds of your students when they begin your course?
(18) Are there any topics that they really struggle with?
(19) Are there any topics that they already know?
(20) What common misconceptions do they have?

Interview II

(21) How do you think that students acquire mathematical knowledge or come to understand ideas in mathematics?
(22) To what extent do you think about your students' learning characteristics before you begin your instruction?
(23) What are some of the instructional strategies or teaching methods you use in your Calculus class?
(24) Do materials or technology play any role in your instructional practice?
(25) How do you define mathematical discourse? What do you believe is its role in the teaching and learning of mathematics?
(26) What are the most important concepts, processes, skills, and results in Calculus for students to learn?
(27) Can you give an example of a real application of Calculus that you presented in class?
(28) Can you give an example of a calculus concept that you presented to students in more than one way (i.e., using more than one form of representation)?
(29) In what ways do you draw upon your personal and-or professional experience in your work as a teacher?
(30) Do you have a vision or image of teaching? If so, could you describe it?
(31) What event(s) or factors have influenced your image of teaching? Why?
(32) What do you believe are the overriding purposes and aims of education?
(33) How do you define learning? What makes learning meaningful?
(34) What are some of the ways that you believe that your students learn, in general, with the greatest success?
(35) Do you provide students with opportunities to learn in this way in your classes? Could you share a specific example?
(36) To what extent do you plan for instruction and what does it typically involve?
(37) What are some of the reasons that you plan?
(38) How would you describe your style of classroom management and organization?
(39) What types of management issues, if any, require the most of your attention?
(40) What is the nature of your research?
(41) What role does your research play in your teaching, if any?

(42) With which mathematical organizations (local, national or international) are you involved? What is the nature of your involvement?
(43) Do you review current research in mathematics education? From which journals or sources?
(44) What do you know about the National Council for Teachers of Mathematics Standards?
(45) Do you believe that the standards apply to college-level instruction? If so, how do they transfer at the college level?
(46) In what direction do you believe that mathematics teaching is moving at the college level?

Interview III

(47) Can you briefly describe the nature of your planning for this lesson?
(48) Describe how you envision this lesson unfolding (i.e., What will you be doing? What will the students be doing?)
(49) What are your goals for this lesson?
(50) Do you expect that students will have difficulty understanding any of the mathematical ideas to be presented in this lesson?
(51) How will you respond if students experience confusion?

Appendix B. Sample of the Line-by-Line Coding of Interview Data

A sample of the line-by-line coding of the interview data.

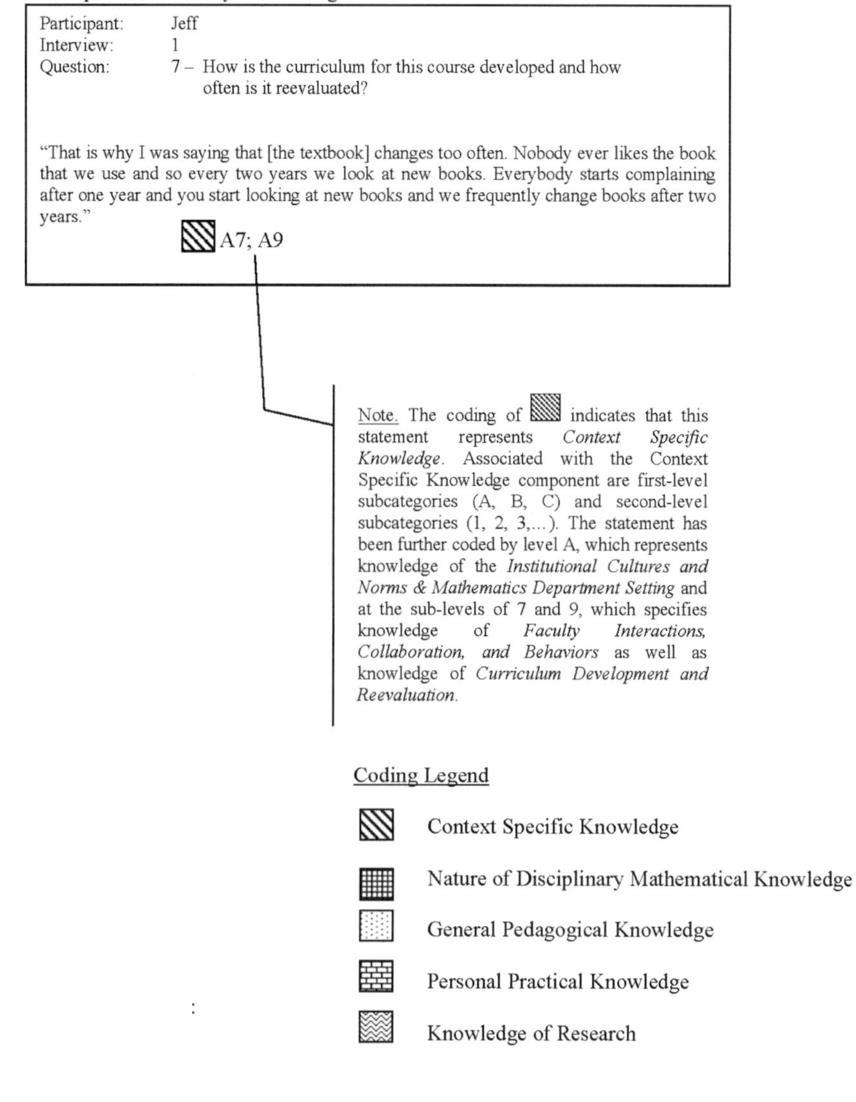

Appendix C. Excerpt I from Calculus Class Observation

Line	Speaker	Dialogue
1	Jennifer (J)	I had a question for you! Calculus was invented or figured
2		out or thought of, do you have any ideas when?
3	Student (S)	Wasn't it by Archimedes?
4	J	No, it wasn't quite that long ago!
5	S	Wasn't it Newton?
6	J	Newton, yeah. So does anybody know when Newton
7		lived?
8	J	...Not as long ago as Archimedes! I am asking you partly
9		because somebody said in the other section... that
10		Calculus had been invented last century.
11	S	Wasn't it about the 1600's?
12	J	Yeah, yeah, and Leibniz too, we should give this guy some
13		credit!...Leibniz lived 1646 to 1717 and Newton 1642 to
14		1727 and you've heard of Newton, right?
15	S	Yeah.
16	J	Have you heard of poor Leibniz?
17	S	Is that the German guy who got published?
18	J	He's a German guy, yeah who um...
19	S	I've heard of the name Leibniz.
20	J	Oh, O.K. So, there was the British guy and the German
21		guy and they invented calculus more or less at the same time.
22		It was an idea whose time had come. The thing about
23		Leibniz... was that he had a nice notation. It is his notation
24		that I wanted to mention. This is $\frac{dy}{dx} = y'$ [for $y = f(x)$].

Appendix D. Excerpt II from Calculus Class Observation

Line	Speaker	Dialogue
1	Jennifer (J)	O.K. Classic calculus related rates problem. There is a
2		ladder leaning up against a wall. Sometimes ladders slip
3		and this one is going to slip. The bottom of a 10 foot
4		ladder is going away from the wall at 2 feet per second.
5		Whenever you have these related rates problems, you
6		should be drawing pictures. How fast is the top going
7		down the wall? Now, don't answer fast. I want you to do
8		it. I want everybody to do it. You can work with somebody,
9		but I want you to set it up and get an answer.
10	Student 1 (S1)	Does the ladder start out standing up straight?
11	J	Oh! That's a good question! Maybe I should have asked
12		you something more! Ah, I should have asked you, "How
13		fast is it going when it is 6 feet from the ground?"... and
14		then the same question at 5 feet. And there is the Student 2
15		question, zero! You will argue with me all day about that
16		one!
17	Student 2 (S2)	How fast is it moving?
18	J	Yeah.
19	S2	At zero feet high?
20	J	Yeah.
21	S2	It isn't moving.
22	J	Keep working.
23	S2	If you look at this like a physics problem, this is pushing
24		down at 9.8 m/s^2.
25	Student 3 (S3)	...Are we supposed to take gravity into consideration?
26	J	No, no, but that's what makes it sort of an artificial
27		problem...
28	S2	O.K.
29	S3	Wait, so do we know how far it is away from the wall?
30		We only know how fast it is moving away from the wall.
31	S2	It is 8 feet away from the wall.
32	S3	How do you know it is 8 feet away from the wall?
33	S2	Pythagorean Theorem.
34	S3	You can't use Pythagorean Theorem if you only have one
35		leg.
36	S2	No, but you know this is 6 feet high and you know the
37		ladder is 10 feet.
38	S3	Oh, O.K.
39	J	Let me butt in here for a minute. You have to start naming
40		something here, right? So, you have to name something,
41		like the distance [points to the diagram with the ladder and
42		refers to the distance from the origin to the bottom of the
43		ladder along the ground]. You have to put variable on them
44		because they change.
45	S2	Oh!
46	J	Like 'x' and 'y'. What is the relationship between them?...

Line	**Speaker**	**Dialogue**
47	Student 4 (S4)	Well, as x increases, y decreases.
48	J	Yes, but a precise geometric definition.
49	Student 5 (S5)	You have a right triangle.
50	J	You have a right triangle, yeah. So, you've got $x^2 + y^2 = 100$ is?
51		Is 100.
52	Student 6 (S6)	Did you do the length of the ladder squared?
53	J	Yes, which was 10 feet.
54	Student 7 (S7)	What does $x^2 + y^2 = 100$ have to do with this?
55	J	It relates, that is why it is called related rates, it relates the
56		rate of change of something you know because you are
57		given that this is going down at 2 feet per second to
58		something that you don't know - how fast is it coming
59		down the wall. O.K.?

Department of Education, Emmanuel College, Boston, MA 02115
E-mail address: sofronki@emmanuel.edu

Neag School of Education and Department of Mathematics, University of Connecticut, Storrs, CT 06269
E-mail address: tom.defranco@uconn.edu

CBMS Issues in Mathematics Education
Volume **16**, 2010

Modeling Students' Conceptions: The Case of Function

Nicolas Balacheff and Nathalie Gaudin

ABSTRACT. We investigate the epistemological complexity of modeling students' knowing of mathematics with the goal of achieving models that acknowledge both the possible lack of coherency and the local efficiency of such knowing. We propose a model of "conception" as a possible tool to address the epistemological complexity we identify. We then provide an illustration of the usefulness of this model by exploring conceptions of "function" as a case in point.

1. From Behavior to Knowing

The possibility to observe students' learning relies heavily on the indications given by students' behaviors and creations, which are hypothesized to be consequences of the knowings they have constructed. Such an appraisal is possible and its results are significant only if one is able to establish a valid relationship between the observed behaviors and the inferred knowings. This relationship between behaviors and knowing is crucial, but also problematic. Its problematic nature has been concealed perhaps as a byproduct of the struggle with behaviorism, but it has always been implicitly present in educational research, at least at the methodological level. Indeed, a knowing cannot be reduced to its associated behaviors, but on the other hand it cannot be diagnosed, understood, or taught without a characterization based on these associated behaviors. We follow the choice made in the translation of Brousseau's work (1997) to use the word knowing as a noun and, as such, to denote a distinction between knowing and knowledge. Knowing refers to students' personal constructs whereas knowledge refers to intellectual constructs recognised by a social body, such as the discipline of mathematics. This distinction corresponds to the distinction made in Roman languages between those words that derive from the Latin "cognoscere" and those words that derive from the Latin "sapere."

The importance of the link between behaviors and knowing was clearly pointed out by Schoenfeld (1987) in his introduction to *Cognitive Science and Mathematics Education.* Schoenfeld described the cognitive science approach in relation to the effort made to describe problem-solving strategies in detail so that they could be taught and reproduced. We would like to focus on and bring into question the level of description of the behaviors and of their outcomes, as tangible expressions

©2010 American Mathematical Society

of the problem-solving strategies that Schoenfeld was interested in teaching. In synthesizing his own research at the time Schoenfeld indicated:

> My intention was to pose the question of problem-solving heuristics from a cognitive science perspective: What level of detail is needed to describe problem-solving strategies so that students can actually use them? (p. 18)

Such inquiry raises for us two essential questions:

(1) On one hand, to what extent would a finer granularity of the description of these strategies guarantee a better reliability of the learning of a problem-solving strategy? Or, rather, given a competency, does there exist a level of granularity of its description that guarantees the efficient learning of such competency?
(2) On the other hand, to what extent does a finer description of problem-solving strategies inform us about the relationships between behaviors and knowings?

In their "microgenetic analysis of one student's evolving understanding of a complex subject domain," Schoenfeld, Smith, and Arcavi (1993) show that the challenge to be taken up is difficult. Two reasons, which are the main lessons we can learn from their seminal study, are at the origin of this difficulty: first, the "highly subjective" character of the empirical data analysis, second what Schoenfeld and his colleagues called the "burden of proof" in showing how behaviors and knowing are related. A way to overcome these difficulties consists of adopting a definition in relation to an explicit characterization likely to reinforce the grounding of the analysis. The present article explores this possibility and provides an illustration. But first, we would like to develop further the issues related to the relationship between behaviors and knowing on which the questions we consider depend.

The question of the relationships between behaviors and knowings is considered fundamental in the theory of didactical situations (Brousseau, 1997). One of the postulates of this theory is that for each knowing that a student could have, there are one or more problem-situations whose demands require certain behaviors from the student that embody that knowing. This fundamental correspondence, established case by case, is justified by the interpretation of problem-situations in terms of games. In the context of this game interpretation, behaviors are seen as expressions of strategies adapted to the representation of the problem-situation attributed to the student (Brousseau, 1997, p. 215). The basis of that postulate is shared by some approaches to cognitive science: Pichot (1994) for example, writes "all behavior implies a knowing" (p. 206). Indeed, this postulate arguably justifies most of our field's research on learning since students' behaviors are the source of the corpus on which we perform our analyses.

Yet the excision of a behavior from the observation of a so-called reality, which could be that of a classroom or a laboratory experiment, requires us to deal at the same time with a methodological and a theoretical problem. An observed behavior is not given by the "reality" but taken out of it as "a result of a decision made by an observer" (Robert, 1992, p. 54).

If a behavior expresses a relationship between a person and her environment, then it depends on the characteristics of the person as well as on the characteristics of her environment. A good example is the case of instruments. While instruments facilitate action – if the user holds the knowing required to use them – they also restrict the action due to their own limitations (Rabardel, 1995; Resnick & Collins, 1994, p. 7). One should notice that these limitations could be related to material constraints as well as to the knowings involved in the design of the instruments.

"Person" and "environment" refer to complex realities, not all of their aspects are relevant to the questions we are considering in this paper. About the person, what is of interest to us is her relationship to a piece of knowledge. Thus we refer to the subject as a kind of projection of the person onto her cognitive dimension. In the same way, we are not interested in the environment in all of its complexity, but only in those features that are relevant to a given piece of knowledge. We call *milieu* such a subset of the environment of a subject; the milieu is a kind of projection of the environment onto its epistemic dimension. In the case of mathematics, the interactions between the subject and the milieu are based on systems of signifiers produced by the subject herself, or by others. We must then extend the classical idea of milieu in order to include symbolic systems and social interaction as means for the production of knowings, as well as the physical milieu to which we generally refer. This is the meaning of Brousseau's (1997) proposal "to define the milieu as the system antagonist to the subject in the learning process" (p. 57).

So, a knowing is neither to be ascribed solely to a subject, nor solely to the milieu. On the contrary, a knowing is a property of the interaction between the subject and the milieu – his or her antagonist system. This interaction is meaningful because its purpose is to fulfill the necessary conditions brought by the situation for the viability of the *subject/milieu system*. By viability we mean that the subject/milieu system has a capacity to return to equilibrium after some perturbations (for example, a contradiction or an uncertainty) the subject was aware of. The knowing may evolve if the perturbations are such that this is necessary, requiring for this purpose the implementation of a learning process. This is, in other words, a formulation of Vergnaud's (1981) postulate that "problems are the source and the criteria of knowings" (p. 220).

Taking into account a remark on a previous version of this article, we note the close relationship between what is stated in this paper and the Piagetian model of knowing. However, there is a difference to be emphasized: Our focus is on the whole system [S (subject) $\leftrightarrow$ M (milieu)] and not on one of its parts. In other words, our concern is not to know how the subject thinks, but to be able to give account of the whole system in a way relevant to a didactical project – the cognitive system we consider is not S but [S $\leftrightarrow$ M].

In what follows, we introduce the classical problem raised by the possible inconsistency in learner behaviors and ways of understanding (Harel, 2006). We present a solution which takes the form of a characterization of students' conceptions and which accounts for their complexity by relating three main features of those conceptions (action, representation, and validation) to the domain in which they are operational. We illustrate the use of this characterization for the case of the concept of function.

2. An Epistemological Problem

2.1. Coherence and Sphere of Practice. The following quotation illustrates the problem of coherence in practice. It comes from the seminal work of the sociologist Pierre Bourdieu (1990) on the practices of Algerian farmers over a year's cycle:

> In the diagram of the calendar, the complete series of the temporal oppositions which are deployed successively by different agents in different situations, and which can never be practically mobilized together because of the necessities of practice never require such a synoptic apprehension but rather discourage it through their urgent demands, are juxtaposed in the simultaneity of a single space. The calendar thus creates, *ex nihilo*, a whole host of relations [...] between reference-points at different levels, which never being brought face to face in practice, are practically compatible even if they are logically contradictory. (p. 83)

Thus the location of farmers' practices in a calendar offers a view of practices that allows the observer to find logical contradictions among them; yet those logical contradictions are not necessarily relevant for the farmer who relates to different practices at different times. Bourdieu's explanation of the paradox of the co-existence of rational thinking on the part of the farmer and of knowings which look contradictory from the observer's point of view, can easily be extended to the case of a single person observed in different situations. The core of this explanation is time on one hand, and, on the other, the diversity of the situations. Time organizes the person's actions sequentially in such a way that the contradictory knowings are equally operational because they appear at different periods of the individual's history: It is reasonable to expect that knowings which are logically contradictory may still ignore each other if they emerge in different situations. The issue of the diversity of situations introduces an element of a different type. This diversity could have several different origins. It may come from a variation of the characteristics of the context (as in the case of problems raised in or out of school), or a variation of the resources (as by making available or not a pocket calculator or a computer), or a variation on the stake of the situation (as in the case of a problem being proposed for an assessment or as homework). The existence of this diversity could be seen as an explanation insofar as one recognizes that a knowing is not of a "universal" nature but that, on the contrary, it is related to a specific and concrete domain of validity. Eventually, this means that the transfer from one situation to another is not an obvious process, even if in the eyes of an informed observer the considered situations look isomorphic. For example, the pupils from a primary mathematics classroom may not use the same procedures to compute the total price of a set of items out of school as they do when they add decimal numbers in order to solve a problem proposed by the teacher. Following Bourdieu, we refer to *spheres of practice* in order to designate these mutually exclusive domains of validity in the history of the subject.

We must insist that although an observer who is able to relate different situations could recognize contradictions, a subject who considers the situations as independent and completely different could ignore these contradictions. However, in the observer's referential system – which is where we are as researchers – these states of the observed subject/milieu system should be labeled in order to show that they relate to the same knowledge, no matter their contradictory nature. So, one often speaks of the subject's knowing of decimal numbers, of continuity of functions, or of line reflection although asserting later that this knowing is not coherent.

To accept the existence of contradictory knowings does not contradict the theoretical principle that these knowings are products of a process of adaptation between the subject and the milieu, ruled by criteria like performance reliability or problem relevance. But this raises educational problems for which solutions have been looked for in different directions. In the following section we review the most significant ones in the case of mathematics.

2.2. A Definition for a Familiar Notion: Conception. In a survey she presented at the 1986 annual meeting of the *American Educational Research Association*, Jere Confrey linked the development of research on misconceptions to the acknowledgement of a failure of teaching. Despite all efforts, she argued, many students held major misconceptions in mathematics and science. Confrey noticed that the mathematics education community had a rather pragmatic approach to this problem: "misconceptions were defined empirically as documented failures of large numbers of students to solve problems which appeared to be related to fundamental concepts" (Confrey, 1986, p. 4).

In a paper published later, Confrey (1990) distinguished different approaches to the question, issued from Piagetian genetic epistemology, scientific epistemology, and the information processing approach. In all of them the child-student is seen as a subject fundamentally different from the adult-expert who appears as the owner of the knowledge of reference. In the case of the child-student, one speaks of "naive theory," "private concepts," "beliefs," or even of the "mathematics of the child" (Confrey 1990 p. 29). But this view does not exclude the recognition of some sort of cognitive legitimacy for these misconceptions:

> [...] a child may not be "seeing" the same set of events as a teacher, researcher or expert. It suggests that many times a child's response is too quickly labeled as erroneous and that if one were to imagine how the child was making sense of the situation, then one would find the errors to be reasoned and supportable. (p. 29)

This remark, made in the case of the scientific epistemology, is in fact valid for all the three mentioned approaches, even for the Piagetian approach as we emphasize it in the following paragraph. Indeed, within the student frame of reference – as opposed to an external one – misconceptions fall under the common rules of knowing:

> a misconception does not require the postulation of an inadequate "picture" of the world; it does require the notion of a successful completion of a number of problems wherein the cycle of problem formulation (expectation), problem-solving (action) and problem reconstruction (re-viewing) are successfully carried out. (p. 29)

In other words: A misconception has a domain of validity, otherwise it would not exist as such. Eventually, there is a very short distance between a misconception and a knowing. The key difference is that for a misconception there exists a refutation known at least to the observer. But even when ascribing the status of a knowing to a misconception – what led most authors to abandon the word misconception – the idea that an intrinsically correct knowledge of reference exists remains a corollary of the initial definition. And yet, such an idea is clearly refuted by our current knowledge of the history of science and mathematics. Let us consider, as an example, the evolution of the concept of function.

In 1938, Bachelard noted, "Le réel n'est jamais 'ce qu'on pourrait croir' mais il est toujours ce qu'on aurait dû penser" (p. 13); that is, "reality is never what one could believe, but is always what one should have thought of." This statement, formulated in the first half of the twentieth century, expressed that knowing is always in progress. If we accept this, then errors witness the inertia of the instrumental power of a knowing that has proved itself by its efficiency in enough situations, but which appears badly adapted in new situations.

A main achievement of research of the 1980s on students' ways of understanding is the recognition that errors are not only the effect of ignorance, of uncertainty, of chance, but the effect of previous knowings which were interesting and successful, but which now are revealed as false or simply not adapted (Brousseau, 1997, p. 82). In one of the first studies within this paradigm, Salin (1976) proposed cognitive characteristics of errors that became essential to the development of the theory of didactical situations. On the one hand, an error is a point of view of a knowing about another knowing (possibly for a subject, the evaluation of an older knowing from the point of view of a new one), and, on the other hand, an error can be identified only if the feedback from the milieu can be "read" by the subject as the indication of a failure (a not satisfied expectation).

The main difference between the previous position and the current one lies in their epistemological meaning. The status of a knowing is different in each case. The first position implies the existence of a *knowing-of-reference general and true* (what encapsulates the common use of the word *knowledge* in English). The second position requires only establishing a relationship between two knowings with the idea of an evolution, without judgment on them. More precisely, from the second perspective a knowing is considered as a set of conceptions which are activated depending on contextual characteristics. These conceptions, whether they appear erroneous, partial, or ill-adapted to an observer, are first of all the results of an optimal adaptation of the subject/milieu system following criteria of relevance and of efficiency to a situation – in other words, an adaptation to a problem-situation. The corollary of this perspective is that a conception has a provisional character: It could be revisited in the course of the adaptation process to produce a new conception. This is especially desirable when the new conception corresponds to content to be taught (Balacheff & Margolinas, 2005).

2.3. A Pragmatic Definition of Conception. The word "conception" has been used for years in research on teaching and learning mathematics. It functions as a tool, but its definition remains implicit; it has not yet been taken as an object of study per se. According to Artigue (1991, p. 266), "conception" refers to a local object; in this sense its epistemological status does not really differ from the one of the word "misconception." Together with Vinner (1983, 1987), we think that there is a need for a better-grounded definition of conception. The expectation is that a better definition will allow us to analyze the differences and commonalties between conceptions in such a way that we will have a better ground to design learning situations.

The first raison d'être of the notion of conception in educational research is the need to conceptualize the specific states of equilibrium achieved by the subject/milieu system satisfying some viability constraints of a situation. These constraints do not address the way the equilibrium is recovered but the criterion of this equilibrium. Following Stewart (1994, pp. 25–26), we would say that these

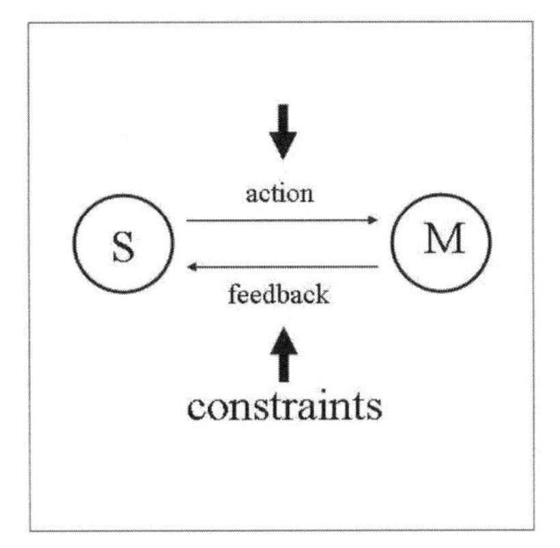

FIGURE 1. Schematic for "conception" in the subject/milieu system

constraints (a) are *proscriptive*, which means that they express necessary conditions to ensure the system's viability and (b) are not prescriptive since they do not tell in detail how an equilibrium must be reached.

We can now propose a definition of "conception" which can be pragmatically and efficiently used in a didactical *problématique*.

> A *conception* is the state of dynamic equilibrium of an action–feedback loop between a subject and a milieu under proscriptive constraints of viability.

From a didactical perspective, we are interested in the nature of the proscriptive constraints that the subject/milieu system must satisfy (see Figure 1). Among these constraints, not known exhaustively, we can mention two that are specific to didactical systems: time constraints and epistemological constraints (Arsac, Balacheff, & Mante, 1992). The former are due to the way schooling is organized (duration of school life, organization of the school year, organization of the lessons, etc.). The latter is due to the existence of a domain of knowledge of reference which underlies the content to be taught and learned and which de facto provides criteria to the acceptability of any learning outcomes.

The role of the teacher, with respect to a given content to be taught, is to organize the encounter between a subject and a milieu so that a conception (acceptable to the didactical intention) can be identified as a result of the evolution of their interactions. Such an encounter is not a trivial event. To create an environment that facilitates it, it is not enough for the student to just be able to read in her environment the milieu relevant to the teaching purpose (Balacheff, 1998). The student also needs to select the relevant features of the environment, to identify the feedback, and to understand it with respect to the intended target of action. To succeed in this task, the teacher must construct a situation that allows the devolution to the students of both the milieu and the relevant relationships (action/feedback) to this milieu. But the didactical intention of such a situation can act as a constraint; this is the case when the student believes in a teacher expectation, that could modify the nature of the subject/milieu system equilibrium and then the nature of the related conception. There resides the basic complexity of didactical systems.

Learning is a process of reconstruction of an equilibrium of the subject/milieu system which has been lost following perturbations of the milieu, or perturbations of the constraints on the system, or even perturbations of the subject herself (modification of her intentions, or as a consequence of a brain disease, etc.). The didactical *problématique* considers the case of perturbations provoked on purpose, with the intention to stimulate learning. The indicator of a perturbation is the gap, recognized by the subject, between the expected result of an action and the actual feedback from the milieu. This means on one hand that the subject is able to recognize the existence of a gap not acceptable with respect to her intention, and, on the other hand, that the milieu can provide identifiable feedback.

Sometimes the subject does not identify a gap whereas we, as observers, see that a gap should have been recognized. We call this unnoticed gap an *error* when it is the symptom of a conception, that is, the symptom of the resilience of a previous equilibrium of the subject/milieu system.

In the following section we propose a model of "conception" complementary to its definition, with the intention of providing an effective tool to concretely represent and analyze the corpora of data that can be constructed from the observation of students' activities. We will then take an example, the case of function.

2.4. From a Definition to a Model of Conception. We model a conception C by a quadruplet (P, R, L, Σ) in which:

P is a set of problems,
R is a set of operators,
L is a representation system,
Σ is a control structure.

Relating this model to the definition proposed in the preceding section, one can see that:

P corresponds to the class of the disequilibria the considered conception is able to recover. In mathematical terms, this is the problems it allows to be solved. In pragmatic terms, we will speak of the *sphere of practice* of the related conception.

R corresponds to the set of operators needed either to perform "concrete" actions on the milieu, or to transform and manipulate linguistic, symbolic or graphical representations.

L describes the linguistic, graphical or symbolic means which support the interaction between the subject and the milieu, either actions or feedback, as well as their outcomes.

Σ is needed to describe the components which support the monitoring of the equilibrium of the [S $\leftrightarrow$ M] system.

The first three components of this quadruplet have been borrowed from the characterization by Vergnaud (1991, p. 145) proposed for a conceptual field (in fact, this definition proposed by Vergnaud was coined at the beginning of the 1980s). We have introduced the fourth component in order to explicitly take into account the dimension of validation which is critical to mathematics and mathematics learning.

The very first question that this model must contend with is that of how it relates to the "reality" of students' ways of understanding as they have been studied and reported until now in our community of researchers. We consider this question for each of the four elements of the quadruplet.

The question of the concrete characterization of the set P of problems is difficult. One option would be to consider the set P to include all the problems that a conception provides efficient tools to solve. This was the option suggested by Vergnaud (1991) for the case of additive structures, but it appears that such an exhaustive description is in most cases out of reach. Another option is to consider P to be a finite set of problems from which other problems can be derived. This is the solution proposed by Brousseau (1997, p. 30). Whereas that would be the most elegant way to proceed, we presently don't know if such a generative set of problems can be constructed in order to describe the sphere of practice of any conception. Instead, the pragmatic position that we can presently implement consists of deriving the description of the set P from a careful task analysis in an empirical way, by building on a continuously developing understanding of the genetic and epistemic complexity of the mathematics considered. This approach can be strengthened by the analysis of historical and actual uses of mathematics (e.g., D'Ambrosio, 1993; Lave, 1988; Nuñes, Carraher, & Schliemann, 1993; Sierpinska, 1989; Thurston, 1994). This approach also reveals the duality of problems and conceptions which Soury-Lavergne (2003) has studied using the characterization we present here as a tool.

The question of the concrete characterization of the set R of operators is more classical. Operators are means to change the relationship between the subject and the milieu; they are the tools for action. Operators could be concrete, allowing the performance of actions on a material milieu, or abstract, allowing the transformation of linguistic, symbolic, or graphical representations. So, an operator could take the form of functionality at the interface of a piece of software, or of a syntactic rule to transform an algebraic expression, or it could even take the form of a theorem in an inference.

The representation system L consists of a repertory of structured sets of signifiers, which may or may not be of a linguistic nature, used at the interface between the subject and the milieu; these support action and feedback, and operations and decisions. Just to mention a few examples: algebraic language, geometric diagrams, natural language, and also interfaces of mathematical software and calculator keys are all examples of structured sets of signifiers. For a given conception, one or more of them can be assembled into a representation system. Whatever it is, depending on the state of the subject/milieu system, the representation system must be adequate to give account of the problems and to allow operators to perform.

We are aware of the difficulty which could be raised by the use of the word "representation," especially when it is read in the light of a psychological *problématique*. We do recognize that any "symbol as representation needs a living person who constructs the representation, or in comprehending reconstructs it" (Furth, 1969, p. 93). But we are here focusing on the system formed by a knowing subject and a milieu, not on any one of them in isolation. Indeed, representation in our sense – which is a semiotic sense – is the basic support of the observed behaviors. We neither mean to reduce the subject or the knowledge to the signifiers. Nor do we reduce the representation systems to a set of signifiers, which we consider in the context of semiotic registers as conceptualized by Duval (2006, p. 106). It is on purpose that our approach is limited here. For more detailed considerations, the reader can see Balacheff and Margolinas (2005).

The last dimension of a conception, the control structure Σ, is made up of all the means needed to make choices, to take decisions, and to express judgment. This dimension is often left implicit although one may realize that the criteria which allow one to decide whether an action is relevant or not, or whether a problem is solved, is a crucial element in understanding a mathematical concept. We would suggest that, in Vergnaud's seminal proposition, the control structure is implied by his reference to theorems-in-action or to inference (1991, pp. 141–142), which are meaningful notions only to the extent that they are associated with the recognition that the subject has procedures to check whether his or her actions are legitimate and correct. After Polya and a long tradition of research on metacognition, Schoenfeld (1985, pp. 97–143) has shown the crucial role of control in problem solving. More recently, Robert (1993) emphasized the role of meta-knowledge, demonstrating the need to treat control structures as such. Indeed this is directly related to a *problématique* of validation, which is intrinsically related to understanding (Balacheff, 1987, p. 160).

It is important to insist that this model of a conception aims at accounting for the subject/milieu system and is not restricted to one of its components, in that the representation system allows the formulation and the use of the operators by the active sender (the subject) as well as the reactive receiver (the milieu). The control structure allows us to express and discuss the means of the subject to decide on the adequacy and validity of an action, as well as the criteria of the milieu for "selecting" a feedback.

3. The Case of Functions

3.1. An Extensively Studied Theme. Instead of giving many examples, which we could explore only superficially in the limited space of the present article, we have chosen to investigate in a manner as precise as possible just one case, namely the case of "functions." The theme of function has been extensively studied. Many bibliographical references are available, all very different from one another. The notion of function is also at the intersection of several mathematical areas (numbers, limit, algebra, etc.) and requires considering several representation systems (graphical representations, symbolic language, etc.). All of that makes functions an important candidate to exemplify our approach.

The following are classical ways to categorize conceptions of function:

Function is a correspondence "law" (a function expresses the correspondence between two sets, an element of the first set being associated with a unique element of the second set).

Function is a symbolic expression (a formula).

Function is a graphical object.

The two first formulations come from Vinner and Dreyfus (1989, pp. 359-360). As the reader may know, these authors consider other conceptions of "function," such as "relation of dependence," "rule" and "operation." Other authors introduce other categories like "ratio and proportion" or "functional dependency" (Sierpinska, 1989) or function-as-processes (Breidenbach, Dubinsky, Hawks, & Nichols, 1992). These categories can be seen as refinements of the more general ones given above. Because of the fragility of the means we have to ascribe a conception to a student, we chose here to remain with the three main categories mentioned above.

The methods used to ascribe a conception to a student have usually included analyzing interviews or questionnaires. These instruments have often asked students whether there exists a function corresponding to a given specification (e.g.,Vinner & Dreyfus, 1989, p. 359), engaged students in modeling some situations (e.g., Breidenbach et al., 1992, p. 279), or even posed students the question "what is a function for you?" (Vinner & Dreyfus, 1989, p. 359). What students produce then is rather difficult to analyze. For example, one may wonder how it is possible to distinguish precisely between the category "dependency relationship between two variables" (a "relation of dependence") and the category "something which relates the value of x to that of y" (a "rule"). The issues pointed at here are, on one hand, that of the way data are collected and its effect on the diagnostic of conceptions, and on the other hand, that of the way in which these conceptions can be described.

The categories we have selected can be seen as invariant in the mathematics education literature, they in fact correspond to the three main representation systems associated with "function" – whether one considers research in mathematics education or research in the history of mathematics. Indeed, it is by historical analysis, using the classical works of several historians (Edwards Jr., 1979; Kleiner, 1989; Kline, 1972; Smith, 1958), that we will introduce a first proposition for modeling conceptions of "function."

3.2. Conceptions of Function from a Historical Perspective. A good starting point to identify the main conceptions of "function" in the course of the history of mathematics is to distinguish them by means of the main system of representation they implemented. One of the most ancient signs of the existence of function is tables and their uses. For example, Ptolemy (in the *Almagest*) knew that positions of planets change with time and compiled astronomical numerical tables (Youschkevitch, 1976, pp. 40–42). Arabian astronomers in the 10th and 11th centuries also used precise tables. Tables go with locating an isolated number using another number (or quantities), and so, the idea of variable is not yet present.

The association of curve with table played a critical role in formulating and solving the problem of determining the trajectories of the planets. Kline (1972), for example, has indicated that Kepler improved the computation of the position of planets essentially by adjusting geometrical curves and astronomical data, without any theoretical reference to explain why he considered the trajectories to be elliptical. The validity of the conjectured trajectories was then essentially related to the precision of the measurement of the planet positions and to the choice of a familiar geometric object, the ellipse, that permitted the description of the universe with simple mathematical laws. Kline also has noted that most of the functions introduced in the 17th century were first studied as curves (p. 338). In fact, curves, as trajectories of moving points, were the main object of study for mathematicians of that time (Kleiner, 1989).

The creation of the symbolism of algebra (by Viète, and later Descartes, Newton, and Leibniz) was decisive for the development of the concept of function. The separation of the study of functions from geometry is credited to Euler. Kleiner (1989, p. 284) emphasized that Euler's (1748) *Introductio in Analysin Infinitorum* offered an entirely algebraic approach without a single picture or drawing. The function was presented as the central object of the analysis. The analytic characterization of functions received a strong formulation by Euler, who stated that a function is an analytical expression formed in any manner from a variable quantity and constants.

Although considering a function as an analytical expression proved to be a powerful tool, it led to contradictions and was inadequate to solve some problems of the 18th century (e.g., the controversy of the vibrating strings). In 1755, Euler formulated a general definition of function expressing the notion of dependence between variable quantities and the notion of causality (Dhombres, 1988, p. 45). The potential of such a definition, together with the difficulties it brought along, stimulated many discussions up to the 20th century (Monna, 1972).

Each of these conceptions can be represented by a quadruplet:

Table conception: $C_T = (P_T, R_T, \text{Table}, \Sigma_T)$,
Curve conception: $C_C = (P_C, R_C, \text{Curve}, \Sigma_C)$,
Analytic conception: $C_A = (P_A, R_A, \text{Algebra}, \Sigma_A)$,
Relation conception: $C_R = (P_R, R_R, L_R, \Sigma_R)$.

The cases of *Table*, *Curve* and *Analytic* conceptions are sufficient to illustrate our purpose, so we will not develop in detail the case of the *Relation* conception. "Table," "Curve," and "Algebra" are used to refer to the corresponding representation systems (which is not the case for the Relation conception) that are characterized by their specific syntactic rules and their own criteria for validity. The question is then to examine C_T, C_C and C_A for whether their defining components – respectively the sets P of problems, the sets R of operators and the control structures Σ – show significant differences depending on the conception they aim to describe. We consider these differences from the point of view of the control structures.

The *Table* conception C_T has essentially empirical foundations; the validity of a table depends on the precision of measurement and of the related computations against the requirements of a given experimental context. In the case of the *Table* conception, which we ascribe to Kepler, the validity must be evaluated against the quality of the interpolations and predictions that the ellipse allowed. Therefore the corresponding control structure Σ_T was fundamentally of an empirical nature, providing the means that allowed the precision of tables to be verified with reference to the observations and to the measurements that had been carried out. However, the input/output table is the first means of representation used, the means by which quite a number of functions were shaped. Kline (1972, p. 338), reminds us that the table of the sine function was known with great precision long before the associated curve became a mathematical object. Then, the validity of the solution of a problem from the corresponding sphere of practice (P_T) did depend in an essential manner on the quality of rather concrete productions and of actions necessary to collect and treat data.

This also applies to the *Curve* conception whose corresponding sphere of practice (P_C), in the beginning of the 18th century, was constituted by the important problem of long distance navigation where coasts were out of sight. Thus, the sets of problems P_T as well as P_C were dominated by practical questions and R_T as well as R_C included – but not exclusively – techniques of measurement, computation, and drawing. We might even suggest that the ellipse of Kepler's first law was a geometrical object, ideally conceived of, but empirically used when constructing curves in order to access objects and to make predictions. A curve was not yet the graphical representation that we acknowledge nowadays as being the graph of a function considered as a relationship between entities (numbers or even quantities).

The *Analytic* conception C_A is of a different nature; it introduces a rupture in the epistemology of functions. A function defined by an analytical expression does not need to refer to an experimental field (either of natural phenomena or of mechanical drawings). It can be studied for itself, as Euler did by presenting functions as the object of study – called "analysis" (e.g., *Introductio in Analysin Infinitorum*). That does not mean that the *problématique* of modeling no longer plays any role; rather, it means that it is no longer central and does not characterize the conception. A purpose of the analysis of the 18th century (and of the 19th and 20th centuries) was the solution of functional equations, which were of great importance in physics (Dhombres, 1988), and the developments into infinite series, which played a central role as operators (R_A) in those solutions. The corresponding control structure Σ_A depends on the specific characteristics of algebra as a representation system and on the operators it allows to implement. Computation of symbolic expressions and mathematical proof are the key tools to decide whether a statement is valid or not. Indeed, symbolic representations are not the only ones to be available and to be used. Following C_A, a function can be associated with a graph, that is, a set of pairs (x, y) in the Cartesian plane (where y is the value of the function for a given x). This possibility often suggests close relationship between C_A and C_C that raises the question of the relationship between graph and curve. While the graph is a possible representation of a function displaying phenomena that algebraic expressions do not easily demonstrate (for example, the intersection of two lines), a curve is rather an evocation of the trajectory of a mobile point or of a geometric object, as Kline (1972, p. 339) expressed it when describing Newton's conception.

The general solution of partial differential equations expressing the vibrations of a finite string, subject to initial conditions, induced Euler to consider arbitrary functions that did not necessarily have analytic representations. The existence of such arbitrary functions was controlled by physical arguments and was related to the various possible initial forms of the string. The emergence of a *Relation* conception, C_R, of function then required the development of new modes of representation L_R and new control structures Σ_R in order to define what such functions could be and in order to work with them without any reference to an analytical representation. These developments took two centuries.

The conceptions of function we have modeled differ from each other in an essential way. The conception of curves as trajectories of a point, ascribed to Newton by Kline, (1972, p. 339) is fundamentally different from the Dirichlet conception of a subset of the Cartesian product of two sets satisfying given constraints (which guarantees a unique image for each element of the source set). The crucial point here is that "function" does not refer to the same object in the two cases but to objects that are different in essence, despite the fact that in modern terms we could mathematically interpret them in the same framework.

Without going too far in the discussion of these points we must notice that each representation system we consider, taken by itself (with its semiotic characteristics), has a different displaying power (which can be defined as the capacity to show or to hide what should be shown). These differences can be better understood by considering (a) the operators that can be implemented and (b) the corresponding control systems.

3.3. Students' Conceptions of Function. We focus on students from the secondary and post secondary levels. These students constitute the bulk of the population studied in the literature. They all have some knowledge of algebra and they all have been exposed to classical elementary functions. A classic in this area, Vinner's (1992) study on students' concept image of function, identified eight components of students' conceptions of function:

> (1) The correspondence which constitutes the function should be systematic, should be established by a rule and the rule itself should have its own regularities.
> (2) A function must be an algebraic term
> (3) A function is identified with one of its graphical or symbolic representations.
> (4) A function should be given by one rule.
> (5) Function can have different rules of correspondence for disjoint domains provided that these domains are regular domains (like half lines or intervals).
> (6) A rule of correspondence which is not an algebraic rule is a function only if the mathematical community officially announced it as a function.
> (7) The graph of a function should be regular and systematic.
> (8) A function is a one-to-one correspondence. (p. 200)

These components of the concept image of function result from investigations carried out by Shlomo Vinner with students in the Jerusalem area at the beginning of the 1908s. Since Vinner published this seminal work in 1983, these features have been confirmed as being largely common to students all over the world (Tall, 1996, pp. 297–301).

As they are presented to the reader, these features are not yet organized into conceptions. In particular, one misses indications about their domain of validity as well as about the way they could be implemented in a problem-solving situation. One can notice that several of them are tightly related to one system of representation, either algebraic or graphical. Actually, representations are good starting points to shed light on the question of the differentiation of conceptions.

For example, from the students' point of view, the idea that a graph-curve should exist in relation to an algebraic expression is central, both representations having to conform to certain constraints. We note here that there is a rich literature on understanding functions and their representations. A common distinction is made between the process and object aspects of functions (Dubinsky & Harel, 1992; Sfard, 1991). These aspects can be related to students' uses of different representations of functions (DeMarois & Tall, 1999; Schwingendorf, Hawks, & Beineke, 1992). The distinction between the global approach and the point-wise approach to functions is critical as well (Bell & Janvier, 1981; Even, 1998), showing in particular that the capacity to deal with (graphical and algebraic) representations of functions or to move from one representation to another is related to the flexibility in using different approaches to functions. Because of limitations of space we do not elaborate on all those aspects here but we limit ourselves to the specific case of the graph/curve distinction.

By the expression "graph-curve" we refer to two different entities that must be distinguished. Although relying on the word "curve" in both cases, this difference was well expressed by Sierpinska (1989) when she introduced the distinction between synthetic and analytic views of curves:

> Curve analytical view: a function is an "abstract" curve in a system of coordinates; this means that it is conceived of points (x, y), where x and y are related to each other somehow (p. 18).
>
> Curve, a synthetic view: [...] function is identified with its representation in the plane; it is a curve viewed in a concrete, synthetic way (p. 17).

This distinction reminds us of the one that is usually made in mathematics between curve and graph: curve refers to a geometrical object and graph to a representation of a function in the graphical representation system (one plots a graph). However, the distinction made by Sierpinska seems not to be exactly this one, since she added that the "relationship (*between x and y, the analytical view*) can be given by an equation. But the curve does not *represent* the relation. Rather, it is represented by the equation." (Sierpinska, 1989, p. 17). This remark draws our attention to the confusion likely to be made by students between the geometric setting and the calculus setting (in the sense of setting proposed by Douady, 1985), which is facilitated by the identity of the diagrams used by both of them.

Thus, we suggest considering two types of student conceptions: the Curve-Algebraic and the Algebraic-Graph. To consider them in a more precise way we describe them by two quadruplets:

Curve-Algebraic conception: $C_{CA} = (P_{CA}, R_{CA}, \text{Curve-Algebraic}, \Sigma_{CA})$,
Algebraic-Graph conception: $C_{AG} = (P_{AG}, R_{AG}, \text{Algebraic-Graphic}, \Sigma_{AG})$.

These two conceptions apparently share the same representation system, algebraic and graphic, but these systems have different degrees of importance in both (this is suggested by the different order in their given names). In the case of the Curve-Algebraic conception, the criterion is that the curve must be related to an algebraic representation (its equation, in Sierpinska's terms). In the case of the Algebraic-Graph conception, the criterion is that the algebraic representation must be associated with a graph which one must be able to *plot*. The distinction between graph and curve is not very easy to make because both rely on graphical representations. This difficulty is very likely to be one of the reasons why some students do not recognize in which setting they are working (and consequently they experience difficulty in knowing what is legitimate to do). It is by looking at the rules that both conceptions require one to abide by, the tools that they allow one to use, and the control structures they provide that we can shape the distinction.

The exploration and modeling of students' conceptions is a rather difficult job to accomplish when one can only rely on the evidence provided by tasks where students are directly asked to answer the question "what is a function?" or where they are invited to decide if graphs or descriptions of correspondences represent functions (e.g., "Does there exist a function all of whose values are equal to each other?" in Vinner & Dreyfus, 1989, Figure 1, p. 359). One can notice two important points about such experiments. First, they show a distance between the answers to the question "what is a function?" and those to the tasks requiring a decision on descriptions of correspondences. This distance confirms the relationship between conceptions and problems. Different tasks (like providing a definition or describing a function) may call for different conceptions. This suggests that students' conceptions are less accessible in statements about a concept than in problem solving situations involving this concept. For this reason, the characterization of conceptions requires the provision of evidence for the relations between conceptions and problems. The second point concerns the type of tasks proposed to students. In none of these tasks did students need to perform actions in order to produce a solution, rather, they had to activate control operators in order to produce a decision. Vinner (1992) himself emphasized the fragility of this type of investigation when he observed a large number of occurrences of what he called "irrelevant reasoning" that he defined as "justification given by the student because he or she assumed it was the right thing to say (and no meaningful thought was involved)" (p. 206).

This observation can be analyzed in light of the very clear description by Castela (1995) of the situation she used to explore students' conceptions of tangent. Castela assumed that the drawings she proposed to students were "straightforward," that their approximate character (they were sketches of functions) did not hide any surprising feature and that the functions represented were what they seemed to be ["Les fonctions représentées sont bien ce qu'elles ont l'air d'être" (p. 21)]. One can then legitimately think that the observed situations may depend heavily on the quality of the *experimental contract*, the (implicit) contract defining the way students should understand the experimental situation.

A researcher may claim that an investigation targeted students' conceptions, but it may be the case that what was observed were students' contingent opinions and not students' conceptions as expected. Students' answers might be a way to fulfill the teacher/observer expectation – this is exactly what Vinner feared. Indeed, how could it be different? Especially in that the graphical representation system of the tasks provided to students by Castela used what we suggest be called *function icons* instead of functions' graphical representations.

In order to understand this point, we invite the reader to consider such tasks from the point of view of the nature of the feedback the students can expect from the environment provided by the situation in which they are involved. They cannot perform any relevant action on the graphical representation since these representations are just sketches, see Figure 2 for Castela's pictures or refer to the tasks proposed in Vinner and Dreyfus (1989).

But, what are the problems for which students' conceptions provide tools allowing them to propose reasonable solutions, at least in the student's eyes? Indeed, the characterization of a conception should not be separated from the characterization of the problem situation that provides evidence for it.

To go beyond the definition and investigate more problem-oriented situations is what Vinner (1992) intended when he asked students to decide on the continuity or on the differentiability of a function. He concluded from his study that for students:

(a) for a "function to be continuous is the same as being defined and to be discontinuous is the same as being undefined at a certain point" (p. 205),
(b) "continuity or discontinuity is related to the graph," for example, "a function is continuous because its graph can be drawn in one stroke" (p. 206), or

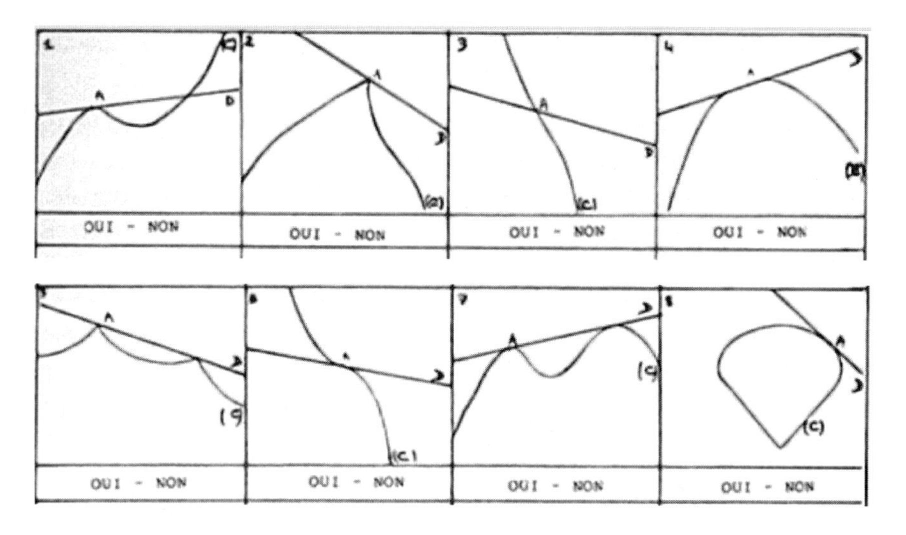

FIGURE 2. Sketches from Castela (1995, p. 17).

(c) "there is a certain reference to the concept of limit" as in "The function is continuous because it tends to a limit for every x" (p. 207).

Whereas the direct question "what is a function?" gave essentially an indication of possible elements of the control structures, we get here an insight about the tools students may have available. Some of these tools clearly relate to a Curve-Algebraic conception. For example, drawing the graph of

$$y = \begin{cases} 1, & x > 0 \\ -1, & x < 0 \end{cases}$$

is a tool which allows us to control the aspect of the graphical object (the graph shows a disruption) and so allows us to conclude that the function $f(x) = \dfrac{|x|}{x}$ is discontinuous (Vinner, 1992, task B1). Other tools relate to an Algebraic-Graph conception, like the ones involving a criterion of limit.

Vinner (1992) did the same with the derivative, identifying (a) correct algebraic characterizations, (b) descriptions of symbolic manipulations to be performed, or (c) correct definitions within the graphical representation system. However, Vinner did not seem to have effective tools for describing the so-called incorrect solutions. This resulted in the use of attributes like "vague," "fuzzy," or "meaningless." Altogether these vague and fuzzy answers represent 46% of the sample of 119 students. We must realize here that students were confronted with graphical representations of functions and asked *to decide* and *to justify* whether they were continuous or not. In the case of the derivative they were directly asked *to define* it, answering the question "what is a derivative?" Again we observe the fragility of the data gathered, insofar as these tasks may well not reveal students' conceptions of derivative but their possible opinion offered in a rather embarrassing situation.

In order to decide or to justify a statement, one has to mobilize one's conception in a different way, more operational, than to define it, say, *in abstracto*. The point here is that what we learn from students' discourse or work must be analyzed against the characteristics of the situation in which this discourse or work is produced. We could say that *conceptions* and *problems* are dual entities (Balacheff, 1995): In order to characterize students' conceptions one should provide them with meaningful problem situations, with enough complexity so that they can engage their conceptions in a significant way, demonstrating for us the tools they use and the nature of the control they involve in the task. Note, we do not pretend that this is enough to solve our diagnostic problem, but that it is a necessary condition.

Artigue (1992) gave us an excellent example of the benefit of coming back to the students' context when she analysed, in the case of the qualitative approach to differential equations, a "false theorem": "if $f(x)$ has a finite limit when x tends toward infinity, its derivative $f'(x)$ tends to 0" (p. 130). She wrote:

> In the field of differential equations, monotonicity conceptions may especially act as obstacles as on one hand, effective predictions are implicitly based on the extra-hypothesis of monotonicity and on the other hand, theorems have to get free of these extra-hypothesis. Let us be more explicit: When sketching solution curves, we draw the simplest one compatible with the identified set of constraints, but in doing this, we add extra constraints concerning the convexity that can be expressed roughly in the following way: convexity has to be the least changing possible or, in algebraic terms, the sign of f'', for a solution f, has to be the most constant possible. So, f' is implicitly the most monotonic possible. (p. 130)

She concludes that the mentioned false theorem,

> ...can be seen as an instantiation of such an extra-hypothesis: if for x large enough, f' is monotonic, then f' has necessarily a limit (finite or not) and the unique limit compatible with an horizontal asymptote is 0. In other words, by adding the condition f' monotonic, this false theorem becomes a true one. (p. 130)

Such an analysis of a tool – which is the actual status of theorems in students' practice – is evidence of the interaction between the graphic and the algebraic representation system and of the role played by the characteristics of the sphere of practice (a role which was recognized by Vinner when he pointed to the phenomenon of compartmentalization).

Let us then come back to the Curve-Algebraic conceptions and the Algebraic-Graph conceptions. We distinguish these conceptions by the type of tools and controls they involve in problem-situations, and their interactions with the representation system (either graphic or symbolic) on which they depend and which makes them tangible.

To state what the set P is in each case remains an open problem for research in mathematics education, even in this heavily investigated domain. One may observe, at this point, that history is of no great help. Actually, students' conceptions are very difficult to analyze against what history teaches us about the evolution of the concept of function. And, indeed we would be very cautious with the idea that the "historical study of the notion of function together with its epistemological analysis helped us to analyze the student's mathematical behavior" (Sierpinska, 1989, p. 2). It is clear that epistemological analysis is an essential tool, but historical analysis may induce a view of the notion of function that hides the role played by the modern school context. Historical analysis will delineate the notion from the mathematical point of view, and from the cognitive point of view we must be prepared to see things in a rather different way. Sierpinska acknowledged that "students' conceptions are not faithful images of the corresponding historical conception" (p. 19). For example, a question one has to consider is that of knowing what could be the essential difference between students' algebraic conceptions and the "corresponding" historical conceptions. It is also striking that tables play a very limited role, if at all, in situations involving functions. If tables are present, it is in relation to concrete situations in which the aim is less one of analyzing a function than one of analyzing data: the function "disappears" behind its use as a tool for data analysis.

Students' spheres of practice are radically different from the ones of the mathematicians that historians consider. The didactical system has first introduced students to "good" functions, mainly playing with two different settings: algebraic and graphical. As Sierpinska (1989) noticed, shapes of graphs of elementary functions can become prototypes of conceptions. Depending on the curriculum they have been exposed to, students have available more or less sophisticated tools in order to analyze some elementary algebraic formulas and to describe the behavior of the corresponding functions. These spheres of practice may be described in detail following a close analysis of textbooks which are available to students; see Mesa (2004, and this volume), for investigation in this direction.

3.4. A Case Study: The Curve-Algebraic and Algebraic-Graph Conceptions. Our intention in this section is to demonstrate how the algebraic and the graphical representation system, the rules-tools, and the controls required in the problem lead to the differentiation of the *Curve-Algebraic* and the *Algebraic-Graph* conceptions of function, taking the case of students from the 12th grade in France. The experimental context was provided by the dynamic geometry computer software environment *Cabri-Geometry II* (distributed by Texas Instruments and hereafter referred to as "Cabri"). Cabri allows one to construct objects and display the dynamic relation between their graphical and algebraic representations (Gaudin, 2002).

The diagram of a parabola in a system of coordinates was presented to the students. Some limitations on the manipulation of the drawing were imposed in order to constrain the action of the students and open better opportunity for their conceptions to be elicited (see below for the specification of Situations 1 and 2). The diagram could be turned around its vertex by dragging the parabola's axis of symmetry. The software would update dynamically the equation associated to the diagram after each of these manipulations (for examples, see Figures 3 to 6). In each situation, we asked students to give the equation of the line tangent to the parabola at point M_0 and to draw this tangent. The students knew how to draw a straight line, using Cabri, whose Cartesian equation was given.

Situation 1: the parabola could be manipulated by moving its axis of symmetry (grabbing a point given on this axis, see Figure 3) so that it could reach a "vertical" position (Figure 5). In this position the parabola is the graph of a quadratic function.

Situation 2: the movements of the diagram were constrained. The parabola could not turn completely around the vertex, hence it could not reach the vertical position as in Situation 1 and students could not get the familiar picture of a quadratic function.

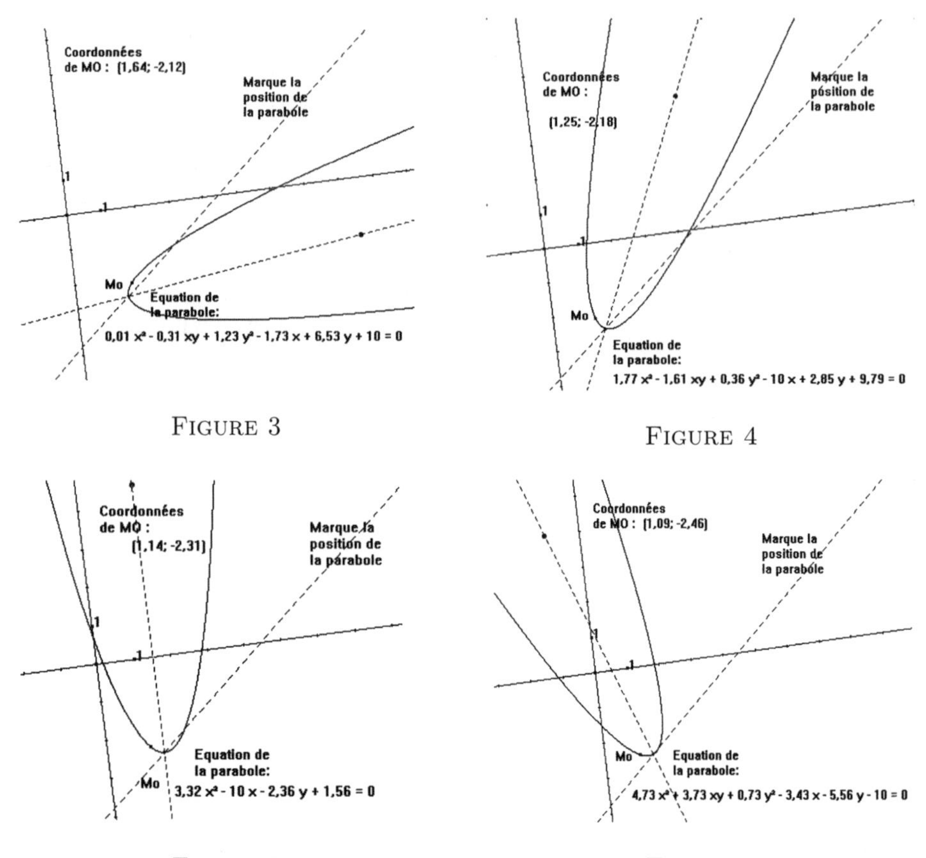

FIGURE 5

FIGURE 6

Note: The French text "Marque la position de la parabole" means that the dashed line was drawn to keep a record of the initial position of the parabola so that the students could always come back to it if they wanted. The other French texts are "Coordonnées de M0" for M_0 coordinates and "Equation de la parabole" for equation of the parabola.

Twelfth grade students were supposed to know:

(a) that a parabola whose equation is $y = ax^2 + bx + c$ is the graph of the function f defined by $f(x) = ax^2 + bx + c$;
(b) that an equation of the tangent to the graph of a function f at point $M_0(x_0, y_0)$ is $y = f'(x_0)(x - x_0) + f(x_0)$, f' being the derivative of f;
(c) some rules of differentiation;
(d) nothing about conics.

In order to frame the analysis of students' problem-solving activity we sketch the strategies that could be expected in both situations.

Expected strategies for Situation 1.

(a) The capacity to control the position of the parabola allows reaching the vertical position to get the graph of a quadratic function f. Its algebraic representation is $f(x) = (1/2.36)(3.32x^2 - 10x + 1.56)$ (see Figure 5).
(b) The property "An equation of the tangent of a graph of a function f at point $M_0(x_0, y_0)$ is $y = f'(x_0)(x - x_0) + f(x_0)$, where f' is the derivative of f" is an operator which processes the algebraic expression of f.

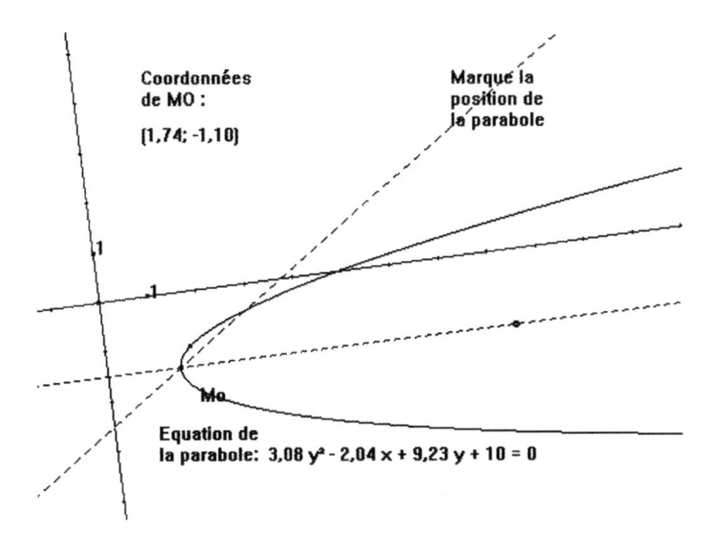

FIGURE 7

(c) One gets the equation of the tangent of the parabola ($y = -1.01x - 1.17$) and draws the tangent with Cabri.

Expected strategies for Situation 2.

(a) The parabola can not reach the vertical position, but in the horizontal position it is the graph of a quadratic function in another system of coordinates defined by the same origin, the point $(0, -1)$ as the unit on the x axis and the point $(1, 0)$ as the unit on the y axis (see Figure 7).

(b) The new system is associated to the change of variable $X = -y$ and $Y = x$. This change is a tool to find the equation $3.08X^2 - 2.04Y - 9.23X + 10 = 0$ of the parabola in the new system and to get the algebraic representation $f(X) = (1/2.04)(3.08X^2 - 9.23X + 10)$ of the function of which the parabola is a graph in the new system.

(c) As in Situation 1, the property "An equation of the tangent of a graph of a function f at point $M_0(x_0, y_0)$ is $y = f'(x_0)(x - x_0) + f(x_0)$, where f' is the derivative of f" is an operator which processes the algebraic expression of f. One gets the equation of the tangent of the parabola ($Y = -1.20X + 3.06$). The change of variable $X = -y$ and $Y = x$ is a tool to find the equation of the tangent in the initial system of coordinates: $1.20y - x + 3.06 = 0$. One draws the tangent with Cabri.

The drawing of a parabola on the screen can be conceptualized as a *graph* or as a *curve*. Recall from Section 3.2, the graph of a function f is the set of points $(x, f(x))$ in a system of coordinates; a curve is a geometrical object which can be drawn independently from the existence of a system of coordinates. Considering the parabola as the graph of a quadratic function is essential in order to associate controls to the tools involved in the above-mentioned strategies.

We will present some aspects of the analysis of the observed behaviors of two pairs of students. André and Rémi were one pair, and Loïc and Sylvain were the other pair. Each pair of students worked together on these problems. We have chosen these two cases because they illustrate in a very good way both conceptions: *Curve-Algebraic* in the case of André and Rémi[1] and *Algebraic-Graph* in the case of Loïc and Sylvain.

From these students' points of view, getting the equation of the tangent of the parabola meant "deriving the equation" (see below) and using the operator "**if** f **is** a function of x **then** the equation of the tangent at the point $M(x_0, f(x_0))$ **is** $y = f'(x_0)(x - x_0) + f(x_0)$." As long as the parabola was not the representation of a graph, students could not perform the rules of differentiation they knew (and so, use the operator and get the equation of the tangent). Their decisions and actions to make the equation in accordance with these rules revealed different controls and tools involved in the problem. By identifying the controls used by André and Rémi, or by Loïc and Sylvain, to establish if an equation could be derived or not in Situation 1, we can discriminate between conceptions.

Situation 1: André and Rémi. For these students, the equation could be derived if its form was $y = \langle\text{an expression in } x\rangle$. Since this was not the case, they tried to change the position of the parabola. They moved it to the vertical position. They chose this position while controlling the equation which, as a result, became simpler and simpler. When y disappeared from the equation the students said "that's good" and when the coefficient of xy decreased, "that's *perfect*." They obtained an equation that they called a "*neat* equation" that could be "derived."

Situation 1: Sylvain and Loïc. This pair of students found the equation on the screen really complicated ("monstrous"). They decided that "the equation was not the representation of a function" because the graph did not pass the vertical line test ("there were two values for one x"). They stated that the equation could not be derived. The control on the derivability was based on the form of the equation which had to be "y=some x, no y." Sylvain anticipated that xy and y^2 should disappear in the equation when the parabola would reach the vertical position, so they decided to get this position.

Situation 1: Comparison. The actions of both student pairs in Situation 1 were the same: moving the parabola to the vertical position and getting the equation of the tangent as described in the expected strategy. But the means of control and the associated representation system differed in an important way. André and Rémi's means of control referred essentially to the algebraic representation system: getting an adequate symbolic writing of the equation. In Sylvain and Loïc's case, they referred to the graphic and algebraic representation system: satisfying the vertical line test to get an equation of the form "y=some x, no y." These controls played an essential role in students' decisions in Situation 2. Noticing that the parabola could not reach the vertical position anymore, both student pairs had to change their strategies.

Situation 2: André and Rémi. In this case, the pair decided to move the parabola to the horizontal position because the equation appeared then to be the simplest one they could expect. They obtained an equation ($3.08y^2 - 2.04x + 9.23y + 10 = 0$) which they described as "nice" and "the best." But this equation was not yet in accordance with the rule of differentiation and students proposed to "change x with y and y with x." Again, their control was essentially algebraic. They were skeptical of the legitimacy of such a change. They proposed to change the system of coordinates into the new one according to the mapping of

[1]For more details, see Gaudin (2005).

variables: $X = -y$ and $Y = x$. They got an expression that they could differentiate "$f(x) = (1/2.04)(3.08X^2 - 9.23X + 10)$" – note that they wrote "x" and not "X" in "$f(x)$." The important point here is the meaning students ascribed to the change of system. They did not relate this change to the other objects of the situation; they used the first coordinate of M_0 in the operator "**if** f **is** a function of x **then** the equation of the tangent at the point $M(x_0, f(x_0))$ **is** $y = f'(x_0)(x - x_0) + f(x_0)$." Consequently, the equation of the tangent was not rewritten in the initial system prior to drawing the tangent (e.g., as described above by the *Expected strategies*), and the equation of the tangent line was not correct. Clearly, the change of system was not associated with a control in the graphical representation system. It was associated with a control on the symbolic writing of the equation conforming to $y = \langle \text{an expression in } x \rangle$.

Situation 2: Sylvain and Loïc. Loïc moved the parabola to the horizontal because this position appeared to him to be "better than any position." Sylvain did not consider this position as a better one to solve the problem. This opinion was consistent with the vertical line test he used in Situation 1 and was confirmed by the presence of y^2 in the equation ("I don't see... how will you cope with the y^2?"). He proposed to change the system of coordinates: the new proposed system being the one in which the parabola satisfied the vertical line test, and so, the one in which the parabola was the graph of a function. Thus, unlike the other student pair, it was a control in the graphical representation system which led to he choice of the new system of coordinates. This control allowed a distinction between graph and curve more effective than getting a nice equation. In this context, "more effective" means that the control was associated to new tools used by the students: reading the coordinates of M_0 in the new system, getting the equation of the tangent in which the parabola is a graph, and then writing this equation in the initial system to draw the tangent.

The identification of different conceptions in these two cases is possible by looking at the controls, the tools, and the associated representation systems the students used to solve the problems. In the present situation, the actions we observed refer to rather different controls and do not define the same settings of work.

Curve-Algebraic conception. The control associated to the operator "**if** f **is** a function of x **then** the equation of the tangent at the point $M(x_0, f(x_0))$ **is** $y = f'(x_0)(x - x_0) + f(x_0)$" is an algebraic control. The parabola is a geometric object designated by an equation and some of its positions are more or less operational from the point of view of this control. Thus, the algebraic tools and transformations, such as a change of variables, do not apply to the objects of the geometric setting (e.g, the parabola, M_0, the tangent straight line). Calculus is reduced to symbolic transformations.

Algebraic-Graph conception. Graphic controls (e.g., recognizing the parabola as a graph, reading the situation in a new system) are related to algebraic tools (e.g., the operator "**if** f **is** a function of x **then** the equation of the tangent at the point $M(x_0, f(x_0))$ **is** $y = f'(x_0)(x - x_0) + f(x_0)$," the change of variables that works on every object of the system). Because of this relation, calculus can be based on variables linked by an equation and/or a graph.

Moreover, it should be noted that drawings and symbolic representations are not used in the same way in these two conceptions. Eventually, they are not components of the same representation systems, despite the similarities of the signs used. In the case of the *Curve-Algebraic* conception, the diagram is the reification of a geometrical object. In the case of the *Algebraic-Graph* conception, the diagram is an object in the graphical system which is considered in close coordination with an object from the algebraic representation system.

4. Conclusion: Conception, Knowing, and Concept

Modeling students' knowing of mathematics is a difficult task, one which we need to achieve in order to be able to better design teaching situations or learning environments. To contribute to the search for a solution we have proposed a definition of the notion of "conception" and a modeling framework. This approach is an attempt to capture the core of a possible characterization of a conception seen as a situated knowing providing the tools and the controls to solve problems from a given sphere of practice. This proposition is intended to overcome the epistemological contradiction of an individual being a rational person but still likely to demonstrate contradictory behaviors from the point of view of an observer. Moreover, it provides an instrument to more precisely account for the possible meaning of students' behaviors and outcomes by describing more precisely how the representations, actions, and controls they use are related and contribute to the problem-solving process.

One may notice that as a result of our definition of conception, we now may have three notions which seem likely to compete in mathematics education discourse, namely: conception, knowing and concept. Our contribution would not improve matters unless we were able to relate these three notions in a clear manner, demonstrating their complementarities. This is possible by defining a knowing as a set of conceptions. A knowing is a set of conceptions ascribed to a subject, whereas a conception represents an instantiation of this knowing by a given situation. The universe of an individual can be described by means of several different spheres of practice, each calling for different conceptions, such as different instances of the individual's knowing. It is only from the point of view of the observer that all these instances relate to the same knowing. Individuals are not necessarily or always aware of the possible relations among their own conceptions. In order to stimulate this awareness, a special event is necessary, which consists of a situation bringing to the fore a problem which forces us to consider two different spheres of practice which are usually mutually exclusive. This is the basis for the design of a learning situation which could stimulate the students feeling of intellectual need[2] to express their conceptions and possibly to reconsider them.

In order to be able to establish links among conceptions being ascribed to the student, the observer needs to translate each system of representation into a single system of representation – in general, the observer's own system of representation. All our analyses as researchers in mathematics education are based on this possibility. This aspect of representations and their use in research is essential, although rarely noticed: we can model a conception, relate and compare conceptions, only if we can ascribe to them representation systems and if we can relate these systems of representation to ours. This imposes constraints on our analysis that could be a kind of anachronism or over-interpretation of learners' behaviors and understanding. We claim that the only way to cope with this inherent difficulty to our task

[2]We borrow here the expression proposed by Harel (2006).

is to make things as explicit as possible. The modeling framework proposed in this paper could contribute to achieving such explicitness.

Considering now the term "concept." We suggest the use of *concept* to refer to the set of *all* the knowings that share the same content of reference. This answers positively the question posed by Rolf Biehler: "Couldn't we say that the meaning of a mathematical concept is the synthesis of all of its uses?" – the question was posed in a working paper he shared in the context of the BACOMET IV project "Meaning in mathematics education" (1996). On the other hand, we are at some distance from Sfard's (1991) proposal that,

> The word "concept" (sometimes replaced by "notion") will be mentioned whenever a mathematical idea is concerned in its "official" form – as a theoretical construct within "the formal universe of ideal knowledge"; the whole cluster of internal representations and associations evoked by the concept – the concept's counterpart in the internal, subjective "universe of human knowing" – will be referred to as "conception." (p. 3)

However, if the view of "concept" could be considered more pragmatic in our presentation, the view of "conception" we have developed is not so far away from the one Sfard proposed. The fact that the two notions of "concept" and "conception" are sufficient for Sfard is consistent with the fact that the level of knowing is necessary only when one wants to reconstruct the totality of the epistemic subject, considering the variety of the situations in which one is engaged. If the situation and what is at stake in it are precise enough, the notion of conception is sufficient. Well defined or not, considered from the "misconception" perspective or from a more general perspective, conception has always played a central role in mathematics education.

Acknowledgments

This paper has a long history. It has benefited from the comments and questions of the participants in the BACOMET IV project "Meaning in mathematics education" (1993-1996) and extensive discussions with participants in the Rutgers RISE project "Understanding students' understanding" (2001). The writing of this paper was finished in San Diego, where it benefited from the stimulating input from Guershon Harel and his group. Finally, we would like to thank especially Patricio Herbst for the final polishing of this text and its careful editing. This version follows a significantly different one available at the following link: `http://hal.ccsd.cnrs.fr/ccsd-00001510`.

References

Arsac, G., Balacheff, N., & Mante, M. (1992). Teacher's role and reproducibility of didactical situations. *Educational Studies in Mathematics*, *23*, 5–29.

Artigue, M. (1991). Épistémologie et didactique. *Recherches en didactique des mathématiques*, *10*(2/3), 241–285.

Artigue, M. (1992). Functions from an algebraic and graphic point of view: Cognitive difficulties and teaching practices. In E. Dubinsky & G. Harel (Eds.), *The concept of function* (pp. 109–132). Washington, DC: Mathematical Association of America.

Bachelard, G. (1938). *La formation de l'esprit scientifique.* Paris: Vrin.

Balacheff, N. (1987). Processus de preuves et situations de validation. *Educational Studies in Mathematics, 18*, 147–176.

Balacheff, N. (1995). Conception, connaissance et concept. In D. Grenier (Ed.), *Didactique et technologies cognitives en mathématiques, séminaires 1994-1995* (pp. 219–244). Grenoble: Université Joseph Fourier.

Balacheff, N. (1998). Construction of meaning and teacher control of learning. In D. J. Tinsley & D. C. Johnson (Eds.), *Information and communication technologies in school mathematics* (pp. 111–120). London: Chapman & Hall.

Balacheff, N., & Margolinas, C. (2005). ckȼ Modèle de connaissances pour le calcul de situations didactiques. In A. Mercier & C. Margolinas (Eds.), *Balises en didactique des mathématiques* (pp. 75–106). Grenoble: La Pensée Sauvage.

Bell, A., & Janvier, C. (1981). The interpretation of graph representing situations. *For the Learning of Mathematics, 1*(2), 34–42.

Bourdieu, P. (1990). *The logic of practice.* Stanford, CA: Stanford University. (English translation of *Le sens pratique.* Paris: Les éditions de Minuit. 1980)

Breidenbach, D., Dubinsky, Hawks, J., & Nichols, D. (1992). Development of the process conception of function. *Educational Studies in Mathematics, 23*, 247–285.

Brousseau, G. (1997). *Theory of didactical situations in mathematics.* Dordrecht: Kluwer.

Castela, C. (1995). Apprendre avec et contre ses connaissances antérieures. *Recherches en didactique des mathématiques, 1*(15), 7–47.

Confrey, J. (1986). *"Misconceptions" across subject matters: Charting the course from a constructivist perspective.* Presentation at the annual meeting of the American Educational Research Association, San Francisco, CA.

Confrey, J. (1990). A review of the research on students conceptions in mathematics, science, and programming. In C. Courtney (Ed.), *Review of research in education* (pp. 3–56). American Educational Research Association.

D'Ambrosio, U. (1993). *Etnomatemática.* São Paulo: Editora Atica.

DeMarois, P., & Tall, D. O. (1999). Function: Organizing principle or cognitive root? In O. Zaslavksy (Ed.), *Proceedings of the 23rd annual conference of the International Group for the Psychology of Mathematics Education* (Vol. 2, pp. 257–264). Haifa, Israel: Technion – Israel Institute of Technology.

Dhombres, J. (1988). Un texte d'Euler sur les fonctions continues et les fonctions discontinues, véritable programme d'organisation de l'analyse au 18ième siècle. *Cahier du Séminaire d'Histoire des Mathématiques, Université Pierre et Marie Curie, Paris, 9*, 23–68.

Douady, R. (1985). The interplay between different settings. Tool-object dialectic in the extension of mathematical ability. In L. Streefland (Ed.), *Proceedings of the 9th annual conference of the International Group for the Psychology of Mathematics Education* (Vol. 2, pp. 33–52). Utrecht: State University of Utrecht.

Dubinsky, E., & Harel, G. (1992). The nature of the process conception of function. In E. Dubinsky & G. Harel (Eds.), *The concept of function* (pp. 195–213). Mathematical Association of America.

Duval, R. (2006). A cognitive analysis of problems of comprehension in a learning of mathematics. *Educational Studies in Mathematics, 61*, 103–131.

Edwards Jr., C. H. (1979). *The historical development of calculus.* Berlin: Springer.

Even, R. (1998). Factors involved in linking representations of functions. *Journal of Mathematical Behavior, 17*(1), 105–121.

Furth, H. G. (1969). *Piaget and knowledge: Theoretical foundations.* Englewood Cliffs, NJ: Prentice-Hall.

Gaudin, N. (2002). Conceptions de fonction et registres de représentation, étude de cas au lycée. *For the Learning of Mathematics, 22*(2), 35–47.

Gaudin, N. (2005). *Place de la validation dans la conceptualisation, le cas du concept de function.* Unpublished doctoral dissertation, Université Jospeh Fourier, Grenoble1, Grenoble. (Retrieved April 12, 2008 from http://tel.archives-ouvertes.fr/tel-00113100/en/)

Harel, G. (2006). The DNR system as a conceptual framework for curriculum development and instruction. In R. Lesh, J. Kaput, & E. Hamilton (Eds.), *Foundations for the future: The need for new mathematical understandings & abilities in the 21st century.* Hillsdale, NJ: Erlbaum.

Kleiner, I. (1989). Evolution of the function concept: A brief survey. *The College Mathematics Journal, 20*(4), 282–300.

Kline, M. (1972). *Mathematical thought from ancient to modern times.* New York: Oxford University Press, US.

Lave, J. (1988). *Cognition into practice.* Cambridge: Cambridge University.

Mesa, V. (2004). Characterizing practices associated with functions in middle school textbooks: An empirical approach. *Educational Studies in Mathematics, 56,* 255-286.

Monna, A. F. (1972). The concept of function in the 19th and 20th centuries, in particular with regard to discussions between Baire, Borel and Lebesgue. *Archive for History of Exact Sciences, 9*(1), 57–84.

Nuñes, T., Carraher, D., & Schliemann, A. (1993). *Street mathematics and school mathematics.* Cambridge: Cambridge University Press.

Pichot, A. (1994). Pour une approche naturaliste de la connaissance. *Lekton, 4,* 199–241.

Rabardel, P. (1995). Qu'est-ce qu'un instrument? *Les Dossiers de l'Ingénierie Éducative, 19,* 61–65.

Resnick, L., & Collins, A. (1994). *Cognition and learning* (Tech. Rep.). University of Pittsburg: Learning Research and Development Center.

Robert, A. (1992). Problèmes méthodologiques en didactique des mathématiques. *Recherches en didactique des mathématiques, 12*(1), 33–58.

Robert, A. (1993). Présentation du point de vue de la didactique des mathématiques sur les métaconnaissances. In M. Baron & A. Robert (Eds.), *Métaconnaissances en IA, en EIAO et en didactique des mathématiques* (pp. 5–18). Paris: Institut Blaise Pascal. (RR LAFORIA 93/18)

Salin, M.-H. (1976). *Le rôle de l'erreur dans l'apprentissage des mathématiques de l'école primaire.* Université de Bordeaux: IREM de Bordeaux.

Schoenfeld, A. H. (1985). *Mathematical problem solving.* Orlando, FL: Academic.

Schoenfeld, A. H. (1987). *Cognitive science and mathematics education.* Hillsdale, NJ: Erlbaum.

Schoenfeld, A. H., Smith, J., & Arcavi, A. (1993). Learning: The microgenetic analysis of one student's evolving understanding of a complex subject matter. In R. Glaser (Ed.), *Advances in instructional psychology* (Vol. 4, pp. 55–175). Mahwah, NJ: Erlbaum.

Schwingendorf, K., Hawks, J., & Beineke, J. (1992). Horizontal and vertical growth of the students' conception of function. In E. Dubinsky & G. Harel (Eds.), *The concept of function* (pp. 133–152). Washington, DC: Mathematical Association of America.

Sfard, A. (1991). On the dual nature of mathematical conceptions: Reflections on processes and objects as different sides of the same coin. *Educational Studies in Mathematics, 22*, 1–36.

Sierpinska, A. (1989). *On 15-17 years old students' conceptions of functions, iteration of functions and attractive fixed points.* Institut de Mathématiques, preprint 454. Varsovie: Académie des Sciences de Pologne.

Smith, D. E. (1958). *History of mathematics.* New York: Dover. (Vol. II, especially Chapter X)

Soury-Lavergne, S. (Ed.). (2003). *Baghera Assessment Project, designing a hybrid and emergent educational society.* Cahier Liebniz 81. Retrieved August 22, 2007, from *Les cahiers du laboratoire Liebniz* web site: `http://www-leibniz.imag.fr/LesCahiers/Cahiers2003.html`. Grenoble, France: Laboratoire Leibniz–IMAG.

Stewart, J. (1994). Un système cognitif sans neurones: Les capacités d'adaptation, d'apprentissage et de mémoire du système immunitaire. *Intellectika, 18*, 15–43.

Tall, D. (1996). Functions and calculus. In A. Bishop (Ed.), *International handbook of mathematics education* (pp. 289–326). Dordrecht: Kluwer.

Thurston, W. P. (1994). On proof and progress in mathematics. *Bulletin of the American Mathematical Society, 30*(2), 161–177.

Vergnaud, G. (1981). Quelques orientations théoriques et méthodologiques des recherches françaises en didactique des mathématiques. *Recherches en didactique des mathématiques, 2*(2), 215–231.

Vergnaud, G. (1991). La théorie des champs conceptuels. *Recherches en didactique des mathématiques, 10*(2/3), 133–169.

Vinner, S. (1983). Concept definition, concept image and the notion of function. *International Journal of Mathematical Education in Science and Technology, 14*, 293–305.

Vinner, S. (1987). Continuous functions – images and reasoning in college students. In J. C. Bergeron, N. Herscovics, & C. Kieran (Eds.), *Proceedings of the 11th annual conference of the International Group for the Psychology of Mathematics Education* (Vol. 3, pp. 177–183). Montréal, Canada: Université de Montréal.

Vinner, S. (1992). The function concept as a prototype for problems in mathematics education. In E. Dubinsky & G. Harel (Eds.), *The concept of function* (pp. 195–213). Washington, DC: Mathematical Association of America.

Vinner, S., & Dreyfus, T. (1989). Images and definition for the concept of function. *Journal for Research in Mathematics Education, 20*(4), 356–366.

Youschkevitch, A. P. (1976). The concept of function up to the middle of the 19th century. *Archives for History of Exact Sciences, 16*(1), 37–85.

Laboratoire Leibniz, IMAG, Grenoble, France
E-mail address: `Nicolas.Balacheff@imag.fr`

Laboratoire Leibniz, IMAG, Grenoble, France
E-mail address: `Nathalie.Gaudin@imag.fr`

CBMS Issues in Mathematics Education
Volume **16**, 2010

Strategies for Controlling the Work in Mathematics Textbooks for Introductory Calculus

Vilma Mesa

ABSTRACT. This study analyzes the availability of strategies for (a) deciding whether an action is relevant when solving a problem, (b) determining that an answer has been found, and (c) establishing that the answer is correct in 80 examples of Initial Value Problems (IVPs) in twelve calculus textbooks intended for first-year undergraduate calculus. Examples in textbooks provided explicit information about deciding what to do to solve the problem and determining the answer more frequently than they discussed establishing that a solution is correct or that it makes sense for the given situation. Strategies geared toward verification that the answer is correct or makes sense correspond to three aspects of verification: plausibility, correctness, and interpretation. Honors textbooks were more explicit than non-honors textbooks. Presenting examples as a collection of steps to solve problems – without consideration of what needs to be done to verify that the answer is correct – might obscure the need for verification in solving problems and suggest to students that reworking the solution is the only verification alternative. Implications for research and practice are discussed.

How, where, and when do students learn to (a) decide what to do when solving a problem, (b) determine that they have found an answer, and (c) establish that the answer is right? These are not trivial questions, as the large body of research on problem solving and metacognition has demonstrated (Brown & Walter, 1990; Kilpatrick, 1987; Nespor, 1990; Schoenfeld, 1985a, 1985b, 1992, 1994; Silver, 1987). These questions are at the core of reasoning and validation practices in mathematics and are closely related to metacognitive processes in mathematical problem solving. I approach these questions using Balacheff's theory of conceptions (Balacheff, 1998; Balacheff & Gaudin, 2003; Balacheff & Margolinas, 2005), which allows me to generate hypotheses about how instructors and students may react in hypothetical situations before observing real-life problem-solving. For generating hypotheses, I concentrate on textbook content, because textbooks' tasks and examples determine, to some extent, the kind of work both teachers and students do in the mathematics classroom (Doyle, 1988; Mesa, 2007; Stein & Smith, 1998). In this paper, I report the results of the analysis of 80 Initial Value Problem (IVP) examples taken from twelve textbooks intended for first-year calculus students used in a wide range of U.S. universities. I start by presenting the theoretical framework that guided the study, making connections to metacognition and validation in mathematics, as well

©2010 American Mathematical Society

as the rationale and the research questions. A description of the methods used and a report of the findings are followed by a discussion of the implications of the study for further research and for instruction.

Theoretical Framework and Prior Research

Balacheff's theory of mathematical conceptions defines a *conception* as the interaction between the cognizant subject and the milieu – those features of the environment that relate to the knowledge at stake (Balacheff & Gaudin, 2003; Balacheff & Gaudin, this volume; Balacheff & Margolinas, 2005). His basic proposition is that students' conceptions of mathematical notions are tied to particular problems in which those conceptions emerge. Thus Newton's conception of function was substantially different than Dirichlet's because each was working with a different phenomenon (Balacheff, 1998). As mathematics develops and we solve new problems, our conceptions get transformed. The combination of all these different conceptions is what constitutes a persons' knowledge (*knowing*) about a particular mathematics notion. Conceptions can be distinguished from each other because they require particular operations, particular systems of representations, and particular *control structures*: the organized set of criteria that "allows one to express the means of the subject to decide on the adequacy and validity of an action, as well as the criteria of the milieu for selecting a feedback" (Balacheff, 1998, p. 10). In operational terms, the control structure allows one to (a) decide what to do in a given situation, (b) determine that an answer has been found, and (c) establish that the answer is correct. Describing the control structure is fundamental because the control structure is closely related to processes of metacognition in problem-solving and to validation in mathematics. Research on these two aspects has been consistent: novice problem solvers do not use metacognitive strategies regularly (Schoenfeld, 1992), and they struggle with validation processes when doing proofs (Alcock & Weber, 2005).

The classical definition of metacognition states that the term refers to two separate but related aspects: "(a) knowledge and beliefs about cognitive phenomena, and (b) the regulation and control of cognitive actions" (Flavell, 1976, as cited in Garofalo and Lester (1985), 1985, p. 163; see also Schoenfeld, 1992, for a comprehensive review on the topic). Knowledge of cognition is categorized depending on whether certain factors – *person factors* like beliefs about self as cognitive being, *task factors* such as knowledge about scope, requirements, or complexity of the task, or *strategy factors* like knowledge of when strategies can be used – influence performance. On the other hand, regulation and control of cognitive actions refers to monitoring the processes in relation to the objectives sought. These categorizations generate four components of metacognition: orientation (strategic behavior to assess and understand a problem), organization (planning of behaviors and choice of actions), execution (regulation of behavior to conform to plans), and verification (evaluation of decisions made and of outcomes of executed plans). Two components, organization and verification, overlap to some extent with Balacheff's notion of control structure.

Determining whether proofs of statements are valid is not only a matter of knowing and using the appropriate format of the proof (Herbst, 2002; Martin & Harel, 1989) but also a matter of establishing the truth of each assertion and having the appropriate data and warrants for supporting each one (Alcock & Weber,

2005; Stephan & Rasmussen, 2002). The literature tells us that undergraduate students generally have significant difficulties validating proofs (Martin & Harel, 1989; Selden & Selden, 1995) but that with appropriate guidance, they can learn to distinguish valid from invalid arguments. Learning to infer warrants in strings of assertions, recognizing what constitutes data that could be used, and explaining why the warrant has authority also overlap with the three elements of Balacheff's control structure as it applies to proving.

Thus, if learning about the control structure is important for mathematical learning, we must investigate the opportunities students have to get acquainted with it. One resource is students' mathematics textbooks. Prior research on textbooks, however, suggests that this structure might not be explicitly presented in students' textbooks. Three studies are pertinent: those by Mesa (2004), Raman (2002), and Lithner (2004).

In an analysis of over 2000 exercises on functions from 8th grade texts from 18 countries, Mesa (2004) found that the control structure included three different types of criteria for verifying the correctness of an answer: *process*, *didactical-contract*, and *content*. Process criteria, found in 55% of the cases, included strategies tied to the procedure involved in solving a problem. Thus, repeating the procedure was necessary, and was suggested explicitly in many cases, as a means to verify that an answer was correct; finding the *same* answer was a warrant for correctness. Didactical-contract criteria, found in 28% of the cases, included strategies that used clues from the presentation of the text that deemed an action and the result appropriate; for example, obtaining an out-of-range ordered pair that could not be plotted in a given Cartesian plane would suggest an error in the solution. Finally, content criteria, found in 17% of the cases, included strategies by which definitions, theorems, or assumptions in the problem needed to be checked to substantiate a claim; thus, obtaining a positive number for a measure of time acted as a warrant that the answer was correct. These results are startling because they illustrate the possibility that part of students' reluctance to use strategies related to the problem itself (i.e., content criteria) to verify the correctness of an answer may be explained by the absence of these strategies in the textbooks. In other words, the opportunities to learn about the control structure are limited in these textbooks.

Raman (2002), in a study about the difficulties that students experience coordinating formal and informal aspects of mathematics, analyzed the role of textbooks in hindering or fostering such coordination. In particular, Raman noted that a pair of students, after working for about 45 minutes on a problem ("Is there a number that is exactly one more than its cube?") were unable to produce a solution until Raman mentioned that the problem was taken from the Intermediate Value Theorem section in their textbook. The students then solved the problem in five minutes. To the question of why was it so difficult to find the solution, a student replied,

> S3: Had you told us to open our books to page 89 and solve problem number 57, we would have done it in 5 min. Because it is, you know, after this section. So we know what we are supposed to do. But this just given like that, we don't know, you know, which part of our knowledge to access. (p. 147)

This episode illustrates the strength of the didactical contract in determining the type of strategies that can be used in a given situation. For the student, the name of the section of the textbook acted as the trigger that helped her decide what to do in the situation, carry out the procedure, and then be sure that the obtained answer was correct. Textbooks can play an important role in helping students to make these decisions, especially if they include specific markers for students to follow, for example, labeling sections of the problem set using the same headings given in the main text.

Using a coding system designed to capture the structure of reasoning present in textbooks, Lithner (2004) demonstrated that about 90% of the exercises in a U.S. freshman calculus textbook were written in such a way that students could use superficial reasoning to solve them without the need to access the meaning of the notions at stake. There is an important parallel between Lithner's definition of reasoning structure and Balacheff's control structure. For Lithner, the reasoning structure has four elements: the *problematic situation*, *strategy choice*, *strategy implementation*, and *conclusion* (p. 406). *Strategy choice* refers to choosing "(in a wide sense: choose, recall, construct, discover, etc.) a strategy that can solve the difficulty... [i.e., it] can be supported by predictive argumentation." *Strategy implementation* refers to "verificative argumentation: Did the strategy solve the difficulty?" and *conclusion* refers to obtaining the result (p. 406). As a consequence, "reasoning" is defined by Lithner as

> The line of thought, the way of thinking, adopted to produce assertions and reach conclusions. The explicit or implicit *argumentation* is the substantiation, the part of the reasoning that aims at convincing oneself, or someone else, that the reasoning is appropriate. (p. 406)

Note that Lithner's predictive argument would correspond to aspects of deciding what needs to be done in a given situation in Balacheff's control structure, the verificative argument would correspond to processes of verification that the answer is indeed correct, and the conclusion could be mapped to deciding whether an answer has been found – that is, recognizing the solution – in Balacheff's theory. The control structure in Balacheff's theory, however, emphasizes the establishment of the criteria that govern the decisions students make, rather than analyzing the actual consequences of the decision or the type of reasoning obtained. In this sense, Balacheff's control structure connects to metacognitive strategies, what Schoenfeld (1985a) called *control*: "Global decisions regarding the selection and implementation of resources and strategies. Planning. Monitoring and assessment. Decision-making. Conscious metacognitive acts" (p. 15).

According to Lithner, in *superficial reasoning*, students may rely on recalling keywords, algorithms, or previous experiences rather than on using *plausible reasoning*, which is based on intrinsic aspects of the content at stake. This classification also parallels Mesa's (2004) findings of the types of control strategies suggested by exercises in middle school textbooks, which were based on process, the didactical contract, or the content. Moreover, both studies report an alarmingly low percentage of exercises that would require verification and reasoning strategies based on the content at stake (10% in Lithner's study, 17% in Mesa's study).

Taken together, these textbook analyses illustrate how textbooks have the potential to shape students' development of verification and reasoning strategies in unproductive ways. It is worth noting here that an important link, instruction, is still understudied. The larger project, from which this paper is derived, was undertaken with the purpose of understanding the link between textbook presentation and instruction regarding the manifestation of the control structure. However, the studies to date that have focused on textbooks have investigated problems or definitions, elements of the textbook that may not be suitable for disclosing what the control structure is. These studies also looked across topics, or across nations, or across levels of instruction. Different results might be possible when looking at a particular level of instruction, within a single educative system, and within a single topic. Therefore, it is important to further probe these findings about textbooks by looking at a specific mathematical notion in which intrinsic mathematical aspects of the content at stake are fundamental for the criteria in the control structure.

In this paper I present results regarding the stability of findings of the analysis of textbook material – specifically, sections in first-year calculus textbooks devoted to initial value problems. First-year calculus was selected for two reasons. First, for many students it is the first university course in which they start to learn important aspects of mathematical work; therefore, an analysis of the opportunity to learn about controlling work in mathematics could be started in this course. Second, calculus reform has emphasized the use of multiple representations and technology; these two tools can become important resources for verification processes in the control structure. Thus, it is possible to expect that they will be manifested in calculus textbooks.

Selecting initial value problems as the content for my analysis was a theoretical decision. Balacheff's theory of conceptions indicates that the control structure is particular to the problems in the subject–milieu system. I chose sections devoted to initial value problems because the topic interweaves integration and differentiation notions with physical and other real-life situations that could offer more opportunities for different types of control strategies to emerge. Thus, the first research question for the study reported here is:

(1) How do textbooks intended for introductory college calculus make their control structures explicit in examples devoted to Initial Value Problems (IVPs)? More specifically, how do these examples indicate
   (a) how to decide what to do to solve the problem,
   (b) how to determine that the answer has been found, and
   (c) how to establish that the answer is correct.

Because textbooks tend to differ depending on their audiences, I anticipated that textbooks intended for honors students would expose the control structure in a different way than would textbooks intended for non-honors students. The second research question was:

(2) Do textbooks intended for honors students differ from textbooks intended for non-honors students regarding their control structures? If so, in what ways?

These two research questions allowed me to: (a) determine the role that textbooks might play in establishing the control structures, (b) trace variations depending on the audience, and (c) suggest hypotheses about ways in which instructors might take advantage of what textbooks present.

Methods

This study applied textual analysis techniques (Bazerman, 2006) to selected sections of a wide variety of textbooks intended for introductory calculus. The purpose of the methodology selected was to mine the *text* (i.e., the content within the textbook) in search of evidence of elements of the control structure. In what follows, I describe the decisions made, together with the rationale, regarding sampling and data analysis.

Sampling. There were two levels of sampling used in the study. They were sampling of textbooks and of text within textbooks.

Textbook Sampling. I chose 14 in-print calculus textbooks intended for introductory calculus at a wide range of postsecondary programs. To select the textbooks, I consulted the syllabi of various higher-education institutions in the U.S. Midwest and obtained information from instructors who were teaching or had taught calculus as a first-year undergraduate course. From over 30 suggestions, I selected the textbooks that were (a) most frequently mentioned and (b) used in post-secondary institutions within 100 miles of my university (to facilitate the continuation of the project). I obtained the latest edition of each of those textbooks. The student's solutions manuals, sold separately, presented the solutions without elaboration and none of the textbooks had a teacher's manual. For this reason, I did not analyze these texts. I visited the institutions' bookstores to make sure the textbooks were offered to the students. The majority of the textbooks were intended for a general audience of undergraduates; some were intended for more specialized audiences, such as math majors or science and engineering majors. Some institutions offered separate tracks of calculus for their students (e.g., applied, regular, honors) and used a different textbook for each. Textbooks intended for an applied or honors sequence in one institution might be used for a regular sequence in others. In the sample of 14 textbooks, five were intended for honors students.

Text Within Textbooks Sampling. All the textbooks followed the most common type of organization, namely: exposition–examples–exercises. In exposition, the text "directs the reader" and will "expound its subject matter in a discursive fashion, maybe use devices such as questions, visual materials, or tasks as assisting concept formation" (E. Love & Pimm, 1996, p. 387). The examples,

> [Are] intended to be 'paradigmatic' or 'generic', offering students a model to be emulated in the exercises which follow. The assumption here is that the student is expected to form a generalisation from the examples which can then be applied in the exercises. ...some books similarly annotate examples, indicating particular points of difficulty and the reasons behind a choice of approach. (p. 387)

In the exercises, the students are encouraged to actively engage with the text, by working through tasks that mimic previous examples or with a more varied collection of problems. Of the three types of texts, I chose to analyze the examples because they were intended to be paradigmatic of the work to be done and would be most likely to contain explicit strategies for controlling that work (Watson & Mason, 2005). Another more pragmatic reason had to do with instructors' beliefs that students rely on examples in order to work out homework problems (Mesa, 2006b).

Thus, I analyzed all examples provided in all sections devoted to initial value problems (IVPs) in these textbooks. The sections were located by finding the entries in the index corresponding to *initial value problems*, *initial condition problems*, or *introduction to differential equations*. Two honors textbooks lacked sections devoted to IVPs; therefore, those textbooks were not included in the analysis. The 12 textbooks analyzed are given in the reference list (Adams, 2003; Apostol, 1967; Goldstein, Lay, & Schneider, 2004; Hughes-Hallett, et al., 2005; Larson, Hostetler, & Edwards, 2006; MacCluer, 2006; Ostebee & Zorn, 2002; Simmons, 1996; R. T. Smith & Minton, 2002; Stewart, 2003; Thomas, Finney, Weir, & Giordano, 2001; Zenor, Slaminka, & Thaxton, 1999).

I examined 24 sections and 80 examples in total. I collected information on the definition of IVP and the contexts in which the examples were presented and on the control structure suggested in the example.

Definition and Context. Knowing what elements belong to the definition of an IVP was crucial for determining the internal logic of the arguments provided. A textbook that contains a definition that includes the order of the equation, for example, may be more likely to include "order of the equation" as part of the considerations in deciding what to do (e.g., in a second order equation, it will be necessary to solve two IVPs). The definitions were usually given in-line with the text (i.e., not under a definition heading or a box), as in the following examples:

> *Example 1.* A *differential equation* is an equation involving an unknown function and one or more of its derivatives. The *order* of such an equation is the order of the highest derivative that occurs in it. ...The arbitrary constant that appears in the general solution of a first-order equation is given a specific numerical value by prescribing, as an *initial condition*, the value of the unknown function $y = y(x)$ at a single value of x, say $y = y_0$ when $x = x_0$. (Simmons, pp. 178-180)
>
> *Example 2.* Sometimes we want to find a particular solution that satisfies certain additional conditions called *initial conditions.* Initial conditions specify the values of a solution and a certain number of its derivatives at some specific value of t, often $t = 0$. (...) The problem of determining a solution to a differential equation that satisfies given initial conditions is called an *initial value problem.* (Goldstein et al., p. 502)
>
> *Example 3.* In many physical problems we need to find the particular solution that satisfies a condition of the form $y(t_0) = y_0$. This is called an initial condition, and the problem of finding a solution of the differential equation that satisfies the **initial condition** is called an **initial-value problem**. (Stewart, p. 626)

I highlighted the text containing these definitions and used the notions included to construct a concept map (Novak, 1998) for IVPs.

Because of the important role that context can play in establishing whether an answer is correct or not, I also categorized the examples depending on whether they were contextualized or not, and created a running list of the applications suggested. When the example provided an IVP that did not make reference to an application, the context was marked as not present.

Control Structure. I selected all the examples that proposed and solved IVPs. When an example used a result that was found in another example (e.g., a general solution for a differential equation or a slope field), that example was included in the analysis. I read the solutions provided, the text preceding and the text following the example, and for each one answered the following three questions:

- How do I know how to solve the problem?
- How do I know that I got an answer?
- How do I know that the answer is correct?

This provided a holistic summary of what type of control structure was present, for which evidence was sought using a more detailed content analysis of the text in the example. I parsed the sentences in the example in order to classify them as belonging to one of four categories.

Category 1. What to do: when the sentence described a decision-making process listing actions that would lead to the solution of the particular problem,

Category 2. Answer found: when the sentence indicated that an answer to the problem had been obtained or described its characteristics without saying whether it was correct,

Category 3. Answer correct: when the sentence described ways in which the answer could be confirmed as making sense therefore deeming it correct or when the sentence verified the correctness of the answer, and

Category 4. Elaboration: when the sentence conveyed reasons or justifications for a given result, or described further implications of that result.

Sentences that did not fit any of these categories were coded as *other* and not considered in the analysis. Table 1 provides examples of the sentence categorization.

Data Analysis. The coding system was tested in two randomly selected textbooks with seven examples ($\sim$ 9% of the sample of examples, 85 sentences, $\sim$ 11% of the total number of sentences) by three other coders who were not familiar with the textbooks. I contrasted my coding to that assigned by the coders generating three contrasting pairs and calculated the Cohen's κ coefficient for each pair. The average was $\kappa = .72$, which is an acceptable level for inter-rater reliability (Landis & Koch, 1977). The remaining sentences and examples were coded using this process.

Definition and Context. I organized the definitions provided in the textbook by the common term being defined (e.g., *differential equation, initial condition, slope field*) and established the connections among those terms by common references; the terms and connections were organized into concept maps that illustrated how the definitions were organized in each textbook. I then produced a summary concept map combining all elements and connections within each individual concept map. I created a running list of all the contexts in which IVPs were presented and tallied them.

Control Structure. The holistic summaries were used to describe the contents of each of the categories; these summaries were useful for determining the extent to which the sentence parsing was capturing the essence of the summaries. The summaries also contained quotes (with page references) that provided evidence for particular categorizations.

TABLE 1. Examples of Sentence Categorization

What to do
"We solve these problems in two steps...: (1) solve the differential equation, and (2) evaluate C"
"We have already seen that the general solution is... The initial condition allows us to detemine the constant C"
"The process of antidifferentiation amounts to solving a simple type of differential equation"
Answer Found
"The (unique) solution is"
"Thus, ... is the desired solution"
"The temperature is decreasing at the rate of 20 degrees per second"
"The result is a curve shown in Fig..."
Answer Correct
"It is easy to check by differentiation that every function of this form is a solution"
"This translates to around 200mph—possible for an old–fashioned cannon"
"Around the height of the tallest human–built structure"
"We can visualize the solution curve by following the flow of the slope field "
Elaboration
"This DE simply asks for an antiderivative f of the function $6x+5$"
"The differential equation (1) says that $y' + 2y$ equals zero for all values of t"
"This DE is more challenging than the previous one."
"First we observe that there is no finite time t at which $f(t)$ will be zero because the exponential e^{-kt} never vanishes."

For each textbook, I tallied the number of examples for which elements of the control structure were evident. For each example, within each textbook, I created a summary table tallying the number of sentences belonging to each category. Simple frequencies are reported and the aggregates are used to contrast the textbooks.

All sentences categorized as *What to do*, *Answer Found*, and *Answer Correct* were further analyzed using a combination of open coding and thematic analysis (Bazerman, 2006). The purpose was to produce descriptions of the kinds of strategies that were provided in the examples, attending to what Balacheff (1998) has described as means to decide adequacy (of both the decisions made regarding processes and of the answers), validity (of both processes and answers), and ways for the milieu to provide feedback (about both adequacy and validity).

Results

The presentation of IVPs in these textbooks followed a consistent pattern: after providing the definition, textbooks then presented a non-contextual example with a relatively simple differential equation involved. The example would give the solution process (integrate and then evaluate the initial condition) accompanied by a figure containing some functions of the family of solutions, with the particular solution to the IVP highlighted in a different color. After the non-contextual example, applications with a more or less known context were provided. Only occasionally were the differential equations in non-contextual examples used in the contextual examples. Some textbooks indicated that most IVPs could not be solved analytically, then suggested Euler's method as a viable process to approximate a solution. Fewer textbooks indicated the origin of IVPs: in real life, what is observable is the *change* of a variable in relation to another variable; thus, the quest of the solver is finding or making explicit, in the best way possible, the nature of the relationship between the two (or more) variables – that is, finding a function whose change will be similar to the one that has been observed.

Figure 1 presents a concept map summarizing terms and connections among those terms associated with IVPs in these textbooks. The ovals and thicker lines indicate concepts and links that were common to all the textbooks studied.

The concept map suggested a basic conceptualization of IVPs in these textbooks: an IVP is a differential equation for which an initial condition is provided which generates a unique solution from among the infinite many that the equation has. In order to solve the problem, an antiderivative must be sought. Looking at more than one textbook, other information regarding the equation (separability, constant solution, slope field) and the solution to the IVP (domain of validity and the necessity of using Euler's method to find an approximation a solution) was garnered. The concept map allows us to anticipate possibilities for the control structure to emerge: (a) solutions can be verified through derivation, (b) slope fields can illustrate the behavior of the situation being modeled, (c) slope fields can also be used to verify a solution, and (d) applications can provide the means for assessing the plausibility of solutions. As we will see later, these elements were, in fact, present and referred to in the control structure of the textbooks analyzed.

Motion – rectilinear, vertical, and near the surface of the earth – was the only topic that was addressed by at least one example in all textbooks; other contexts frequently mentioned in the examples included growth/decay problems (24% of the examples), Newton's cooling or gravitational laws (17%), electric circuits (12%), and mixture problems (5%). Honors textbooks, on average, had more contextualized examples than non-honors textbooks (85% vs. 36%).

Control Structure. The initial reading of IVP examples revealed that, in general, there was substantial information about how to deal with particular problems and about recognizing an answer, but very few suggestions for how to establish that the solution was correct or that it made sense for the situation. Only four examples showed students how to test whether a certain function was a solution to an IVP, and it was illustrated through derivation. Only 20% of the examples (16 out of 80) made all three aspects explicit. The following excerpts illustrate such an example:

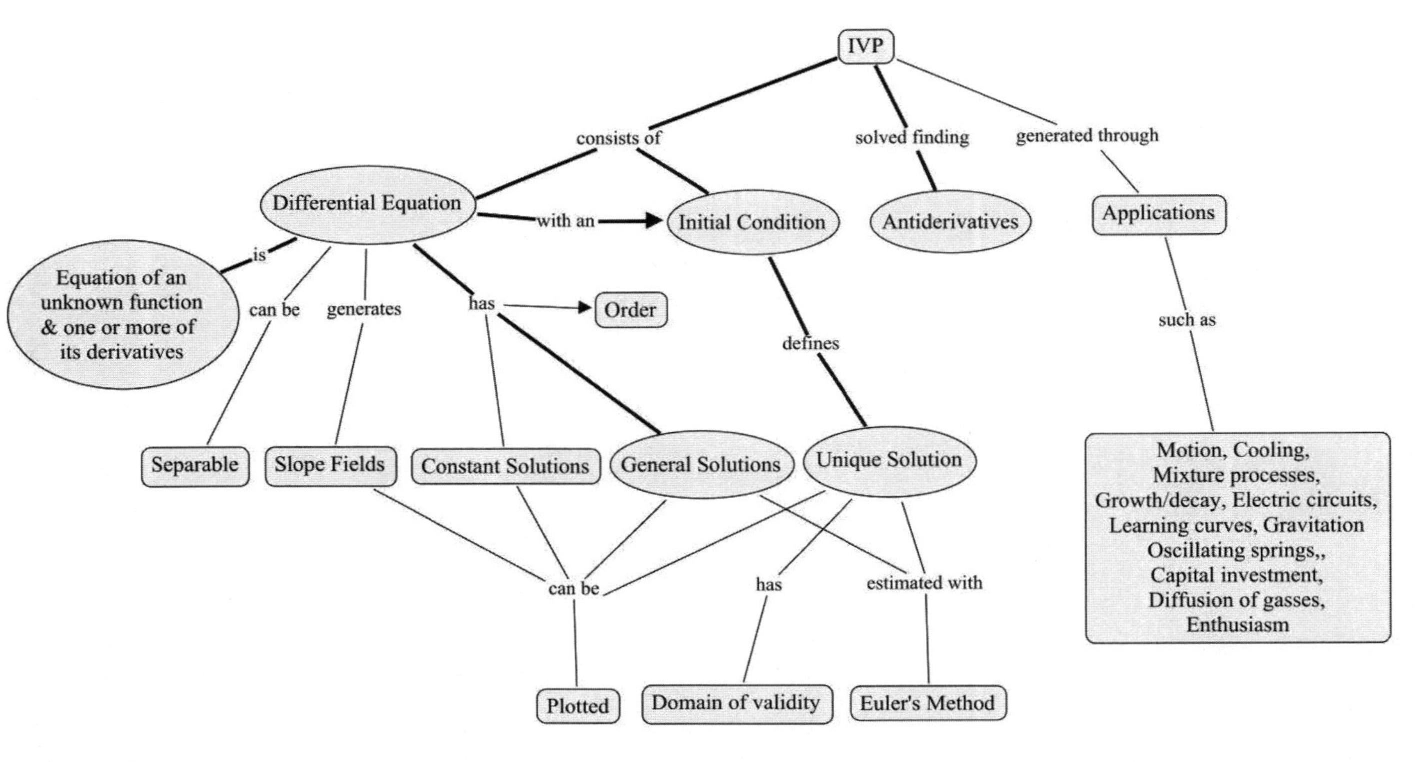

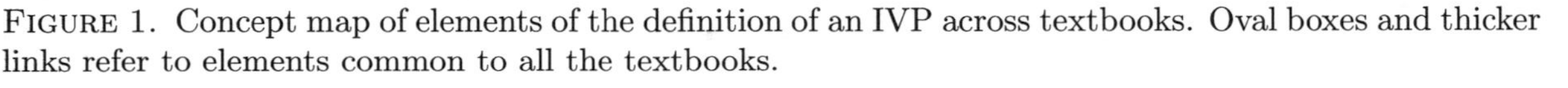

FIGURE 1. Concept map of elements of the definition of an IVP across textbooks. Oval boxes and thicker links refer to elements common to all the textbooks.

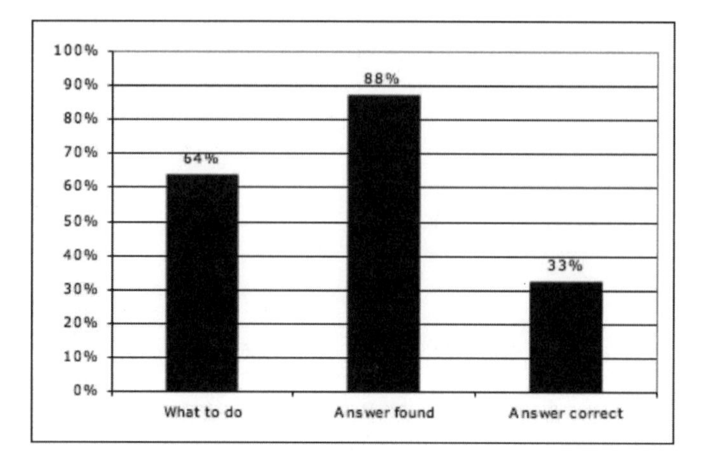

FIGURE 2. Percentage of examples ($n = 80$) in which elements of the control structure were made explicit across textbooks.

Example 4. Find an expression for the current in a circuit where the resistance is 12Ω, the inductance is $4H$, a battery gives a constant voltage $60V$, and the switch is turned on when $t = 0$. What is the limiting value? [An accompanying figure shows the circuit; the equation comes from an earlier example]
Solution: [a number of substitutions are made into the differential equation that models the current in the circuit given]... *we recognize this equation as being separable, and we solve it as follows:* [integration is carried out to yield a general solution for I]... Since I(0) = 0, we have ... [substitution in I] *and the solution is* $I(t) = 5 - 5e^{-3t}$ ... The *limiting current, in amperes, is*... 5. Figure 6 shows how the solution in Example 4 (the current) approaches its limiting value. Comparison with Figure 11 in Section 10.2 *shows that we were able to draw a fairly accurate solution* curve from the direction field. (Stewart, 2003, p. 639-640, emphasis added)

The highlighted sentences are those in which elements that define the control structure were made explicit: what needed to be done (*equation [is] separable, solve it as follows*), that the answer had been found (*the solution is, the limiting current is*), and that it made sense (contrast with both the slope field plotted in a previous problem and with the graph of the current).

These observations were corroborated with the sentence categorization. I compared these data at the example level (strategies present or not) and at the sentence level (frequency observed). Figure 2 presents the percentage of examples across textbooks in which at least one element of the control structure was made explicit – that is, examples in which authors explicitly indicated how to decide what needed to be done, or how to recognize the answer, or how to establish its correctness.

Figure 2 indicates that about 6 out of 10 examples included information describing how to approach the solution of the given problem, 9 out of the 10 examples highlighted when the answer had been found, and 3 out of the 10 examples provided information that assessed the correctness of the answer or its appropriateness for

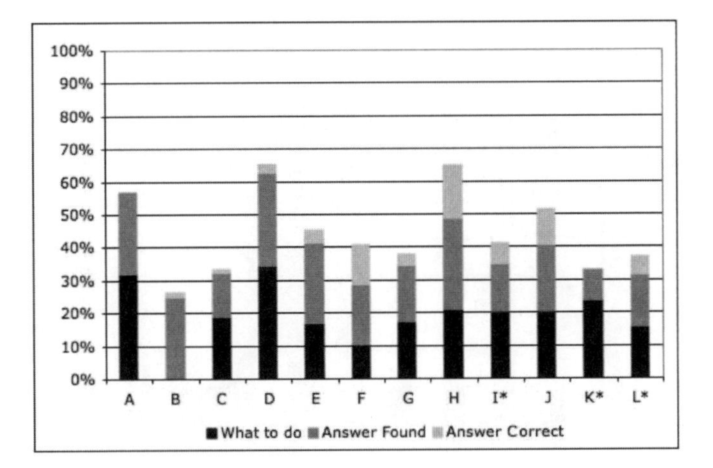

FIGURE 3. Percentage of sentences ($n = 763$) within textbooks in which elements of the control structure were made explicit. A starred letter indicates an honors textbook.

the situation. Thus, taken together, the examples in these textbooks were more explicit about how to solve the problems and about recognizing the answers than about establishing that the answers found were indeed correct.

Within textbooks, however, the situation changed. The examples in some textbooks provided more information regarding the control structure than in others and the strategies tend to carry over content across the textbook – that is, they were part of the writing style. Looking within textbooks may yield more practical implications than looking at the textbooks as a set because instructors and students tend to use a single textbook in their class work. Figure 3 presents the proportion of sentences in examples that were devoted to the different elements of the control structure, organized by textbook.Though textbooks are denoted with a letter in Figure 3, the letter was randomly assigned to avoid text identification.

Figure 3 illustrates that textbooks differed from each other in terms of the manifestation of the control structure. In general, the pattern observed across textbooks was found here: examples in these textbooks were more explicit regarding how to solve the problem or highlighting that the answer had been found than about establishing that an answer was correct. Notice that textbook H had about the same number of sentences for each element, making it the most balanced of all the textbooks analyzed. All textbooks explicitly showed how to recognize that an answer had been found, and most explicitly suggested ways to solve the problem. It is important to note that the coding system accounted for only explicit language used to highlight decision-making processes regarding the solutions. In the following example, no sentences were coded as *What to do* because the presentation consisted only of the steps taken, the actual actions leading to solving the problem:

Example. Find the function $f(x)$ whose derivative is $f'(x) = 6x^2 - 1$ for all real x and for which $f(2) = 10$.
Solution: Since $f'(x) = 6x^2 - 1$, we have $f(x) = \int(6x^2 - 1)dx = 2x^3 - x + C$ for some constant C. Since $f(2) = 10$ we have $10 = f(2) = 16 - 2 + C$. Thus $C = -4$ and $f(x) = 2x^3 - x - 4$.

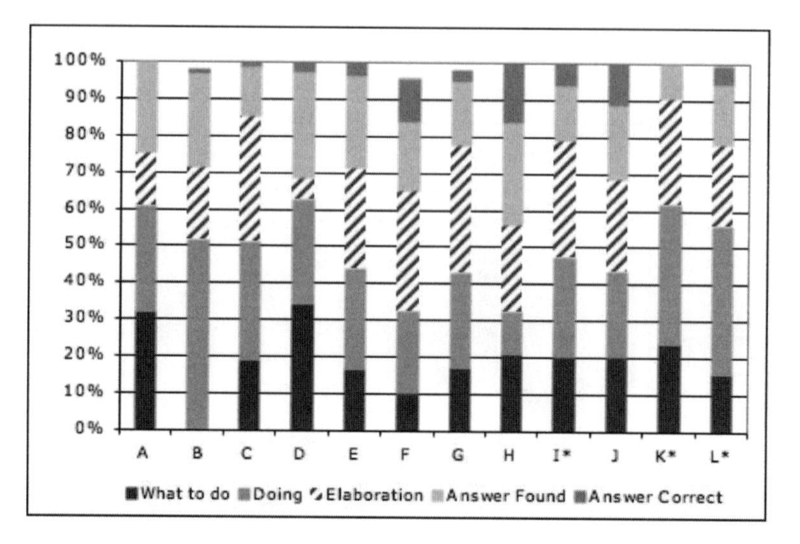

FIGURE 4. Percentage of sentences within textbooks corresponding to actions, elaborations, and control strategies.

(*By direct calculation we can verify that* $f'(x) = 6x^2 - 1$ and $f(2) = 10$). (Adams, Example 3, emphasis added)

The example tells what to do by doing the steps. This approach was fairly typical across all the textbooks observed and led me to include a new category, *Doing*, to categorize sentences of this type. Notice that this example includes an explicit verification sentence: by taking the derivative of the function that was found, one can obtain the original expression; also substitution of 2 in the function found yields the initial condition. The conclusion is, then, that the function found must be correct. However, as seen in Figure 3, verification statements were not very common.

A number of sentences that had relevant information for justifying the solution process also made up an important portion of the content of the examples. Figure 4 shows the percentage of sentences coded in each book considering both the *Doing* and the *Elaboration* sentences. Observe that for most of the cases, the five types of codes accounted for most of the content in the examples. *Doing* and *Elaboration* sentences accounted for about half of the sentences in most of the books (min 35%, max 71%).

The inclusion of *Doing* sentences, the sentences in which actual operations are carried out, showed that examples function as spaces for illustrating procedures. The demonstration is such that the reader is the spectator, similar to an apprentice, who observes the master carrying out the work and has to reproduce that work later on. In some textbooks the apprentice receives extra guidance from the master, with the *Elaboration* sentences serving the purpose of connecting ideas, justifying steps, or translating into similar language what has been or needs to be done. *Doing* and *Elaboration* sentences carry out the most important functions that examples are meant to fulfill; they provide the meat of the action for the apprentice to see how things work (Watson & Mason, 2005).

Examples in the honors textbooks analyzed tended to be longer than those in other textbooks, to offer more explicit control strategies, to deal with fewer constraints and more general cases, and to mostly elaborate ideas, sometimes not following the pattern of problem-then-solution observed in all the non-honors textbooks (i.e., the example might consist of a discussion of a situation, for which no explicit questions were stated). Seven of the 16 examples that made explicit the three elements of the control structure were found in honors textbooks.[1] The following excerpts illustrate some of these points.

Excerpt 1.

> *Radioactive Decay.* Although various radioactive elements show marked differences in their rates of decay, they all seem to share a common property–the rate at which a given substance decomposes at any instant is proportional to the amount present at that instant. If we denote by $y = f(t)$ the amount present at time t, the derivative $y' = f'(t)$ represents the rate of change of y at time t, and the "law of decay" states that $y' = -ky$, where k is a positive constant (called the 'decay constant') whose actual value depends on the particular element that is decomposing. The minus sign comes in because y decreases as t increases, and hence y' is always negative. The differential equation $y' = -ky$ is the mathematical model used for problems concerning radioactive decay. Every solution $y = f(t)$ of this differential equation has the form $f(t) = f(0)e^{-kt}$. Therefore, *to determine the amount present at time t, we need to know* the initial amount $f(0)$ and the value of the decay constant k. It is interesting to see *what information can be deduced from* [the general solution], without knowing the exact value of $f(0)$ or k. First we observe that there is no finite time t at which $f(t)$ will be zero... therefore, *it is not useful to study* the "total lifetime" of a radioactive substance. (Apostol, Example 1, emphasis added)

Excerpt 2.

> A 0.1 kg ball [...] is thrown at an angle of 60 deg with the horizontal on a level field, with an initial speed of $50m/sec$. See Figure [it shows the ball's initial velocity and the forces acting on it]. The constant of resistance due to the atmosphere is known to be $k = 0.02$. How high will the ball go, and how far away will it land?
>
> Solution: [some initial substitutions]...
>
> *Proceeding exactly as we did in the previous example*, we get [$v_x(t)$, and $v_y(t)$, maximum altitude, time, $x(t)$].
>
> Note that, even if the ball were going over a cliff and t could become quite large, *its range would be less than* $125m$. ... *To find the range, we must first find* the time $t \neq 0$ when $y(t) = 0$. (*Why can't we assume that it takes the same amount of time to get down as it took to get to the top of its trajectory?*)... Notice

---

[1] Numerical data is available. I chose not to present a summary table with this information, because the different sample sizes may lead to unwarranted conclusions about the differences between the textbooks.

> *how much longer the ball takes to come down than it took going up. Why?* (Zenor et al., Examples 5 and 9, emphasis added).

Excerpt 3.

> Observe how the [...] problem above tracks exactly what some [...] call the "scientific method":
> *Step 1:* model the phenomenon as a differential equation.
> *Step 2:* Solve the differential equation
> *Step 3:* Impose the given data
> *Step 4:* Interpret the results.
> (MacCluer, Example 15, emphasis in original)

Excerpt 1 was taken from an example that did not pose a question to be answered; among the concepts that the example illustrates is that the "half-life is the same for every sample of a given material." Excerpt 2, on the other hand, was from an example on motion. It uses vectors for speed and position, which allows the problem to ask for the range of the motion and it also considers the effects of air resistance. None of the non-honors textbooks assumed air resistance (although one mentioned that the results would be "slightly" different if air resistance were considered). Finally, Excerpt 3 looks for extracting a general "method" illustrated by the previous example. Thus, the example is not only to be used as reference —a standard case that can be used to check understanding—but also as a model illustrating the general method behind the process (Michener, 1978).

Content of the Control Structure. In showing students how to decide what to do when solving an IVP, authors of these textbooks offered two options, depending on whether the differential equation could be solved with methods already introduced or not. That antiderivation is to be used is made clear in the presentation. The decision about how to antdifferentiate (or whether one should attempt it) depended on what the authors had included in the textbook. In principle, Euler's method could be used all the time, but the textbooks that include the method made a point of using it only when analytical strategies cannot be applied. Because this option was not available in all textbooks (see Figure 1), this decision was not always required. Similarly, when the textbooks introduced a section about separable equations, the decision about how to antidifferentiate involved recognition that the equation was separable. Without this section, that decision was not required either. When neither of these sections was included, the equations were simple enough that authors could refer to previous theorems to justify the antiderivation processes.

This approach to antiderivation suggests that deciding what needs to be done is in general driven by the didactical contract. Thus, when the section was labeled "Euler's method," the equation was solved using the approximation, even when the equation could have been solved analytically. In such cases, the analytical solution worked as a template against which the accuracy of the Euler's approximation could be measured.

Once a family of solution functions was found, the reader might be directed toward making sure that that family of functions fit the original differential equation (a verification step that will be discussed below). The next step consisted of using the initial conditions to fix the function, and therefore find the unique solution to the IVP. No decisions seemed to be needed for this step; rather, it is a matter

TABLE 2. Description and Frequency of the Content Strategies

Strategy	Frequency
Establishing the goodness of fit: A given answer matches a previously known fact	18
Using the definition or a theorem: Use of analytical procedures to establish that the function found is a solution to the differential equation provided	15
Assessing the meaning of numerical values: Draw attention to the magnitude of the values in the situation	6
Analyzing Equations: Draw attention to the form of the expression of the solution	3
Issuing warnings: Draw attention to potential problems	3
Total	45

of substituting values, of applying a process. The outcome of this process is the answer. In showing students how to decide whether the answer is right, however, the textbooks seemed to offer more differentiated strategies.

The analysis of the 45 Answer Correct sentences from 23 examples generated five types of strategies that would be classified as content, following Mesa's (2004) study: establishing the goodness of fit, using the definition or a theorem, assessing the meaning of numerical values, analyzing equations, and issuing warnings.

Establishing the goodness of fit was the most common of the strategies (18 out of 45 sentences) and indicated the cases in which the authors showed how a given answer matched a previously known fact. For example, having found the solution to an IVP analytically, authors might use Euler's method to find an approximation and "use [the] exact solution as a basis for comparison for the performance of the Euler's method." The essence of the strategy was to rule out gross errors and to show that one could be certain the solution was accurate: "to make this correspondence more apparent, we have drawn a graph of the approximate solution superimposed on the direction field in [the figure]." This strategy used numbers, graphs, or symbolic expressions, and would be seen in both contextualized and non-contextualized situations. In one instance, the reader was encouraged to conduct an experiment to corroborate the findings.

Using the definition or a theorem was also common (15 out of 45 sentences) and corresponded to the use of analytical procedures to establish that the function found was indeed a solution to the differential equation provided: "(1) To decide whether the function $y = e^{2t}$ is a solution, substitute it into the differential equation: $d^2y/dt^2 + 4y = 2(2e^{2t}) + 4e^{2t} = 8e^{2t}$. (2) Since $8e^{2t}$ is not identically zero, $y = e^{2t}$ is not a solution." Theorems were invoked to justify the correctness of a statement: "By theorem 9 (...), these are the only possible derivatives."

Assessing the meaning of numerical values occurred in 6 of the 45 cases. In these sentences, authors called the reader's attention to the magnitude of the values in the situation. For example, one noted that $495m$, the height to which a cannonball rose, was "around the height of the tallest human-built structure." Another emphasized how the numerical values obtained were to be interpreted in the context provided: "note that, even if the ball were going over a cliff and t could

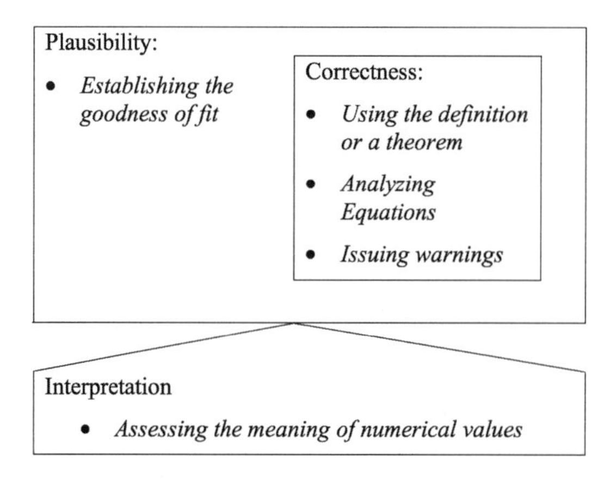

FIGURE 5. Three aspects of the process of verification.

become quite large, its range would be less than $125m$." Graphical representations were used in some cases.

Analyzing equations, seen in 3 of the 45 sentences, corresponded to instances in which the reader's attention was drawn to the form of the expression of the solution to establish the validity of the statement. For example, "this [$v = -32(t-4)$] tells us that for $t < 4$, the velocity is positive, so the stone is moving upward."

Finally, in 3 of the 45 sentences, authors *Issued warnings* that were meant to draw the reader's attention to potential problems. For instance, one textbook included the sentence, "we should be careful about the domain of our solution," suggesting that a mistake could be generated if the warning was not followed. A summary is presented in Table 2.

The five strategies seemed to address three different aspects of the process of verification, namely plausibility (whether what is obtained is possible), correctness (whether what is obtained is true), and interpretation (that what is obtained makes sense). In this reduced set of sentences, it was difficult to notice possible connections among these aspects, some of which could be conjectured. In some cases, it may not be possible to establish the correctness of a solution, and plausibility may be the best confirmation of accuracy that one can get (as when an IVP cannot be solved analytically).

By definition, "plausibility" suggests that there "appears to be merit for acceptance" (Plausible, 1989) and that possible discrepancies can be ignored. Correctness is grounded on the assumption that there are legitimate means for demonstrating or proving that a statement is true and that those means are valid – that is, that the process "contains premises from which the conclusion may logically be derived" (Valid, 2000). Thus, even though all correct statements may be plausible, not all plausible statements might be correct. Interpretation occurred only in contextualized examples, because context was what determined the meaning to be attributed to particular numbers. In this way, interpretation may be seen to support both plausibility and correctness (see Figure 5).

In summary, the types of strategies illustrated in these examples for deciding about what to do to solve a problem, determining that an answer has been found, and establishing that the answer is correct appeared to be grounded differently.

"What to do" decisions seemed to be grounded mostly on the didactical contract that suggested the process to be followed. "Determining that an answer has been found" appeared to be a non-issue: the answer is either the function or the values that the situation asked for. Finally, "establishing whether the answer is correct or makes sense" seemed to be the only aspect that was grounded in the content at stake, and the strategies given in the textbooks seemed to address interrelated aspects of verification, plausibility, correctness, and interpretation.

Discussion

This study was conducted to determine (a) the ways in which textbooks intended for introductory college calculus were explicit about the control structure in examples devoted to Initial Value Problems (IVPs) and (b) the ways in which textbooks intended for honors students differ from textbooks intended for non-honors students regarding the control structure. I organize the discussion by these two topics.

The Control Structure for IVPs in Introductory Calculus Textbooks. Balacheff's theory of conceptions states that conceptions can be distinguished from each other because different problems require different operations, different systems of representations, and different control structures. In the case of IVPs undertaken in this study, and in the special case of what textbooks offer, it is interesting to note that how IVPs emerge—from real-life situations in which we can only trace the change of a variable in relation to another variable—is not prominently discussed across all textbooks. The concept map compiled from the content across the textbooks seems to emphasize the abstract nature of IVPs; even though these problems have originated from real-life situations, they have been conceptualized as models, and their treatment in textbooks emphasizes dealing with those models (antiderivating and finding a unique solution). This treatment seems related to the control structure that emerges.

The analysis of the sentences referring to the different aspects of the control structure, which are meant to speak about adequacy, validity, and the way in which the milieu provides feedback, revealed that these textbooks offer more opportunities to learn about what needs to be done to solve the IVPs and for recognizing the answer, than opportunities for making sure the answer is correct. The emphasis on *doing* and justifying the doing (*elaboration*) seems to reinforce the technical aspects of solving IVPs. Decisions about what needs to be done are tied to the didactical contract, and obtaining answers seems to be a consequence of doing a procedure, which is assumed to be correct by virtue of being an illustration. In the few sentences in which verification was carried out, the five identified strategies took advantage of the content at stake. But, as the content map and the coding reflect, these strategies did not appear in all textbooks analyzed.

Before discussing whether this situation could be different, especially regarding verification strategies, I will offer conjectures about why this is so. A look at calculus textbooks from different eras (e.g., De Morgan, 1842; Lacroix, 1797, 1837; Lamb, 1924; C. E. Love, 1916; C. E. Love & Rainville, 1962; Perry, 1897; R. H. Smith, 1897) and at reviews of some of those calculus textbooks shows that several changes have occurred since the first calculus textbook was written by the Marquis of L'Hopital in 1696 (Eves, 1990, p. 426). Until 1890, the textbooks appeared to be mostly compendia of all that was known at the time about calculus. The early

1900s has been recognized as a period that profoundly affected the teaching of mathematics at the collegiate level (Kilpatrick, 1992), and which resulted in a new type of textbook available for calculus (Mesa, 2006a). Textbooks became sources by which instruction at universities could be delivered to larger numbers of students, who were not only interested in mathematics but also in engineering, physics, or chemistry. At the time, there were criticisms about the lack of verification in some textbooks. For example, Ransom (1917), in a review of a Clyde E. Love's *Differential and Integral Calculus* (1916), noted that in spite of C. E. Love's claim made in the preface that "A feature of the book is its insistence on the importance of checking the results of the exercises,"

> The reviewer found no attention given to the general question of checking, and except under double integration (where more than one order of work is often possible) very few problems call for solutions in two ways... No rough methods of checking are suggested, such as sketching derivative or integral curves or other graphical devices, nor is there any hint of checking limits by computing neighboring values, nor of checking differentials by computing small increments, nor of checking integrals by Simpson's rule. (Ransom, 1917, p. 175)

Textbooks intended for the applied disciplines contained step-by-step directions for solving problems, had long lists of exercises for practice, avoided theory when not related to practice, and limited the presentation to what was considered the most useful part of the subject. As in C. E. Love's textbook, there were few instances of verification strategies. It was not until the late 1970s and the calculus reform movement that attention to multiple representations and to technology use, among other things, created new opportunities for students to verify their work.

Although very few of the textbooks analyzed in this study claimed to be reform oriented, most of them included some reform features such as examples and exercises to be solved with technology (graphing calculators or computer algebra systems) and more references to graphical representations. These features make them different from earlier editions of the same textbooks. However, even though these features could be used for verifying answers, they were not prominently referred to in the textbooks.

Through the analysis of these textbooks, we see that even with a topic that provides many opportunities for establishing the correctness of answers, these opportunities are not made explicit. Perhaps the textbook writing tradition that emerged in the earlier 1900s still persists. Perhaps what E. Love and Pimm (1996) called the authority of the textbook is also playing a role: the textbook does not have mistakes, therefore what is presented is correct. Or perhaps carrying out the intentions of providing alternative solutions or rough methods of checking is difficult: how many alternatives can be proposed without making the textbook longer or cumbersome to read? Textbooks authors might also have wondered whether an introductory textbook was the place for such work. To respond to these questions, we need to weigh considerations of the potential benefit for students.

Research on students' use of verification strategies has suggested that what students can do in particular situations is limited. Eizenberg and Zaslavski (2004), in their study of students solving combinatorics problems, found five types of verification strategy: *reworking the solution, using a different solution method and comparing answers*, *adding justifications to the solution*, *evaluating the reasonability of the answer*, and *modifying some components of the solution*. *Reworking the solution*, *using a different solutions method*, and *modifying some components of the solution* map into what Mesa (2004) categorized as process-oriented strategies, as they depend on the procedures learned. The other two strategies identified by Eizenberg and Zaslavski (2004), *adding justifications* and *evaluating the reasonability of the answer*, could be categorized as content strategies because they require the content at stake to make decisions about the correctness of the solution. Eizenberg and Zaslavski found that some students were capable of producing verification strategies, but that this happened more frequently *when they were told* that their answers were incorrect; additionally, they found that:

> Many of the students who made attempts to verify their incorrect solutions, whether out of their own initiative or in response to the interviewer's prompts, *were not able to come up with efficient verification strategies* and were thus neither able to detect an error nor to correct their solution (p. 33, emphasis added).

Of the five strategies in their study, *reworking the solution* was the most frequently used and the most inefficient. It is a remarkable coincidence that just as textbooks emphasize decisions about how to deal with problems based mostly on procedures, they also emphasize the use of procedures with regard to verification strategies.

In the present study, the five verification strategies found were grounded in the content, and they referred to three different aspects of verification, namely: plausibility, correctness, and interpretation. But they appeared in a relatively reduced number of sentences (45 out of 763) in about one fourth of all the examples analyzed. Meanwhile, *all* examples were very explicit about the process to follow to solve an IVP. I argue that presenting examples as a collection of steps that suggest what needs to be done to solve the problem, without consideration of what needs to be done to verify that the answer is correct, leads, in the best case, to students who rely mostly on process-oriented strategies to assess whether their answers make sense, and, in the worst case, to students who do not attempt to verify the correctness of the solutions in the first place. This should be an area of concern for those who design and use textbooks.

As suggested by the Eizenberg and Zaslavski (2004) study, adding justification to the solution is useful. We see that in this study a number of textbooks provided elaboration sentences that explicitly justified intermediate steps. This is to be taken as a good sign, as an indication that it is feasible to include information that could potentially be used for justifying the correctness of the solution. It should be possible, then, to include statements that explicitly address why the justifications allow us to make sure that the answer is correct. The link between justification and verification needs to be made more explicit. Making this link more explicit might also be useful as students learn to prove.

Content strategies obviously depend on the content at stake. In this study, the availability of slope fields, of technology that can plot the differential equations, of a considerable amount of knowledge about functions and their behavior (these sections appeared after differentiation and after integration), and of a real-life context, all make this topic ideal for studying the uses of these devices in verification of the solution. As could be expected from Balacheff's theory of conceptions, different problems lead to different conceptions for a notion; in this case, it was possible to illustrate how aspects of the didactical contract, traditions regarding the content of examples that emphasize steps to follow, and the content itself interact to create a control structure that is, in many ways, dependent upon the types of problems considered here. It was crucial to find that the few strategies provided refer to different aspects of verification, but it is disappointing that in so few instances textbooks capitalized on the rich content associated with IVPs. Possibly, textbooks are written under the assumption that during class these opportunities will be realized. Having observed a wide range of instructors teaching introductory courses in different settings makes me doubt whether this assumption is often borne out.

The Control Structure in Honors Textbooks. The honors and non-honors textbooks differ. The prefaces of these textbooks indicate that it is assumed that the student who uses an honors textbook, or who is in an honors class, will read the textbook. Besides reading the textbook, the student is supposed to continually produce data and infer warrants so that the assertions made can be validated. As seen in one of the excerpts, questions such as "Why?" are introduced in the narrative, presumably as a reminder to the student that verification work is needed.

The discussion of why these textbooks are different is matter of a different analysis. The need for honors textbooks is predicated upon the idea of the honors student, one who is a markedly different person from the "average" ability student. However, for some scholars "ability" as a measure of innate capabilities or talent is a fictitious construct because mathematics performance, which is used to establish ability, is a consequence of structural practices that construct and reify the differences between high and low performers, therefore making the differences appear as "natural" (Dowling, 1998; Walshaw, 2001; Zevenbergen, 2005). I note here that instructors' possible assumption that non-honors students, in general, do not read the text in their textbook needs to be established empirically. Some of my data suggests that this might not be the case.

We can ask whether the differences that we observed in the two types of textbooks should be maintained, in particular, when we look at the control structure. I believe that there is some value in incorporating some of the strategies that were observed in the honors textbooks in the presentation of the non-honors textbooks. Here are a few examples:

- Pointing out that the number of initial conditions needed to find a unique solution depends on the order of the differential equation.
- Explaining the meaning and existence of constant solutions, an important characteristic of differential equations (it can be seen as a particular IVP) that can be addressed in tandem with the discussion of slope fields.
- Showing why the values obtained for time, distance, or height make sense in the proposed situation. In this study only one non-honors textbook explicitly addressed this issue, and another invited the reader to verify the reasonableness of the values by replicating the experiment.
- Discussing the kind of values that could be expected for a particular variable: IVPs grounded in a given context have the hypothetical advantage over abstract IVPs of providing other ways to verify answers and check procedures. The implicit assumption that most of the quantities are positive is pervasive. Beyond that, it would be important to know whether the quantities can range from 0 to infinity or are bounded.

- Stating that finding solutions is very difficult whereas verifying whether a solution has been found is easy. The statement was made in only one of the non-honors textbooks, but it was made in all three honors textbooks. The contrast between verifying and finding is an important one, yet it is overlooked in most non-honors textbooks.
- Taking advantage of the slope fields to explain why the general and particular solutions are reasonable for the differential equation and IVP proposed. This was done in only 2 of the 9 non-honors textbooks.
- Making explicit the nature of the modeling involved in generating the differential equation and in finding a solution that would fit the IVP. The context was prominent in making decisions about the solution.

Conclusion

I am aware that these findings are limited by my selection of content and textbooks to analyze. The sections analyzed belonged to textbooks devoted to introducing the tools of calculus, rather than to presenting a theory of differential equations. We could not expect to have a complete treatment in a textbook that is by necessity introductory. However, I believe that what I found in the IVP sections is symptomatic of what happens across all textbooks. I chose this particular content because it affords the most potential for the control structure to emerge. The finding of strategies that address different aspects of the verification process corroborates that this was a good choice. However, given the few instances in which these strategies were made explicit across these textbooks, I wonder: Where should students learn to control their work? Further investigation is needed to determine whether and how these strategies should be made available during instruction and how the students who read these examples interpret and use these strategies.

The structure disclosed here does not look at regulatory aspects of metacognition that would be evident in an actual problem-solving session. Would students exposed to the different strategies described here exhibit them in a live problem-solving situation?

Assuming that students use the textbooks for doing exercises and that they read the examples in the text when in need, the study suggests that students will learn and will be able to recognize when they have found a solution, no matter what textbook they use. The study suggests, also, that depending on what textbook is used, students will be more or less aware of the need to verify that the solution is correct or that it makes sense. I believe that we should address this shortcoming. If examples in mathematics are to continue playing a crucial role for learning and understanding (Michener, 1978; Watson & Mason, 2005), they should illustrate not only the processes associated with solving problems, but also strategies for verifying that the solution is correct and that it makes sense. Doing so may provide a step in the right direction to improve students' understanding of the nature of validation and verification in mathematics.

Acknowledgments

A paper based on this study was presented at the third International Conference of the Teaching of Mathematics at the Undergraduate Level (ICTM-3) in Istanbul, July 4, 2006. The Horace H. Rackham School of Graduate Studies at the University of Michigan supported this research. I thank Charalambos Charalambous, Amy

Jeppsen, and Jenny Sealy for their assistance with several aspects of this work and Nancy Songer, Annemarie Palincsar, and the reviewers for insightful comments to earlier versions of the manuscript.

References

Adams, R. A. (2003). *Calculus: A complete course* (5th ed.). Toronto: Pearson.

Alcock, L., & Weber, K. (2005). Proof validation in real analysis: Inferring and checking warrants. *Journal of Mathematical Behavior*, *24*, 125–134.

Apostol, T. (1967). *Calculus* (Vol. 1). Hoboken, NJ: John Wiley & Sons.

Balacheff, N. (1998). *Meaning: A property of the learner-milieu system* (Unpublished manuscript).

Balacheff, N., & Gaudin, N. (2010). Modeling students' conceptions: The case of function In F. Hitt, D. Holton, & P. Thompson (Eds.), *Research in collegiate mathematics education. VII* (pp. 207–234). Providence, RI: American Mathematical Society.

Balacheff, N., & Gaudin, N. (2003). Baghera Assessment Project. In S. Soury-Lavergne (Ed.), *Baghera Assessment Project: Designing an hybrid and emergent educational society.* Cahier Liebniz 81. Retrieved April 14, 2007, from *Les cahiers du laboratoire Liebniz* web site: `http://www-leibniz.imag.fr/LesCahiers/Cahiers2003.html`. Grenoble, France: Laboratoire Leibniz–IMAG.

Balacheff, N., & Margolinas, C. (2005). ck¢ Modèle de connaissances pour le calcul de situations didactiques. In A. Mercier & C. Margolinas (Eds.), *Balises en didactique des mathématiques* (pp. 75–106). Grenoble: La Pensée Sauvage.

Bazerman, C. (2006). Analyzing the multidimensionality of texts in education. In J. L. Green, G. Camilli, & P. B. Elmore (Eds.), *Complementary methods in education research* (pp. 77–94). Mahwah, NJ: Erlbaum.

Brown, S. I., & Walter, M. I. (1990). *The art of problem posing* (2nd ed.). Hillsdale, NJ: Erlbaum.

De Morgan, A. (1842). *The differential and integral calculus, containing differentiation, integration, development, series, differential equations, differences, summation, equations of differences, calculus of variations, definite integrals,– with applications to algebra, plane geometry, solid geometry, and mechanics. Also, elementary illustrations of the differential and integral calculus*. London: R. Baldwing & Cradock.

Dowling, P. (1998). *The sociology of mathematics education: Mathematical myths/pedagogic texts* (Vol. 7). London: Falmer Press.

Doyle, W. (1988). Work in mathematics classes: The context of students' thinking during instruction. *Educational Psychologist*, *23*, 167–180.

Eizenberg, M., & Zaslavski, O. (2004). Students' verification strategies for combinatorial problems. *Mathematical Thinking and Learning*, *6*, 15–36.

Eves, H. (1990). *An introduction to the history of mathematics* (6th ed.). Orlando, FL: Harcourt Brace Jovanovich.

Flavell, J. H. (1976). Metacognitive aspects of problem solving. In L. Resnick (Ed.), *The nature of intelligence* (pp. 231–236). Hillsdale, NJ: Erlbaum.

Garofalo, J., & Lester, F. K. (1985). Metacognition, cognitive monitoring, and mathematical performance. *Journal for Research in Mathematics Education, 16*, 163–176.

Goldstein, L. J., Lay, D. C., & Schneider, D. I. (2004). *Calculus and its applications* (10th ed.). Upper Saddle River, NJ: Pearson.

Herbst, P. (2002). Establishing a custom of proving in American school geometry: The evolution of the two-column proof in the early twentieth century. *Educational Studies in Mathematics*, *49*, 283–312.

Hughes-Hallett, D., Gleason, A., McCallum, W. G., Lomen, D. O., Lovelock, D., Tecosky-Feldman, J., & et al. (2005). *Calculus single variable* (5th ed.). Hoboken, NJ: John Wiley & Sons.

Kilpatrick, J. (1987). Where do good problems come from? In A. H. Schoenfeld (Ed.), *Cognitive science and mathematics education* (pp. 123–148). Hillsdale, NJ: Erlbaum.

Kilpatrick, J. (1992). A history of research in mathematics education. In D. Grouws (Ed.), *Handbook of research on mathematics teaching and learning* (pp. 3–38). New York: Macmillan.

Lacroix, S. F. (1797). *Traité du calcul différentiel et du calcul intégral* [A course in differential and integral calculus]. Paris: Chez J. B. M Duprate.

Lacroix, S. F. (1837). *Traité élémentaire de calcul différentiel et de calcul intégral* [An elementary course in differential and integral calculus]. Paris: Bachelier.

Lamb, H. (1924). *An elementary course of infinitesimal calculus*. Cambridge, UK: Cambridge University Press.

Landis, J. R., & Koch, G. G. (1977). The measurement of observer agreement for categorical data. *Biometrics*, *33*, 159–174.

Larson, R., Hostetler, R. P., & Edwards, B. H. (2006). *Calculus with analytic geometry* (8th ed.). Boston: Houghton Mifflin.

Lithner, J. (2004). Mathematical reasoning in calculus textbook exercises. *Journal of Mathematical Behavior*, *23*, 405–427.

Love, C. E. (1916). *Differential and integral calculus*. New York: Macmillan.

Love, C. E., & Rainville, E. D. (1962). *Differential and integral calculus* (6th ed.). New York: Macmillan.

Love, E., & Pimm, D. (1996). 'This is so': A text on texts. In A. J. Bishop, K. Clements, C. Keitel, J. Kilpatrick, & C. Laborde (Eds.), *International handbook of mathematics education* (Vol. 1, pp. 371–409). Dordrecht: Kluwer.

MacCluer, C. R. (2006). *Honors calculus*. Princeton, NJ: Princeton University Press.

Martin, W. G., & Harel, G. (1989). Proof frames of pre-service mathematics teachers. *Journal for Research in Mathematics Education*, *20*, 41–51.

Mesa, V. (2004). Characterizing practices associated with functions in middle school textbooks. *Educational Studies in Mathematics*, *56*, 255–286.

Mesa, V. (2006a). *Calculus teaching in the US* (Unpublished raw data). University of Michigan.

Mesa, V. (2006b). *Understanding the role of resources in developing collegiate teaching expertise: The case of mathematics textbooks* (Unpublished raw data). University of Michigan.

Mesa, V. (2007, February). *Insights from instructors using textbooks for teaching mathematics.* Paper presented at the 10th annual conference on Research in Undergraduate Mathematics Education, San Diego, CA.

Michener, E. R. (1978). Understanding understanding mathematics. *Cognitive Science*, *2*, 361–383.

Nespor, J. (1990). The jackhammer: A case study of undergraduate physics problem solving in its social setting. *Qualitative Studies in Education*, *3*, 139–155.

Novak, J. D. (1998). *Learning, creating, and using knowledge*. Mahwah, NJ: Erlbaum.

Ostebee, A., & Zorn, P. (2002). *Calculus from graphical, numerical, and symbolic points of view*. Belmont, CA: Brooks/Cole.

Perry, J. (1897). *The calculus for engineers*. New York: Edward Arnold.

Plausible (1989). *Oxford English Dictionary*. Oxford: Oxford University Press.

Raman, M. (2002). Coordinating informal and formal aspects of mathematics: Student behavior and textbook messages. *Journal of Mathematical Behavior, 21*, 135–150.

Ransom, W. R. (1917). Differential and integral calculus by Clyde E. Love. *American Mathematical Monthly*, *24*, 173–175.

Schoenfeld, A. H. (1985a). *Mathematical problem solving*. Orlando, FL: Academic.

Schoenfeld, A. H. (1985b). Metacognitive and epistemological issues in mathematical problem solving. In E. A. Silver (Ed.), *Teaching and learning mathematical problem solving. Multiple research perspectives* (pp. 361–379). Hillsdale, NJ: Erlbaum.

Schoenfeld, A. H. (1992). Learning to think mathematically: Problem solving, metacognition, and sense making in mathematics. In D. A. Grouws (Ed.), *Handbook of research on mathematics teaching and learning* (pp. 334–370). New York: Macmillan.

Schoenfeld, A. H. (1994). *Mathematical thinking and problem solving*. Hillsdale, NJ: Erlbaum.

Selden, J., & Selden, A. (1995). Unpacking the logic of mathematical statements. *Educational Studies in Mathematics*, *29*, 123–151.

Silver, E. A. (1987). Foundations of cognitive theory and research for mathematics problem-solving. In A. H. Schoenfeld (Ed.), *Cognitive science and mathematics education* (pp. 33–60). Hillsdale, NJ: Erlbaum.

Simmons, G. F. (1996). *Calculus with analytic geometry* (2nd ed.). New York: McGraw-Hill.

Smith, R. H. (1897). *The calculus for engineers and physicists: Integration and differentiation with applications to technical problems, with classified reference tables of integrals and methods of integration*. London: C. Griffin & Co.

Smith, R. T., & Minton, R. B. (2002). *Calculus* (2nd ed.). New York: McGraw Hill.

Stein, M. K., & Smith, M. S. (1998). Mathematical tasks as a framework for reflection: From research to practice. *Mathematics Teaching in the Middle School*, *3*, 268–275.

Stephan, M., & Rasmussen, C. L. (2002). Classroom mathematical practices in differential equations. *Journal of Mathematical Behavior*, *21*, 459–490.

Stewart, J. (2003). *Calculus with early transcendentals* (5th ed.). Belmont, CA: Brooks/Cole.

Thomas, G. B., Finney, R. L., Weir, M. D., & Giordano, F. R. (2001). *Thomas' calculus* (10th ed.). Boston: Addison Wesley.

Valid (2000). *American Heritage Dictionary of the English Language*. Boston: Houghton Mifflin.

Walshaw, M. (2001). A Foucauldian gaze on gender research: What do you do when confronted with the tunnel at the end of the light? *Journal for Research in Mathematics Education*, *32*, 471–492.

Watson, A., & Mason, J. (2005). *Mathematics as a constructive activity*. Mahwah, NJ: Erlbaum.

Zenor, P., Slaminka, E. E., & Thaxton, D. (1999). *Calculus with early vectors*. Upper Saddle River, NJ: Prentice Hall.

Zevenbergen, R. (2005). The construction of mathematical habitus: Implications of ability grouping in the middle years. *Journal of Curriculum Studies*, *37*, 607–619.

University of Michigan 1360F SEB, 610 East University Ann Arbor, MI, 48109-1259
E-mail address: vmesa@umich.edu

Titles in This Series

16 **Fernando Hitt, Derek Holton, and Patrick W. Thompson, Editors,** Research in collegiate mathematics education. VII, 2010

15 **Robert E. Reys and John A. Dossey, Editors,** U.S. doctorates in mathematics education: Developing stewards of the discipline, 2008

14 **Ki Hyoung Ko and Deane Arganbright, Editors,** Enhancing university mathematics, 2007

13 **Fernando Hitt, Guershon Harel, and Annie Selden, Editors,** Research in collegiate mathematics education. VI, 2006

12 **Annie Selden, Ed Dubinsky, Guershon Harel, and Fernando Hitt, Editors,** Research in collegiate mathematics education. V, 2003

11 **Conference Board of the Mathematical Sciences,** The mathematical education of teachers, 2001

10 **Solomon Friedberg et al.,** Teaching mathematics in colleges and universities: Case studies for today's classroom. Available in student and faculty editions, 2001

9 **Robert Reys and Jeremy Kilpatrick, Editors,** One field, many paths: U. S. doctoral programs in mathematics education, 2001

8 **Ed Dubinsky, Alan H. Schoenfeld, and Jim Kaput, Editors,** Research in collegiate mathematics education. IV, 2001

7 **Alan H. Schoenfeld, Jim Kaput, and Ed Dubinsky, Editors,** Research in collegiate mathematics education. III, 1998

6 **Jim Kaput, Alan H. Schoenfeld, and Ed Dubinsky, Editors,** Research in collegiate mathematics education. II, 1996

5 **Naomi D. Fisher, Harvey B. Keynes, and Philip D. Wagreich, Editors,** Changing the culture: Mathematics education in the research community, 1995

4 **Ed Dubinsky, Alan H. Schoenfeld, and Jim Kaput, Editors,** Research in collegiate mathematics education. I, 1994

3 **Naomi D. Fisher, Harvey B. Keynes, and Philip D. Wagreich, Editors,** Mathematicians and education reform 1990–1991, 1993

2 **Naomi D. Fisher, Harvey B. Keynes, and Philip D. Wagreich, Editors,** Mathematicians and education reform 1989–1990, 1991

1 **Naomi D. Fisher, Harvey B. Keynes, and Philip D. Wagreich, Editors,** Mathematicians and education reform: Proceedings of the July 6–8, 1988 workshop, 1990